Dieter Vollath
Nanomaterials

Related Titles

Ajayan, P. M., Schadler, L. S., Braun, P. V.,
Keblinski, P.

**Nanocomposite Science and
Technology**

2009
ISBN: 978-3-527-31248-1

Schmid, G. (ed.)

Nanotechnology

Volume 1: Principles and Fundamentals

2008
ISBN: 978-3-527-31732-5

Krug, H. (ed.)

Nanotechnology

Volume 2: Environmental Aspects

2008
ISBN: 978-3-527-31735-6

Waser, R. (ed.)

Nanotechnology

Volume 3: Information Technology I

2008
ISBN: 978-3-527-31738-7

Waser, R. (ed.)

Nanotechnology

Volume 4: Information Technology II

2008
ISBN: 978-3-527-31737-0

Köhler, M., Fritzsche, W.

Nanotechnology

An Introduction to Nanostructuring
Techniques

2007
ISBN: 978-3-527-31871-1

Rao, C. N. R., Müller, A., Cheetham, A. K.
(eds.)

Nanomaterials Chemistry

Recent Developments and New
Directions

2007
ISBN: 978-3-527-31664-9

Borisenko, V. E., Ossicini, S.

**What is What in the
Nanoworld**

A Handbook on Nanoscience and
Nanotechnology

2004
ISBN: 978-3-527-40493-3

Schmid, G. (ed.)

Nanoparticles

From Theory to Application

2004
ISBN: 978-3-527-30507-0

Dieter Vollath

Nanomaterials

An Introduction to Synthesis, Properties and Application

WILEY-VCH

WILEY-VCH Verlag GmbH & Co. KGaA

The Author

Prof. Dr. Dieter Vollath
NanoConsulting
Primelweg 3
76297 Stutensee
Germany

Cover picture:
Photograph of ferrofluid, a liquid which becomes
magnetic when subjected to magnetic fields.
Ferrofluids are composed of nanoscale ferromagnetic
particles.

The picture was kindly supplied by Ferrotec GmbH,
Germany.

■ All books published by Wiley-VCH are carefully
produced. Nevertheless, authors, editors, and
publisher do not warrant the information contained
in these books, including this book, to be free of
errors. Readers are advised to keep in mind that
statements, data, illustrations, procedural details or
other items may inadvertently be inaccurate.

Library of Congress Card No.:
applied for

British Library Cataloguing-in-Publication Data
A catalogue record for this book is available from the
British Library.

**Bibliographic information published by
the Deutsche Nationalbibliothek**
Die Deutsche Nationalbibliothek lists this
publication in the Deutsche Nationalbibliografie;
detailed bibliographic data are available in the
Internet at http://dnb.d-nb.de.

Typesetting   Thomson Digital, Noida, India
Printing   Strauss GmbH, Mörlenbach
Binding   Litges & Dopf Buchbinderei GmbH,
Heppenheim

Printed in the Federal Republic of Germany
Printed on acid-free paper

ISBN: 978-3-527-31531-4

Contents

Preface   *IX*

1   **Introduction**   *1*
1.1   Nanomaterials and Nanocomposites   *5*
1.2   Elementary Consequences of Small Particle Size   *11*
1.2.1   Surface of Nanoparticles   *11*
1.2.2   Thermal Phenomena   *12*
1.2.3   Diffusion Scaling Law   *13*
               References   *19*

2   **Surfaces in Nanomaterials**   *21*
2.1   General Considerations   *21*
2.2   Surface Energy   *23*
2.3   Some Technical Consequences of Surface Energy   *33*
               References   *40*

3   **Phase Transformations of Nanoparticles**   *41*
3.1   Thermodynamics of Nanoparticles   *41*
3.2   Heat Capacity of Nanoparticles   *42*
3.3   Phase Transformations of Nanoparticles   *45*
3.4   Phase Transformation and Coagulation   *54*
3.5   Structures of Nanoparticles   *55*
3.6   A Closer Look at Nanoparticle Melting   *60*
3.7   Structural Fluctuations   *64*
               References   *69*

4   **Gas-Phase Synthesis of Nanoparticles**   *71*
4.1   Fundamental Considerations   *71*
4.2   Inert Gas Condensation Process   *78*

Nanomaterials: An Introduction to Synthesis, Properties and Application. Dieter Vollath
Copyright © 2008 WILEY-VCH Verlag GmbH & Co. KGaA, Weinheim
ISBN: 978-3-527-31531-4

4.3 Physical and Chemical Vapor Synthesis Processes *79*
4.4 Laser Ablation Process *83*
4.5 The Microwave Plasma Process *86*
4.6 Flame Aerosol Process *92*
4.7 Synthesis of Coated Particles *103*
References *108*

5 Magnetic Properties of Nanoparticles *109*
5.1 Magnetic Materials *109*
5.2 Superparamagnetic Materials *113*
5.3 Susceptibility and Related Phenomena in Superparamagnets *125*
5.4 Applications of Superparamagnetic Materials *132*
5.5 Exchange-Coupled Magnetic Nanomaterials *136*
References *143*

6 Optical Properties of Nanoparticles *145*
6.1 General Remarks *145*
6.2 Adjustment of the Index of Refraction *145*
6.3 Optical Properties Related to Quantum Confinement *149*
6.4 Quantum Dots and Other Lumophores *161*
6.5 Metallic and Semiconducting Nanoparticles in Transparent Matrices *169*
6.6 Special Luminescent Nanocomposites *180*
6.7 Electroluminescence *188*
6.8 Photochromic and Electrochromic Materials *194*
6.8.1 General Considerations *194*
6.8.2 Photochromic Materials *195*
6.8.3 Electrochromic Materials *200*
6.9 Magneto-optic Applications *204*
References *207*

7 Electrical Properties of Nanoparticles *211*
7.1 Fundamentals of Electrical Conductivity in Nanotubes and Nanorods *211*
7.2 Carbon Nanotubes *216*
7.3 Photoconductivity of Nanorods *222*
7.4 Electrical Conductivity of Nanocomposites *225*
References *230*

8 Mechanical Properties of Nanoparticles *233*
8.1 General Considerations *233*
8.2 Bulk Metallic and Ceramic Materials *236*
8.2.1 Influence of Porosity *236*
8.2.2 Influence of Grain Size *238*
8.2.3 Superplasticity *251*

8.3   Filled Polymer Composites   *253*
8.3.1   Particle-Filled Polymers   *253*
8.3.2   Polymer-Based Nanocomposites Filled with Platelets   *257*
8.3.3   Carbon Nanotube-Based Composites   *262*
         References   *266*

9   **Nanofluids**   *267*
9.1   Definition   *267*
9.1.1   Nanofluids for Improved Heat Transfer   *267*
9.2   Ferrofluids   *270*
9.2.1   General Considerations   *270*
9.2.2   Properties of Ferrofluids   *271*
9.2.3   Applications of Ferrofluids   *272*
         References   *277*

10   **Nanotubes, Nanorods, and Nanoplates**   *279*
10.1   Introduction   *279*
10.1.1   Conditions for the Formation of Rods and Plates   *283*
10.1.2   Layered Structures   *285*
10.1.3   One-Dimensional Crystals   *286*
10.2   Nanostructures Related to Compounds with Layered Structures   *288*
10.2.1   Carbon Nanotubes   *288*
10.2.2   Nanotubes and Nanorods from Materials other than Carbon   *300*
10.2.3   Synthesis of Nanotubes and Nanorods   *303*
         References   *311*

11   **Characterization of Nanomaterials**   *313*
11.1   Global Methods for Characterization   *313*
11.1.1   Specific Surface Area   *313*
11.2   X-Ray and Electron Diffraction   *319*
11.3   Electron Microscopy   *327*
11.3.1   General Considerations   *327*
11.3.2   Interaction of the Electron Beam and Specimen   *331*
11.3.3   Localized Chemical Analysis in the Electron Microscope   *334*
11.3.4   Scanning Transmission Electron Microscopy using a High-Angle
         Annular Dark-Field (HAADF) Detector   *340*
         References   *342*

Index   *343*

Preface

This book is an elementary introduction to nanomaterials; one may consider it as an approach to a textbook. It is based on the course on nanomaterials for engineers that I give at the Graz University of Technology and on the courses that NanoConsulting organizes for participants from industry and academia. I want to provoke your curiosity. The reader should feel invited to learn more about nanomaterials, to use nanomaterials and to go later on beyond the content of this book. However, even when it is thought as introduction, reading this book requires some basic knowledge in physics and chemistry. I have tried to describe the mechanisms ruling the properties of nanoparticles in a simplified way. Therefore, specialists in different fields will feel uncomfortable, but I saw no other way to describe the mechanism leading to the fascinating properties of nanoparticles for a broader audience.

I am fully aware of the fact that the selection of examples from literature is, to some extent, unfair against those who discovered these new phenomena. However, in most cases, where a new phenomenon was described for the first time, the effect is just shown in principle. Later papers had only a chance, when they showed these phenomena very clear. Therefore, in view of a textbook the latter papers are preferred. I really apologize this selection of literature.

Many exiting phenomena and processes are connected with nanoparticles. However, the size of this book is limited, therefore, I had to make a selection of topics presented in this book. Unavoidably, such a selection is not at least influenced by personal experience and preferences. I really apologize if a reader does not find information on a field, important for him or his company.

I hope the readers will find this book inspiring and motivating to go deeper into this fascinating field of science and technology.

It is an obligation for me to thank my family, in special my wife Renate, for her steady support during the time when I wrote this book and her enduring understanding for my passion for science. Furthermore, I have to thank Mrs. Dr. D. V. Szabó, Forschungszentrum Karlsruhe, and Prof. Dr. F. D. Fischer, University of Leoben, for reading some chapters and giving important advices. Not at least, I have to thank W. H. Down for careful language editing.

April 2008   *Dieter Vollath*

Nanomaterials: An Introduction to Synthesis, Properties and Application. Dieter Vollath
Copyright © 2008 WILEY-VCH Verlag GmbH & Co. KGaA, Weinheim
ISBN: 978-3-527-31531-4

1
Introduction

Today, everybody is talking about nanomaterials and, indeed, very many publications, books, and journals are devoted to this topic. Usually, such publications are directed towards specialists such as physicists and chemists, and the "classic" materials scientist encounters increasing problems in understanding the situation. Moreover, those people who are interested in the subject but who have no specific education in any of these fields have virtually no chance of understanding the development of this technology. It is the aim of this book to fill this gap. The book will focus on the special phenomena related to nanomaterials and attempt to provide explanations which avoid – as far as possible – any highly theoretical and quantum mechanical descriptions. The difficulties with nanomaterials arise from the fact that, in contrast to conventional materials, a profound knowledge of materials science is not sufficient. The cartoon shown in Figure 1.1 shows that nanomaterials lie at the intersection of materials science, physics, chemistry, and – for many of the most interesting applications – also of biology and medicine.

However, this situation is less complicated than it first appears to the observer, as the number of additional facts introduced to materials science is not that large. Nonetheless, the user of nanomaterials must accept that the properties of the latter materials demand a deeper insight into their physics and chemistry. Whereas, for conventional materials the interface to biotechnology and medicine is related directly to the application, the situation is different in nanotechnology, where biological molecules such as proteins or DNA are also used as building blocks for applications outside biology and medicine.

So, the first question to be asked is, "What are nanomaterials?" There are two definitions. The first – and broadest – definition states that nanomaterials are materials where the sizes of the individual building blocks are less than 100 nm, at least in one dimension. This definition is well suited for many research proposals, where nanomaterials have a high priority. The second definition is much more restrictive, and states that nanomaterials have properties which depend inherently on the small grain size and, as nanomaterials are usually quite expensive, such a restrictive definition makes more sense. The main difference between nanotechnology and conventional technologies is that the "bottom-up" approach (see below) is preferred in nanotechnology, whereas conventional technologies usually use the

Nanomaterials: An Introduction to Synthesis, Properties and Application. Dieter Vollath
Copyright © 2008 WILEY-VCH Verlag GmbH & Co. KGaA, Weinheim
ISBN: 978-3-527-31531-4

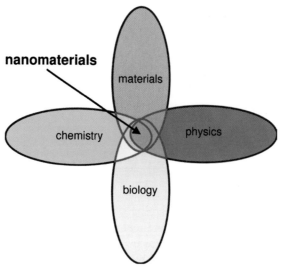

Figure 1.1 A basic understanding of physics and chemistry, and some knowledge of materials science, is necessary to understand the properties and behavior of nanomaterials. As many applications are connected with biology and medicine, some knowledge of these areas is also required.

"top-down" approach. The difference between these two approaches can be explained simply by using an example of powder production, where the chemical synthesis represents the bottom-up approach while the crushing and milling of chunks represents the equivalent top-down process.

On examining these technologies more closely, the expression "top-down" means starting from large pieces of material and producing the intended structure by mechanical or chemical methods. This situation is shown schematically in Figure 1.2. As long as the structures are within a range of sizes that are accessible by either

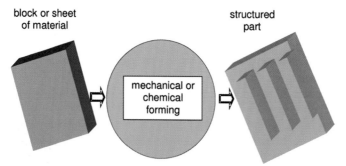

Figure 1.2 Conventional goods are produced via top-down processes, starting from bulk materials. The intended product is obtained by the application of mechanical and/or chemical processes.

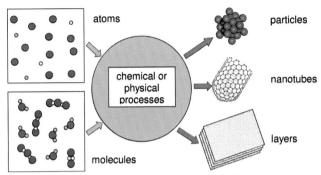

Figure 1.3 Nanotechnologies are usually connected to bottom-up processes, and are characterized by the use of atoms or molecules as building blocks. Bottom-up processes result in particles, nanotubes, nanorods, thin films, or layered structures.

mechanical tools or photolithographic processes, then top-down processes have an unmatched flexibility in their application.

The situation is different in "bottom-up" processes, in which atoms or molecules are used as the building blocks to produce nanoparticles, nanotubes or nanorods, or thin films or layered structures. According to their dimensionality, these features are also referred to as zero-, one-, or two-dimensional nanostructures (see Figure 1.3). Figure 1.3 also demonstrates the building of particles, layers, nanotubes, or nanorods from atoms (ions) or molecules. Although such processes provide tremendous freedom among the resultant products, the number of possible structures to be obtained is comparatively small. In order to obtain ordered structures, bottom-up processes (as described above) must be supplemented by the self-organization of individual particles.

Often, top-down technologies are described as being "subtractive", in contrast to the "additive" technologies which describe bottom-up processes. The crucial problem is no longer to produce these elements of nanotechnology; rather, it is their incorporation into technical parts. The size ranges of classical top-down technologies compared to bottom-up technologies are shown graphically in Figure 1.4. Clearly, there is a broad range of overlapping where improved top-down technologies, such as electron beam or X-ray lithography, enter the size range typical of nanotechnologies. Currently, these improved top-down technologies are penetrating into increasing numbers of fields of application.

For industrial applications, the most important question is the product's price in relation to its properties. In most cases, nanomaterials and products utilizing nanomaterials are significantly more expensive than conventional products. In the case of nanomaterials, the increase in price is sometimes more pronounced than the improvement in properties, and therefore economically interesting applications of nanomaterials are often found only in areas where specific properties are demanded that are beyond the reach of conventional materials. Hence, as long as the use of nanomaterials with new properties provides the solution to a problem which cannot

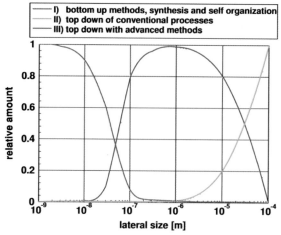

Figure 1.4 The estimated lateral limits of different structuring processes. Clearly, the size range of bottom-up and conventional top-down processes is limited. New, advanced top-down processes expand the size range of their conventional counterparts, and enter the size range typical of bottom-up processes.

be solved with conventional materials, the price becomes much less important. Another point is that, as the applications of nanomaterials using improved properties are in direct competition to well-established conventional technologies, they will encounter a fierce price competition, and this may lead to major problems for a young and expensive technology to overcome. Indeed, it is often observed that marginal profit margins in the production or application of nanomaterials with improved properties may result in severe financial difficulties for newly founded companies. In general, the economically successful application of nanomaterials requires only a small amount of material as compared to conventional technologies; hence, one is selling "knowledge" rather than "tons" (see Table 1.1). Finally, only those materials which exhibit new properties leading to novel applications, beyond the reach of conventional materials, promise interesting economic results.

Table 1.1 The relationship between the properties of a new product and prices, quantities, and expected profit.

Properties	Price		Quantity		Profits
	Low	High	Small	Large	
Improved	X	–	–	X	Questionable
New	–	X	X	–	Potentially high

Note: only those products with new properties promise potentially high profits.

1.1
Nanomaterials and Nanocomposites

Nanomaterials may be zero-dimensional (e.g., nanoparticles), one-dimensional (e.g., nanorods or nanotubes), or two-dimensional (usually realized as thin films or stacks of thin films). As a typical example, an electron micrograph of zirconia powder (a zero-dimensional object) is shown in Figure 1.5.

The particles depicted in Figure 1.5 show a size of ca. 7 nm, and also a very narrow range of sizes. This is an important point, as many of the properties of nanomaterials are size-dependent. In contrast, many applications do not require such sophistication, and therefore cheaper materials with a broader particle size distribution (see Figure 1.6a) would be sufficient. The material depicted in Figure 1.6a, which contains particles ranging in size from 5 to more than 50 nm, would be perfectly suited for applications such as pigments or ultra-violet (UV) absorbers.

A further interesting class of particles may be described as fractal clusters of extreme small particles. Typical examples of this type of material are most of the amorphous silica particles (known as "white soot") and amorphous Fe_2O_3 particles, the latter being used as catalysts (see Figure 1.6b).

Apart from properties related to grain boundaries, the special properties of nanomaterials are those of single isolated particles that are altered, or even lost, in the case of particle interaction. Therefore, most of the basic considerations are related to isolated nanoparticles as the interaction of two or more particles may cause significant changes in the properties. For technical applications, this proved to be negative, and consequently nanocomposites of the core/shell type with a second phase acting as distance holder were developed. The necessary distance depends on the phenomenon to be suppressed; it may be smaller, in case of the tunneling of electrons between particles, but larger in the case of dipole–dipole interaction. Nanocomposites – as described in this chapter – are composite materials with at

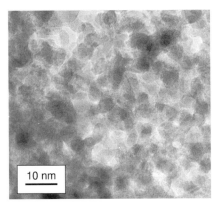

10 nm

Figure 1.5 An electron micrograph of zirconia, ZrO_2, powder. This material has a very narrow distribution of grain size; this is important as the properties of nanomaterials depend on grain size. (Reprinted with permission from [1]; Copyright: Imperial College Press 2002).

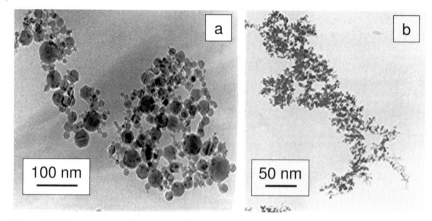

Figure 1.6 The two types of nanoparticulate Fe$_2$O$_3$ powder. (a) Industrially produced [2] nanomaterial with a broad particle size distribution; this is typically used as a pigment or for UV protection. (b) Nanoparticulate powder consisting of fractal clusters of amorphous (ca. 3 nm) particles [3]. As this material has an extremely high surface area, catalysis is its most important field of application. (TEM micrographs reprinted with permission from Nanophase Technologies Corporation, Romeoville, IL, USA).

least one phase exhibiting the special properties of a nanomaterial. In general, random arrangements of nanoparticles in the composite are assumed.

The three most important types of nanocomposites are illustrated schematically in Figure 1.7. The types differ in the dimensionality of the second phase, which may be zero-dimensional (i.e., isolated nanoparticles), one-dimensional (i.e., consisting of

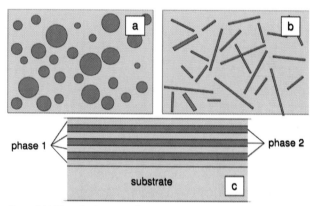

Figure 1.7 Three basic types of nanocomposite. (a) Composite consisting of zero-dimensional particles in a matrix; ideally, the individual particles do not touch each other. (b) One-dimensional nanocomposite consisting of nanotubes or nanorods distributed in a second matrix. (c) Two-dimensional nanocomposite built from stacks of thin films made of two or more different materials.

nanotubes or nanorods), or two-dimensional (i.e., existing as stacks or layers). Composites with platelets might also be thought of as second phase. In most cases, such composites are close to a zero-dimensional state; some of those with a polymer matrix possess exciting mechanical and thermal properties and are used to a wide extent in the automotive industry.

In general, nanosized platelets are energetically not favorable, and therefore not often observed. However, this type of nanocomposite using polymer matrices may be realized using delaminated layered silicates (these nanocomposites are discussed in connection with their mechanical properties in Section 8.3.2). In addition to the composites shown in Figure 1.7, nanocomposites with regular well-ordered structures may also be observed (see Figure 1.8). In general, this type of composite is created via a self-organization processes. The successful realization of such processes require particles that are almost identical in size.

The oldest, and most important, type of nanocomposite is that which has more or less spherical nanoparticles. An example is the well-known gold-ruby-glass, which consists of a glass matrix with gold nanoparticles as the second phase. This material was first produced by the Assyrians in the seventh century BC, and reinvented by Kunkel in Leipzig in the 17th century. It is interesting to note that the composition used by the Assyrians was virtually identical to that used today. This well-known gold-ruby-glass needed a modification of nanocomposites containing a second phase of spherical nanoparticles. In many cases, as the matrix and the particles exhibit mutual solubility, a diffusion barrier is required to stabilize the nanoparticles; such an arrangement is shown in Figure 1.9. In the case of gold-ruby-glass, the diffusion barrier consists of tin oxide. In colloid chemistry, this principle of stabilization is often referred to as a "colloid stabilizer".

A typical electron micrograph of a near-ideal nanocomposite, a distribution of zirconia nanoparticles within an alumina matrix, is shown in Figure 1.10. Here, the material was sintered and the starting material alumina-coated zirconia powder; the particles remained clearly separated.

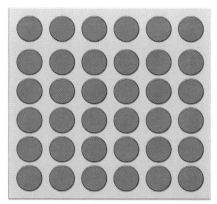

Figure 1.8 A perfectly ordered zero-dimensional nanocomposite; this type of composite is generally made via a self-organization processes.

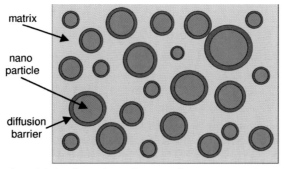

matrix

nano
particle

diffusion
barrier

Figure 1.9 An advanced zero-dimensional nanocomposite. Here, a diffusion barrier surrounds each particle. This type of material is required if the nanoparticle and matrix are mutually soluble.

Composites with nanotubes or nanorods (in most cases, a polymer matrix and long carbon nanotubes) are used for reinforcement or to introduce electric conductivity to the polymer.

When producing nanocomposites, the central problem is to obtain a perfect distribution of the two phases; however, processes based on mechanical blending never lead to homogeneous products on the nanometer scale. Likewise, synthesizing the two phases separately and blending them during the stage of particle formation never leads to the intended result. In both cases, the probability that two or more particles are in contact with each other is very high, and normally in such a mixture the aim is to obtain a relatively high concentration of "active" particles, carrying the physical property of interest. Assuming, in the simplest case, particles of equal size, the probability p_n that n particles with the volume concentration c are touching each other is $P_n = c^n$. Then, assuming a concentration of 0.30, the probability of two touching particles is 0.09, and for three particles it is 0.027. The necessary perfect

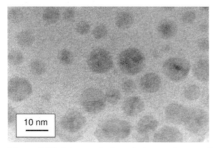

10 nm

Figure 1.10 A transmission electron micrograph of a zero-dimensional nanocomposite, showing zirconia particles embedded in an alumina matrix. The specimen was produced from zirconia particles coated with alumina. The image was taken from an ion beam-thinned sample. There is a high probability that these particles do not touch each other as they are in different planes (reprinted with kind permission from [4]).

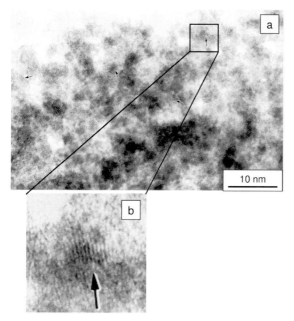

Figure 1.11 Electron micrograph of a nanocomposite consisting an amorphous alumina matrix and precipitated crystallized zirconia particles. (a) Within the amorphous alumina, the crystallized zirconia precipitations are indicated by arrows. (b) One of the precipitations shown at a higher magnification. The precipitation sizes range between 1.5 and 3 nm; such precipitation occurs because zirconia is insoluble in alumina at room temperature.

distribution of two phases is obtained only by coating the particles of the active phase with the distance holder phase. In general, this can be achieved by either of the two following approaches:

- Synthesis of a metastable solution and precipitation of the second phase by reducing the temperature [5]. A typical example is shown in Figure 1.11a , which shows amorphous alumina particles within which zirconia precipitation is realized. As the concentration of zirconia in the original mixture was very low, the size of these precipitates is small (<3 nm). Arrows indicate the position of few of these precipitates. One of the precipitates is depicted at higher magnification in Figure 1.11b, where the lines visible in the interior of the particle represent the lattice planes. This is one of the most elegant processes for synthesizing ceramic/ceramic nanocomposites as it leads to extremely small particles, although the concentration of the precipitated phase may be low (in certain cases, this may be a significant disadvantage).

- The most successful development in the direction of nanocomposites was that of coated particles, as both the kernel and coating material are distributed homogeneously on a nanometer scale. The particles produced in a first reaction step are coated with the distance holder phase in a second reaction step. Two typical examples of coated nanoparticles are shown in Figure 1.12. In Figure 1.12a, a ceramic–polymer composite is shown in which the core consists of iron oxide,

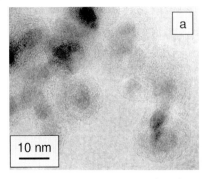
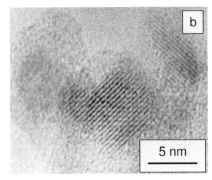

Figure 1.12 Nanocomposite particles. Electron micrographs depicting two types of coated particle. (a) The particles consist of a γ-Fe₂O₃ core and are coated with PMMA (reproduced with kind permission from [6]). (b) Crystallized zirconia particles coated with amorphous alumina (reproduced with kind permission from [7]).

γ-Fe₂O₃, and the coating of polymethylmethacrylate (PMMA). The second example, a ceramic–ceramic composite, uses a second ceramic phase for coating; here, the core consists of crystallized zirconia and the coating of amorphous alumina. It is a necessary prerequisite for this type of coated particle that there is no mutual solubility between the compounds used for the core and the coating. Figure 1.12b shows three alumina-coated zirconia particles, where the center particle originates from the coagulation of two zirconia particles. As the process of coagulation was incomplete, concave areas of the zirconia core were visible. However, during the coating process these concave areas were filled with alumina, such that finally the coated particle had only convex surfaces. This led to a minimization of the surface energy, which is an important principle in nanomaterials.

The properties of a densified solid may also be adjusted gradually with the thickness of the coating. Depending on the requirements of the system in question, the coating material may be either ceramic or polymer. In addition, by coating nanoparticles with second and third layers, the following improvements are obtained:

- The distribution of the two phases is homogeneous on a nanometer scale.

- The kernels are arranged at a well-defined distance; therefore, the interaction of the particles is controlled.

- The kernel and one or more different coatings may have different properties; this allows a combination of properties in one particle that would never exist together in nature. In addition, by selecting a proper polymer for the outermost coating it is possible to adjust the interaction with the surrounding medium; for example, hydrophilic or hydrophobic coatings may be selected.

- During densification (i.e., sintering) the growth of the kernels is thwarted, provided that the core and coating show no mutual solubility. An example of this is shown in Figure 1.10.

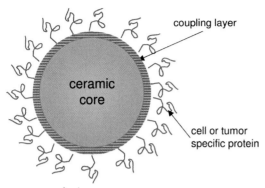

Figure 1.13 A nanocomposite particle for application in biology or medicine. The ceramic core may be magnetic or luminescent. The cell- or tumor-specific proteins at the surface, which are necessary for application, require a coupling layer as typically they cannot be attached directly to the ceramic surface.

These arguments confirm that coated nanoparticles represent the most advanced type of nanocomposite because they allow: (i) different properties to be combined in one particle; and (ii) exactly adjusted distances to be inserted between directly adjacent particles in the case of densified bodies.

Today, coated particles are widely used in biology and medicine, although for this it may be necessary to add proteins or other biological molecules at the surface of the particles. Such molecules are attached via specific linking molecules and accommodated in the outermost coupling layer. A biologically functionalized particle is shown schematically in Figure 1.13, where the ceramic core is usually either magnetic or luminescent. Recent developments in the combination of these two properties have utilized a multishell design of the particles. In the design depicted in Figure 1.13, the coupling layer may consist of an appropriate polymer or a type of glucose, although in many cases hydroxylated silica is also effective. Biological molecules such as proteins or enzymes may then be attached at the surface of the coupling layer.

1.2
Elementary Consequences of Small Particle Size

Before discussing the properties of nanomaterials, it may be advantageous to describe some examples demonstrating the elementary consequences of the small size of nanoparticles.

1.2.1
Surface of Nanoparticles

The first and most important consequence of a small particle size its huge surface area, and in order to obtain an impression of the importance of this geometric variable, the surface over volume ratio should be discussed. So, assuming spherical particles, the surface a of one particle with diameter D is $a = \pi D^2$, and the corresponding volume v is $v = \frac{\pi}{6} D^3$. Therefore, one obtains for the surface/volume ratio

$$R = \frac{a}{v} = \frac{6}{D} \tag{1.1}$$

This ratio is inversely proportional to the particle size and, as a consequence, the surface increases with decreasing particle size. The same is valid for the surface per mol A, a quantity which is of extreme importance in thermodynamic considerations.

$$A = na = \frac{M}{\rho \frac{\pi D^3}{6}} \pi D^2 = \frac{6M}{\rho D} \tag{1.2}$$

In Equation (1.2), n is the number of particles per mol, M the molecular weight, and ρ the density of the material. Similar to the surface over volume ratio, the area per mol increases inversely in proportion to the particle diameter; hence, huge values of area are achieved for particles that are only a few nanometers in diameter.

It should be noted that as the surface is such an important topic for nanoparticles, Chapter 2 of this book has been devoted to surface and surface-related problems.

1.2.2
Thermal Phenomena

Each isolated object – in this case a nanoparticle – has a thermal energy of kT (k is the *Boltzmann* constant and T temperature). First, let us assume a property of the particle which depends for example on the volume v of the particle; the energy of this property may be $u(v)$. Then, provided that the volume is sufficiently small such that the condition

$$u(v) < kT \tag{1.3}$$

is fulfilled, one may expect thermal instability. As an example, one may ask for the particle size where thermal energy is large enough to lift the particle. In the simplest case, one estimates the energy necessary to lift a particle of density ρ over the elevation $x \cdot u(v) = \rho v x = kT$. Assuming a zirconia particle with a density of $5.6 \times 10^3 \, \text{kg m}^{-3}$, at room temperature the thermal energy would lift a particle of diameter 1100 nm to a height equal to the particle diameter, D. If one asks how high might a particle of 5-nm diameter jump, these simple calculations indicate a value of more than 1 m. Clearly, although these games with numbers do not have physical reality, they do show that nanoparticles are not fixed but rather are moving about on the surface. By performing electron microscopy, this dynamic becomes reality and, provided that the particles and carbon film on the carrier mesh are clean, the specimen particles can be seen to move around on the carbon film. On occasion, however, this effect may cause major problems during electron microscopy studies.

Although the thermal instability shown here demonstrates only one of the consequences of smallness, when examining the other physical properties then an

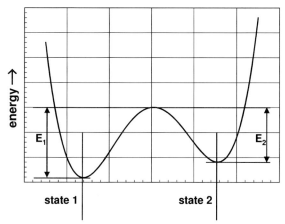

Figure 1.14 A graphical representation of the energy barrier, showing the energy necessary to jump from state 1 to state 2, and vice-versa.

important change in the behavior can be realized. Details of the most important phenomenon within this group – superparamagnetism – are provided in Chapter 5. In the case of superparamagnetism, the vector of magnetization fluctuates between different "easy" directions of magnetization, and these fluctuations may also be observed in connection with the crystallization of nanoparticles. In a more generalized manner, thermal instabilities leading to fluctuations may be characterized graphically, as shown in Figure 1.14.

Provided that the thermal energy kT is greater than the energies E_1 and E_2, the system fluctuates between both energetically possible states 1 and 2. Certainly, it does not make any difference to these considerations if E_1 and E_2 are equal, or more than two different states are accessible with thermal energy at temperature T.

The second example describes the temperature increase by the absorption of light quanta. Again, a zirconia particle with density $\rho = 5.6 \times 10^3 \, kg \, m^{-3}$, a heat capacity $c_p = 56.2 \, J \, mol^{-1} \, K^{-1}$ equivalent to $c_p = 457 \, J \, kg^{-1} \, K^{-1}$ and, in this case, a particle diameter of 3 nm is assumed. After the absorption of one photon with a wavelength, λ, of 300 nm, the temperature increase ΔT is calculated from $c_p \rho v \Delta T = h v = h \frac{c}{\lambda}$ (c is the velocity of light and h is *Planck*'s constant) to 18 K. Being an astonishingly large value, this temperature increase must be considered when interpreting optical spectra of nanomaterials with poor quantum efficiency or composites with highly UV-absorbing kernels.

1.2.3
Diffusion Scaling Law

Diffusion is controlled by the two laws defined by *Fick*. The solutions of these equations, which are important for nanotechnology, imply that the mean square diffusion path of the atoms $\bar{x}^2$ is proportional to $D't$, where D' is the diffusion coefficient and t the time. The following expression will be used in further considerations:

$$\bar{X}^2 \propto D't \tag{1.4}$$

Equation (1.4) has major consequences, but in order to simplify any further discussion it is assumed that $\bar{X}^2$ is proportional to the squared particle size. Conventional materials usually have grain sizes of around 10 μm, and it is well known that at elevated temperatures these materials require homogenization times of the order of many hours. When considering materials with grain sizes of around 10 nm (which is 1/1000 of the conventional grain size), then according to Equation (1.3) the time for homogenization is reduced by a factor of $(10^3)^2 = 10^6$. Hence, an homogenization time of hours is reduced to one of milliseconds; the homogenization occurs instantaneously. Indeed, this phenomenon is often referred to as "instantaneous alloying". It might also be said that "...each reaction that is thermally activated will happen nearly instantaneously", and therefore it is not possible to produce nonequilibrium systems (which are well known for conventional materials) at elevated temperature. Whilst this is an important point in the case of high-temperature, gas-phase synthetic processes, there are even more consequences with respect to synthesis at lower temperatures or the long-term stability of nonequilibrium systems at room temperature. The diffusion coefficient has a temperature dependency of $D' = D'_0\exp(-Q/RT)$, with the activation energy Q, the gas constant R, and the temperature T. However, on returning to the previous example, for a material with 10 μm grain size, we can assume a homogenization time of 1000 s at a temperature of 1000 K, and two different activation energies of 200 kJ mol^{-1} (which is typical for metals) and 300 kJ mol^{-1} (which is characteristic for oxide ceramics). The homogenization times for the 10-μm and 5-nm particles are compared in Table 1.2. In terms of temperature, 1000 K for gas-phase synthesis, 700 K for microwave plasma synthesis at reduced temperature, and 400 K as a storage temperature with respect to long-term stability, were selected. The results of these estimations are listed in Table 1.2.

The data provided in Table 1.2 indicate that, under the usual temperatures for gas-phase synthesis (1000 K and higher), there is no chance of obtaining any nonequilibrium structures. However, when considering microwave plasma processes,

Table 1.2 Relative homogenization time (s) for 5-nm nanoparticles at activation energies of 200 and 300 kJ mol^{-1} compared to 10-μm material at 1000 K.[a]

Particle size	Activation energy (kJ mol^{-1})	Temperature (K)		
		1000	700	400
10 μm	300	10^3	5.0×10^9	2.8×10^{26}
	200	10^3	2.9×10^7	4.3×10^{18}
5 nm	300	2.4×10^{-4}	1.3×10^3	7.0×10^{19}
	200	2.4×10^{-4}	7.3×10^0	1.1×10^{12}

[a]Assumed homogenization time = 1000 s.

where the temperatures rarely exceed 700 K, there is a good chance of obtaining nonequilibrium structures, or combinations of such materials. A temperature of 400 K represents storage and synthesis in liquids, and at this temperature the 5-nm particles are stable; from the point of thermal stability, it should be straightforward to synthesize nonequilibrium structures. However, according to Gleiter, diffusion coefficients up to 20 orders of magnitude larger than those for single crystals of conventional size were occasionally observed for nanomaterials [8]. Diffusion coefficients of such magnitude do not allow the synthesis and storage of nonequilibrium nanoparticles under any conditions. It should be noted that the above discussion is valid only in cases where transformation from the nonequilibrium to the stable state is not related to the release of free energy.

The possibility of near-instant diffusion through nanoparticles has been exploited technically, the most important example being the gas sensor. This is based on the principle that changes in electric conductivity are caused by changes in the stoichiometry of oxides, variations of which are often observed for transition metals. The general design of such a sensor is shown in Figure 1.15.

This type of gas sensor is set up on a conductive substrate on a carrier plate, and the surface of the conductive layer covered completely with the oxide sensor nanoparticles. Typically, for this application, nanoparticles of TiO_2, SnO_2, Fe_2O_3 are used. A further conductive cover layer is then applied on top of the oxide particle; it is important that this uppermost layer is permeable to gases. A change in the oxygen potential in the surrounding atmosphere causes a change in the stoichiometry of the oxide particles, which means that the oxygen/metal ratio is changed. It is important that this process is reversible, as the oxides are selected to show a large change in their electric conductivity as they change stoichiometry. The response of a sensor made from conventional material with grains in the micrometer size range, compared to a sensor using nanomaterials, is shown in Figure 1.16. Clearly, the response of the nanoparticle sensor is faster and the signal better but, according to Equation (1.4), one might expect

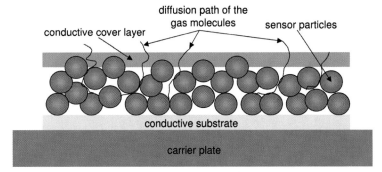

Figure 1.15 The general layout of a gas sensor based on nanoparticles. The sensor comprises a layer of sensing nanoparticles placed on a conductive substrate, and the whole system is covered with a gas-permeable electrode. Time controlling is via diffusion in the open pore network; the influence of bulk diffusion through the grains is negligible.

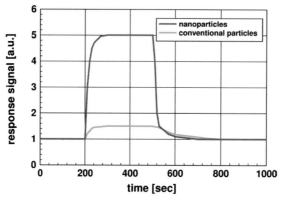

Figure 1.16 The comparative response over time of two gas sensors utilizing a conventional material with grain size either in the micrometer or nanometer range [9].

an even faster response. In a sensor using nanoparticles (see Figure 1.16) the time constant depends primarily on the diffusion of the gas molecules in the open-pore network and through the conducting cover layer.

The details of a gas sensor which was developed following the design principle shown in Figure 1.15 is illustrated in Figure 1.17 [10]. Here, the top electrode was a sputtered porous gold layer, and a titania thick film was used as the sensing material.

A further design for a gas sensor applying platinum bars as electrical contacts is shown in Figure 1.18. Although this design avoids the response-delaying conductive surface layer, the electrical path through the sensing particles is significantly longer. However, it would be relatively straightforward to implement this design in a chip. An experimental sensor using the design principles explained above is shown in Figure 1.19 [11]; this design uses SnO_2 as the sensing material, while the contacts and contact leads are made from platinum.

The response of this sensor is heavily dependent on the size of the SnO_2 particles used as the sensing material, there being a clear increase in the sensitivity of detection for carbon monoxide, CO, with decreasing grain size (see Figure 1.20). Such behavior may occur for either of two reasons: (i) that there is a reduced diffusion time,

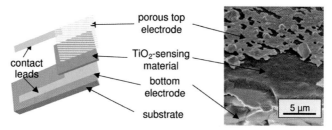

Figure 1.17 A gas sensor following the design principle as shown in Figure 1.15 [10]. The titania-sensing particles are placed on a gold electrode, and the top electrode is gas-permeable (this is clearly visible in the image at the right-hand side of the figure).

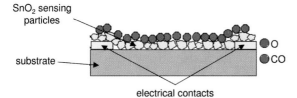

SnO₂ sensing
particles

substrate

electrical contacts

Figure 1.18 A sensor design applying platinum bars as electrical contacts. The sensing nanoparticles (e.g., SnO_2) are located between these contacts. The molecules to be detected (in this example oxygen and carbon monoxide) are shaded dark and light gray, respectively. (Note: The molecules and nanoparticles are not drawn to the same scale.)

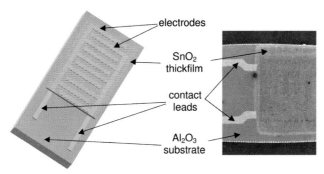

electrodes

SnO₂
thickfilm

contact
leads

Al₂O₃
substrate

Figure 1.19 A gas sensor in which a SnO_2 thick film made from nanoparticles is applied as the sensing element [11].

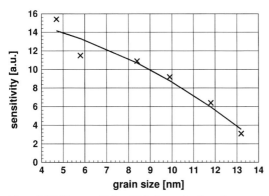

Figure 1.20 The sensitivity of CO determination of a gas sensor designed according to Figure 1.6. A significant increase in sensitivity is achieved with decreasing grain size [11].

Figure 1.21 The structure of a SnO_2 thick-film layer (note the open structure here) (Barunovic and Hahn, with kind permission [11]).

according to Equation (1.3); and (ii) that there is an enlarged surface, thereby accelerating exchange with the surrounding atmosphere.

For the successful operation of a thick-film sensor, it is a necessary prerequisite that the sensing layer be prepared from nanoparticles consisting of a highly porous structure that allows a relatively rapid diffusion of the gas to be sensed. A scanning electron microscopy image of the characteristic structure of such a SnO_2 thick-film layer is shown in Figure 1.21; the high porosity of the sensing thick-film layer, which is required to facilitate rapid diffusion of the gas species, is clearly visible.

Sensors based on this design are well suited for implementation in technical systems, and the structure of electrical contacts at the surface of a chip and integration into a technical system is shown in Figure 1.22 [12]. This design uses, for example, Pt/SnO_2 particles as the sensor for oxygen partial pressure, with the electrical conductivity of the sensor layer increasing with increasing CO concentration at the surface. Such a system consists of many sensing cells, as depicted in Figure 1.22a. This provides two possibilities: (i) by detecting the same signal in more than one cell, there is a possibility of improving the signal-to-noise ratio; (ii) the cells can be covered with a diffusion layer of varying composition and thickness; after calibration, this design allows an additional determination of the gas species.

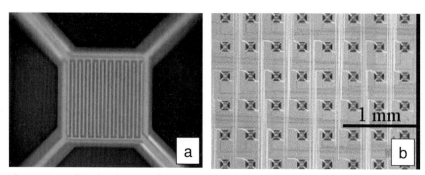

Figure 1.22 Technical realization of a gas sensor according to a design as depicted in Figure 1.18 [12]. (a) The sensing element on a chip. (b) An array of sensing elements; these arrays also allow identification of the gas species (Reprinted with permission from NIST Boulder Laboratories, Semoncik [12]).

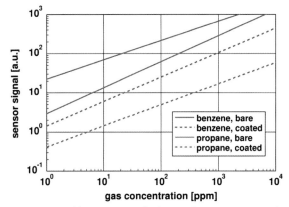

Figure 1.23 Calibration curves for bare and 10 nm SiO$_2$-coated gas sensors using SnO$_2$ to prepare the sensing nanoparticles. As the influence of the coating is dependent on the gas species, the nature, concentrations and/or relative proportions of the two species can be determined [13].

As mentioned above, it is possible to cover each sensing elements with a diffusion barrier of different thickness and composed of silica or alumina. Depending on the molecule's size, the time response for different elements depends on the thickness of the surface coating. After empirical calibration, such a design is capable of providing not only the oxygen potential but also information on the gas species. The integration of many sensor chips on one substrate (as shown in Figure 1.22b) opens the gate for further far-reaching possibilities, especially if the individual sensing elements are coated with a second material of varying thickness [13,15], or if the sensing elements are maintained at different temperatures [14,15]. A typical example of the influence of a coating at the surface of the sensor is shown in Figure 1.23, where the sensor signal is plotted against the concentration of the gas to be determined (in this case, benzene and propane). Because of the different sizes of these two molecules, the coating has an individual influence on the signal, and the subsequent use of some mathematics allows the gas species and its concentration to be determined. However, this approach is clearly valid only for those species where the calibration curves already exist.

References

1 Vollath, D. and Szabó, D.V. (2002) in: *Innovative Processing of Films and Nanocrystalline Powders*, (ed. K-.L. Choi), Imperial College Press, London, UK, pp. 219–251.

2 Nanophase. Nanophase Technologies Corporation, Romeoville, IL. www. nanophase.com 2007.

3 MACH I, Inc. , King of Prussia, PA. www.machichemicals.com 2007.

4 Vollath, D. and Szabó, D.V. (1999) *J. Nanoparticle Res.*, **1**, 235–242.

5 Vollath, D. and Sickafus, K.E. (1992) unpublished results.

6 Vollath, D., Szabó, D.V. and Fuchs, J. (1999) *Nanostructured Mater*, **12**, 433–438.

7 Vollath, D. and Szabó, D.V. (1994) *Nanostructured Mater*, **4**, 927–938.

8 Schumacher, S., Birringer, R., Strauß, R. and Gleiter, H. (1989) *Acta Metall.*, **37**, 2485–2488.

9 ww.boulder.nist.gov/div853/Publication% 20files/NIST_BCC_Nano_Hooker_2002. pdf.

10 Cho, Y.S. and Hahn, H. (2003) Technical University Darmstadt, Germany private communication.

11 Barunovic, R. and Hahn, H. (2003) Technical University Darmstadt, Germany private communication.

12 Semoncik, S. (2007) NIST private communication.

13 Althainz, P., Dahlke, A., Frietsch-Klarhof, M., Goschnick, J. and Ache, H.J. (1995) *Sensors and Actuators B.*, **24–25**, 366–369.

14 Althainz, P., Goschnick, J., Ehrmann, S. and Ache, H.J. (1996) *Sensors and Actuators B.*, **33**, 72–76.

15 Semoncik, S., Cavicchi, R.E., Wheeler, C., Tiffong, J.E., Walton, R.M., Svehle, J.S., Panchapakesau, B. and De Voe, D.L. (2007) *Sensors and Actuators B.*, **77**, 579–591.

2
Surfaces in Nanomaterials

2.1
General Considerations

In nanomaterials, the surface forms a sharp interface between a particle and its surrounding atmosphere, or between a precipitated phase and the parent phase. These are free surfaces in the case of particulate materials, or grain boundaries in bulk material. Nanomaterials have large surfaces, a fact which can be demonstrated by using spherical particles as examples. As mentioned above in the Introduction, nanoparticles demonstrate a large ratio of surface area (a) to volume (v). The surface area-to-volume ratio, $R = \frac{a}{v} = \frac{6}{D}$, is inversely proportional to the particle diameter D, and this assumes a mathematical surface. Realistically, however, it must be assumed that the surface has a certain thickness, and that the surface is partly influencing the volume. Based on many physical properties, it is known that the region of a particle, which is influenced by the surface, has a thickness (δ) between 0.5 and 1.5 nm. Therefore, a modified, dimensionless ratio R^* must be defined as:

$$R^* = \frac{D^3 - (D - 2\delta)^3}{D^3} = 1 - \left(\frac{(D - 2\delta)}{D}\right)^3 \tag{2.1}$$

This ratio, for an assumed surface thickness of 0.5 and 1.0 nm, is shown graphically in Figure 2.1. On examining this figure, it is clear that in the case of a 5-nm particle, 49% or 78%, respectively, of the volume belongs to the surface or, more precisely, to the surface-influenced volume.

As surface is related to energy, the amount of surface energy per particle, $u_{surface}$, is equal to γa, where γ is the specific surface energy and a the surface area of one particle. In this context, the geometric surface area of the particle is calculated from $a = \pi D^2$. In this book, the physical values related to one particle are denoted by lower-case letters, while those related to 1 mole are denoted by upper-case letters. For thermodynamic considerations, the surface energy per mol of material is the essential quantity. Hence, if N is the number of particles per mol, one obtains $N \gamma a = \frac{M}{\rho v} \gamma a = \gamma A$ (where ρ is the density of the material, M the molar weight, D the

Nanomaterials: An Introduction to Synthesis, Properties and Application. Dieter Vollath
Copyright © 2008 WILEY-VCH Verlag GmbH & Co. KGaA, Weinheim
ISBN: 978-3-527-31531-4

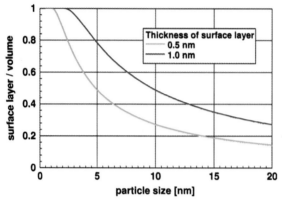

Figure 2.1 The relationship between particle volume and the surface layer. The thickness of the surface layer was assumed to be 0.5 or 1.0 nm.

particle diameter, and A the surface area of 1 mol). Finally, one obtains for the surface energy per mol:

$$U_{\text{surface}} = \frac{6M\gamma}{\rho} \frac{1}{D} \tag{2.2}$$

Equation (2.2) states that the surface energy per mol increases with $1/D$, and in some cases – especially those related to very small particles – this may have dramatic consequences.

The same considerations are valid for polycrystalline materials, where the volume related to the grain boundaries increases as the grain size decreases. In contrast to the well-ordered crystalline areas, the atoms or ions in the grain boundaries are, in a first approximation, arranged randomly. The famous picture of Gleiter [1] representing the arrangement of grains and grain boundaries, is shown in Figure 2.2.

Figure 2.2 Grain boundaries in polycrystalline material with grains in the nanometer range. A large portion of the material is associated with the surface (Reprinted with permission from Elsevier [1]).

2.2
Surface Energy

The origin of surface energy is explained by a model which assumes that particles are produced by breaking a large solid piece of material into smaller parts. In order to achieve this, it is necessary to cut the bonds between the neighboring atoms. (In this simplified explanation, the term "atom" is used equally to describe atoms, ions, and molecules.) Between each two atoms in the lattice the energy of bonding, u, is active (see Figure 2.3).

In order to separate one bond, energy u (symbolized as arrows in Figure 2.3) is required; therefore, to break a large piece of material into smaller pieces, energy nu is required, where n is the number of broken bonds at the surface. After breaking, two new surfaces emerge; consequently, for each broken bond of the new surface, energy $u/2$ is required. It follows, therefore, that the total energy required to remove one particle from a larger piece of material is $n_s\, u/2$, where n_s is the number of atoms at the surface of the particle. The number of broken bonds per unit area N is used to estimate the contribution γ_0 of the broken bonds to the surface energy.

$$\gamma_0 = Nu/2. \tag{2.3a}$$

Within the interior of a particle, an atom, or ion is held in a mechanical equilibrium by binding forces, which fix the atoms in their lattice positions. These forces are indicated by arrows in Figure 2.4, from which it is clear that those atoms at the surface have lost their bonds to the outside.

Due to the reduced number of neighbors, at each surface of atom, a force f acts perpendicular to the surface. At a plane surface (to be mathematically exact, the surface of a plane infinite half space), this does not cause any hydrostatic pressure in the material, but rather leads to stress in the surface plane; surface stress $\bar{\sigma} = f/a$, where a is the area occupied by one atom of the surface. Consequently, a surface stress which deforms the surface will result in surface stretching, and this allows the surfaces of particles to be modeled as an elastic material skin. According to Gurtin *et al.* [2,3] and Fischer *et al.* [4], this provides an additional contribution to the surface-free energy γ as a function of the surface stretching ε_s (much like the

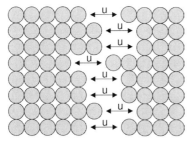

Figure 2.3 The creation of new surfaces (e.g., by breaking a larger portion into smaller pieces) requires energy u for each bond to be broken.

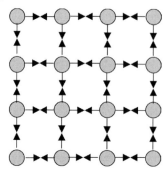

Figure 2.4 Forces acting between atoms or ions at lattice positions. Note that atoms at the surface are attracted into the interior of the particle, as they have a reduced number of neighbors. This does not lead to a pressure comparable with a hydrostatic pressure; rather, it leads to stress in the surface (the surface stress).

stretching of a rubber skin) and the surface stress $\bar{\sigma}$. Consequently, the surface energy is described by the relationship:

$$\gamma = \gamma_0 + \gamma_s(\varepsilon_s) \tag{2.3b}$$

where γ_s is the contribution of the surface stress to the surface energy. The surface stress $\bar{\sigma}$ and ε_s, the corresponding stretch, are assumed to be constant in any direction of the particle's tangent plane. It follows that

$$\bar{\sigma} = \gamma + \frac{\partial \gamma_s}{\partial \varepsilon_s} \tag{2.3c}$$

In the case of liquids, the second term of Equation (2.3c) vanishes as $\gamma_s = 0$. This often raises confusion between γ and $\bar{\sigma}$, especially as both have the same dimension. In order to estimate thermal effects, as for example during the coagulation of two particles, the sum value γ from Equation (2.3b) must be used. For a spherical particle of limited size and with a radius of curvature R at the surface, the situation is different. Due to the curvature, and in connection with the surface stress, a hydrostatic pressure within the particle, and which is comparable to that stemming from a gas or a liquid at the outside, comes into action. To calculate the hydrostatic pressure caused by surface stress, $\bar{\sigma}$ must be applied, the pressure being given by $p = 4\bar{\sigma}/D$.

Even when the situation at the surface can be described by quite plausible physical and exact mathematical models, the experimental situation is poor. To date, no data have been reported for the surface energy discriminating between γ, γ_s, and $\bar{\sigma}$, and therefore it is necessary to use published values of the surface energy γ for all applications. Based on the considerations above, it is clear that the determination of surface energy by measuring interface stress is insufficient as these methods deliver only γ_s, whereas calorimetric measurements (e.g., connected to grain growth) result in a value for $\gamma = \gamma_0 + \gamma_s$. Lastly, only these values are useful for thermodynamic considerations.

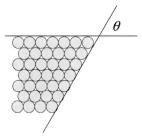

Figure 2.5 The angle θ between an arbitrary crystallographic plane and the reference plane must be taken into account when calculating the surface energy.

A more general situation is depicted in Figure 2.5, where the angle between two planes at the surface is assumed to differ from 90°. It may now also be considered how this configuration influences the surface energy.

Figure 2.5 illustrates an additional fact, namely that the energy related to the surface depends on the crystallographic orientation, while the number of broken bonds per surface unit depends on the orientation. In a cubic system, the surface energy related to different crystallographic planes can easily be calculated. If the angle between a reference plane and a second plane is termed θ (see Figure 2.5), then the surface energy of this second plane is given by:

$$\gamma_\theta = \frac{u}{2a}(\cos\theta + \sin|\theta|). \tag{2.4}$$

The dependency of the surface energy as a function of the angle θ is displayed in Figure 2.6.

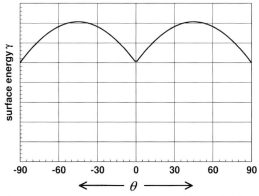

Figure 2.6 Surface energy as a function of the angle θ from a reference plane. As a function of the crystallographic orientation θ, the number of broken bonds per surface unit is different. In a cubic system, the anisotropic surface energy of the different crystallographic planes may be calculated using Equation (2.4).

In case of more anisotropic lattices, the relationships are more complicated as directional bonds are also present. In order to minimize the surface energy, these directed bonds raise crystallization in rods or platelets, while surface-active substances can also influence the surface energy. From a technical aspect, this is used in the production of one- or two-dimensional particles such as needles or plates.

In the case of oxides, it is advantageous to examine the surface is greater detail. Depending on the nature of the terminating ion, which, in most cases, is oxygen, (termination by hydrogen or a hydroxyl group is also possible) the surface energy of the different crystallographic planes is also changing. Excellent reviews of this subject have been produced by Barnard *et al.* [5,6]. As the termination changes the surface energy of dissimilar crystallographic planes in different ways, facetted particles appear with crystallographic planes, leading to a minimum surface energy. However, in experimental procedures, small particles are usually spherical (or close to being spherical) due to the vapor pressure, increasing with curvature $1/R$ (where R is the radius of the edge; see Section 2.3). Therefore, sharp edges or tips – which energetically are unfavorable – are removed by evaporation and condensation processes. However, particles of materials with an extremely low vapor pressure may be facetted, even when produced by high-temperature processes. An example of facetted particles, ceria (CeO_2) is shown in Figure 2.7.

An example of how surface energy has a major influence on the behavior of particles, in relation to particle synthesis, may be of benefit here, whereby the question might be asked as to what is the consequence of the coagulation of two particles. For reasons of simplicity, it is assumed that both coagulating particles are spherical and equal in size, and that the new particle is also assumed to be spherical. The difference in surface energy between the surface of these particles and that of the coagulated particle is:

$$\Delta u = \gamma \Delta a = \gamma(2\pi D^2 - \pi D^2_{\text{coagulated}}) = \gamma\pi D^2(2 - 2^{2/3}) \tag{2.5}$$

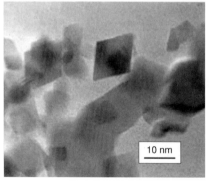

Figure 2.7 Facetted ceria (CeO_2) nanoparticles [7] (TEM micrograph reprinted with permission from Nanophase Technologies Corporation, Romeoville, IL, USA). Particles of materials with an extremely low vapor pressure may be facetted, even when produced by high-temperature processes.

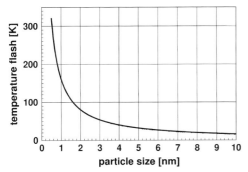

Figure 2.8 Temperature flash after adiabatic coagulation of two ZrO$_2$ nanoparticles of equal size.

The reduction in surface energy leads, due to dissipation, to an increase in temperature, ΔT:

$$\Delta T = \frac{\Delta u}{\rho c_p} \frac{6}{\pi D^3} = \frac{\gamma}{\rho c_p} \frac{0.413}{D} \tag{2.6}$$

If zirconia particles are assumed to have density $\rho = 5.6\,\mathrm{g\,cm^{-3}}$, surface energy $\gamma = 1\,\mathrm{J\,m^{-2}}$, and heat capacity $c_p = 56.2\,\mathrm{J\,mol^{-1}\,K^{-1}}$, this will cause an increase in temperature during the adiabatic coagulation process (see Figure 2.8). (For reasons of simplicity, the materials' data are those of conventional materials; the surface energy value is roughly approximated.)

Based on data in Figure 2.8, it can be seen that via the exchange of surface energy a remarkable temperature flash occurs during the coagulation of two equal-sized particles. It is this temperature flash which makes coagulation possible, as the rise in temperature causes the mobility of the atoms to be increased. The strong decrease in temperature flash with increasing particle size explains the occurrence of odd-shaped particles in the size range above 3 or 4 nm. This situation is not purely theoretical; rather, the coagulation of nanoparticles is a phenomenon which makes the production of small particles difficult, as they tend to agglomerate when they come into contact with each other. The process of agglomeration may also be observed in the electron microscope; a series of excellent electron micrographs showing coagulation between two gold particles are shown in Figure 2.9 [8]. The sequence starts with two particles, with one oriented such that lattice fringes are visible. The particles are moving, as indicated by the change in the lattice fringes. When the particles touch each other, they rotate until their orientation is equal, at which moment the coagulation begins as the larger particles engulf their smaller counterparts. For such a process to occur, significant thermal mobility of the atoms is essential, while the required energy is provided via a reduction of the surface.

This example of coagulation is related directly to grain growth during sintering where, in general, a dramatic grain growth is observed. Such growth leads to a reduction in surface energy and hence to a reduction in free energy. It is possible to estimate the energy released during the sintering of nanoparticles in connection with

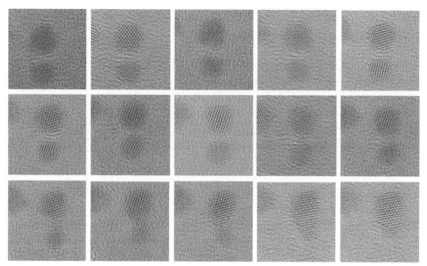

Figure 2.9 A series of electron micrographs depicting the coagulation of two gold particles. The orientation of the lattice fringes is changing from frame to frame, indicating movement of the particles. During the process of coagulation, a grain boundary is not formed; rather, the orientation of the two particles is aligned [8] (reprinted with kind permission from Jorge Antonio Ascencio Gutierrez).

grain growth. Assuming spherical particles with an initial grain size D and (after growth) a final grain size of D_{final}, then due to the reduction of the surface the energy Q per mol is released. Q is given by

$$Q = \gamma(nA - A_{final}) \frac{M}{\rho} \frac{1}{V_{final}} \quad \text{with} \quad n = \frac{v_{final}}{v} = \frac{D_{final}^3}{D^3}$$

leading to

$$Q = \frac{M}{\rho} \frac{6\gamma}{D_{final}} \left(\frac{D_{final}}{D} - 1 \right) \tag{2.7}$$

As long as $D_{final}/D \gg 1$, the surface energy released during grain growth is proportional to the inverse particle size. The energy released per mol, again using ZrO_2 as an example, is shown in Figure 2.10.

Here, the curves were calculated for final grain sizes of 50, 100, and 200 nm, but if the initial grain size is less than ca. 20 nm an amount of released surface energy, which is in the range of the free enthalpy for the tetragonal–monoclinic phase transformation, is realized. It is also of interest to note that, at least for relatively small initial grain sizes, the energy released is almost independent of the final grain size. This effect makes the calorimetric measurement of surface energy very insensitive when compared to the more or less broad distribution of final grain size. In addition, the energy released (which is of the order of a few kJ mol^{-1}) can easily be measured using conventional calorimetric methods. Consequently, it is advantageous to

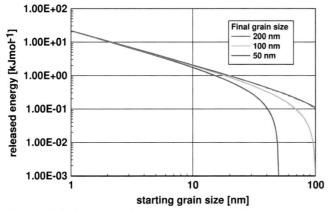

Figure 2.10 Surface energy release during grain growth. Provided that the starting grain size is sufficiently small, the energy released is almost independent of the final grain size.

determine surface energies by measuring the energy released during grain growth. A typical application for ZrO_2 was reported by Navrotsky [9,10].

In order to demonstrate the relative amount and importance of surface energy, the free enthalpy of formation, ΔG_{ZrO_2}, and the free enthalpy for the monoclinic–tetragonal $\Delta G_{monoc\text{-}tetr}$ transformation in comparison to the surface energy, is shown graphically in Figure 2.11, for the case of zirconia. Again, a value of $1\,J\,m^{-2}$ was assumed for the surface energy.

It is clear from the data shown in Figure 2.11 that for particles smaller than 2 nm, the surface energy is comparable to the energy of formation. The free enthalpy of the monoclinic–tetragonal transformation is significantly less than the surface energy,

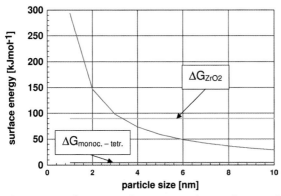

Figure 2.11 Surface energy of zirconia particles as a function of grain size. The free enthalpy of formation ($\Delta G_{Zr}O_2$) and the free enthalpy of the monoclinic–tetragonal phase transformation ($\Delta G_{monoc\text{-}tetr.}$) is plotted for comparison.

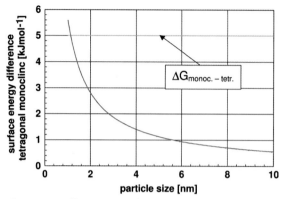

Figure 2.12 Differences in surface energy of the monoclinic and tetragonal phases as a function of particle size in comparison to the free enthalpy of the monoclinic–tetragonal phase transformation, $\Delta G_{monoc\text{-}tetr}$

although in the latter case only the surface change during the phase transformation should be considered (see Figure 2.12).

The small difference in surface area in relation to volume change (ca. 4%) during phase transformation leads to a change in the surface energy that is comparable with the free enthalpy of transformation. Thus, it is clear that the particle size has a significant influence on phase transformation. (This phenomenon is described in detail in Section 3.3). To date, the most important studies of the influence of particle size on phase transformations relate to the melting of metals and to monoclinic–tetragonal phase transformations in zirconia.

When considering isolated particles, it is important to take care of the hydrostatic pressure caused by surface stress in the particles. Such hydrostatic pressure, p, is a function of the curvature and surface stress and, in the simplest case of spherical particles,

$$p = 4\bar{\sigma}/D \tag{2.8}$$

is valid. The hydrostatic pressure caused by surface energy within a nanoparticle is depicted in Figure 2.13. As values for surface energy and surface stress are poorly known for ceramic materials, a value of $1\,N\,m^{-1}$ ($= 1\,J\,m^{-2}$) is often selected, although in general the difference from the unknown true value may be significant.

The hydrostatic pressure in a spherical particle with a diameter of 5 nm and a surface energy of $1\,N\,m^{-1}$ is (according to Figure 2.13) relatively high at $4 \times 10^8\,Pa$ ($= 4 \times 10^3\,bar$). Certainly, such a high hydrostatic pressure in nanoparticles has a major influence on any phase transformation connected to volume change. Phase transformations connected to volume change are pressure-sensitive, which means that the temperature of transformations depends on the external pressure. Thus, it is obvious that the particle size influences phase transformations. As explained in Section 3.3, the most significant influence is observed at the melting point of metal nanoparticles.

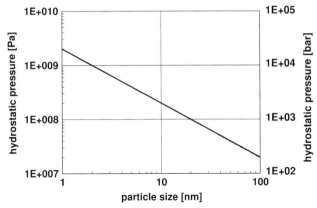

Figure 2.13 Hydrostatic pressure in nanoparticles as a function of particle size. The surface stress $\bar{\sigma}$ was assumed to be $1\,\mathrm{N\,m^{-1}}$.

This hydrostatic pressure p in a particle (as depicted in Figure 2.13) causes a hydrostatic stress σ and a strain ε constant in the particle. The strain energy per particle is $\frac{1}{2}\varepsilon\sigma V_{\text{particle}}$ or $\frac{1}{2}\varepsilon\sigma\frac{M}{\rho}$, the strain energy per mol. By setting $\varepsilon = \sigma/K$ and $\sigma = p = 4\bar{\sigma}/D$ (K is the bulk modulus or $K = \frac{E}{3(1-2v)} = 1/\kappa$, where E is *Young's* modulus, v the *Poisson* number, and κ compressibility), one obtains for the strain energy of small spherical particles:

$$Q_{\text{strain}} = \frac{1}{2K}\left(\frac{4\gamma}{D}\right)^2\frac{M}{\rho} = \frac{(3(1-2v))}{2E}\left(\frac{4\bar{\sigma}}{D}\right)^2\frac{M}{\rho} \tag{2.9}$$

In contrast to other formulae describing the influence of surface phenomena on thermodynamic quanta, the strain energy depends inversely on the square of the particle size, and therefore a significant influence for very small particles is expected. The strain energy for small particles of aluminum and zirconia as a function of the particle size is shown graphically in Figure 2.14. However, when comparing the data from Figure 2.14 to those for the surface energy depicted in Figure 2.11, it is realized that the contribution of the strain energy is small.

In order to calculate the data for Figure 2.14, a bulk modulus K of 200 GPa for zirconia and 76 GPa for aluminum, was assumed. In the case of aluminum, the strain energy is almost meaningless as it is significantly smaller than the heat of fusion for bulk materials. The situation is different for zirconia, however, where the strain energy for particle with a diameter below a few nanometers is more than 10% of the free enthalpy for the phase transformation monoclinic–tetragonal.

The hydrostatic pressure in the particles, caused by the surface stress, deforms the particle and, as might be expected, this phenomenon leads – in the case of metallic nanoparticles – to particle contraction. However, this contraction is so small that it can only be measured using high-precision X-ray lattice constant measurements. In the case of ceramic particles, this reduction is often superimposed by other phenomena, leading to a lattice expansion.

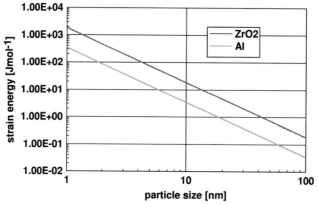

Figure 2.14 Strain energy of aluminum and zirconia nanoparticles calculated according to Equation (2.9).

Considering the experimental problems encountered when determining experimentally the lattice constant with high precision in the case of small particles, this phenomenon is well documented. As an example, the lattice contraction of gold [11] and palladium [12] is shown in Figure 2.15, and in both the cases a significant reduction in the lattice constant was observed. According to Qi and Wang [13], this lattice contraction can be described by

$$\frac{\Delta a}{a} = \frac{1}{1 + C\alpha^{0.5}} \tag{2.10}$$

where a is the lattice constant and α is the ratio between particle surface and the surface of a sphere with equal volume. Lastly, α is a function of the particle shape,

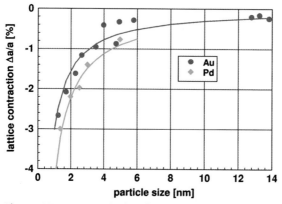

Figure 2.15 Experimental values for the lattice constant of gold [11] and palladium [12] nanoparticles. Due to hydrostatic pressure originating from surface tension, decreasing lattice parameters are observed with decreasing particle size.

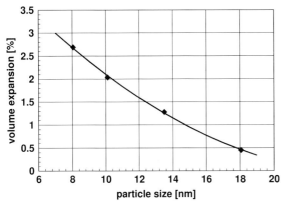

Figure 2.16 Volume expansion of γ-Fe$_2$O$_3$ nanoparticles [14] as a function of particle size. In oxides, in contrast to metallic particles, a volume increase is observed with decreasing particle size. This is a consequence of electrostatic repulsion due to the termination of metal cations by anions with electric charges of equal sign at the surface.

describing the deviation from a sphere. A detailed analysis shows that the palladium particles in Figure 2.15 are almost spherical, whereas the gold particles having an α-value of 3.09 are disk-shaped and have a diameter/thickness ratio of approximately 10.

In the case of ceramic oxide particles, the lattice behaves differently. The data in Figure 2.16 depict the dependency of the unit cell volume of γ-Fe$_2$O$_3$ as a function of the particle size [14]. It is remarkable to realize that, in contrast to metals, the lattice expands with decreasing particle size. This phenomenon is explained by a change in the lattice structure with decreasing particle size. The starting point for this explanation is the observation that in most cases, the cations at the surface of an oxide are terminated by oxygen ions, and therefore the surface is covered with oxygen ions bearing negative charge. As these negatively charged ions repel each other, the lattice is expanded [15].

2.3
Some Technical Consequences of Surface Energy

From the *Clausius–Clapeyron* law, it is possible to derive the vapor pressure of a particle as a function of the diameter D. This formula, which is known as the *Kelvin* formula (sometimes also called the *Thomson* formula), connects the vapor pressure with surface energy and particle size

$$\ln\left(\frac{p}{p_\infty}\right) = \frac{4\gamma V_m}{DRT} \quad \text{or} \quad p = p_\infty \exp\left(\frac{4\gamma V_m}{DRT}\right) \tag{2.11a}$$

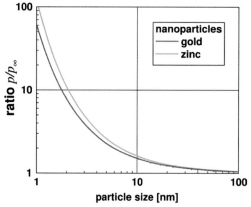

Figure 2.17 Vapor pressure ratio of a nanoparticle, p, in relation to that of a flat plane, p_∞. Note the drastic increase in ratio at small particle sizes.

where p_∞ is the vapor pressure over a flat plane, V_m is the molar volume, R the gas constant, and T the temperature. Assuming constant temperature, the vapor pressure over a curved surface shows the proportionality

$$p \propto \exp\left(\frac{1}{D}\right) \tag{2.11b}$$

Equations (2.11a) and (2.11b) state that the vapor pressure in equilibrium with a particle of diameter D increases drastically with decreasing particle diameter. This is demonstrated in Figure 2.17, using zinc and gold as examples.

The graph in Figure 2.17 displays the ratio of the vapor pressure for nanosized droplets at the melting point of the bulk material over the vapor pressure of a flat surface. It is interesting to realize that the difference between the metals with very different properties $(\gamma_{Au} = 1.13\,\mathrm{J\,m}^{-2}, \gamma_{Zn} = 0.77\,\mathrm{J\,m}^{-2})$ does not vary by much, although it is important to recognize the severe increase in vapor pressure over droplets below approximately 3 nm. The values for the surface energy are taken from the reports of Miedema and colleagues [16,17], and provide a consistent set of values for the surface energy of solid and liquid metals.

The strong increase in vapor pressure for small particles has important technical consequences, three of which are briefly explained in the following:

- When considering the formation of particles in a gas-phase reaction, it is clear that the nuclei must have a minimum size in order to avoid evaporation before they have the chance to grow by the condensation of further material. Therefore, it is clear that in nature, heterogeneous nucleation is preferred over homogeneous nucleation. For particle sizes close to zero, the vapor pressure is extremely large, and the low probability of homogeneous nucleation is well demonstrated. This explains also why in gas-phase reactions it is easier to produce small particles of materials with a low vapor pressure as compared to materials with a high vapor pressure, because there is a low probability for nucleation. Alternatively, this

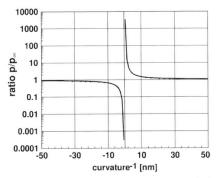

Figure 2.18 Vapor pressure ratio of a curved plane, *p*, over that of a flat plane, p_∞, for zinc. A positive curvature is related to convex surfaces, and a negative curvature to concave surfaces.

provides a good opportunity to produce small particles for materials with an extremely low vapor pressure (e.g., the refractory oxides such as ZrO_2, HfO_2, etc.).

- The next consequence is related to the particle shape. For nanoparticles consisting of a material with low vapor pressure, there is a greater opportunity to obtain facetted particles, whereas nanoparticles of materials with a higher vapor pressure would crystallize in a more spherical shape. This point was stressed above, in connection with Figure 2.7.

- The third example is related to sintering. At this point it is important to note that in a more general sense, the expression $2/D$ in Equation (2.11) may be replaced by $1/R$, where R is the radius and $1/R$ the curvature. The curvature may be either positive, convex surfaces or negative, concave surfaces. In Figure 2.18, the ratio of the vapor pressures is displayed as a function of the inverse curvature for zinc nanoparticles. It is remarkable that outside the range of nanoparticles this ratio is close to one, whereas for small particles or for narrow necks or wedges between two grains the function becomes extremely large, for small values. The difference in vapor pressure between ranges with positive and negative curvature leads to the formation of necks during sintering.

At the particle surface the curvature is positive, whereas at the point of contact of two touching particles the curvature is negative. For a negative curvature, the vapor pressure is decreasing and therefore bodies which consist of pressed nanoparticles begin to sinter very quickly, as the small particles have a high vapor pressure. The evaporated material is deposited at the points of contact of the particles, where the curvature is negative. In total, this process of material transport by enhanced evaporation and condensation leads to an early start of the sintering process at comparatively low temperatures; the situation is depicted schematically in Figure 2.19.

The situation in Figure 2.19 is represented in the micrograph shown in Figure 2.20, which is an example of such a sintering neck between two alumina particles. In the electron micrographs, it can be seen clearly that the material transported by evaporation and condensation processes to the contact point of the two particles.

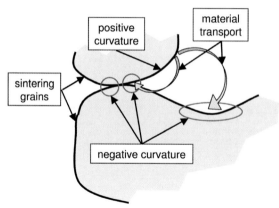

Figure 2.19 Two sintering grains. The diagram indicates ranges with positive and negative curvature, and material transport associated with the curvature-dependent vapor pressure.

Besides the first step of sintering, direct applications of surface energy are rare, although attempts have been made to exploit the energy exchange during coagulation for technical use. One prominent example is the proposal of Regan for a nanomotor based on coagulation processes [18,19]. The basic idea of a "motor" based on coagulation is relatively simple as it utilizes the fact that metallic atoms are migrating

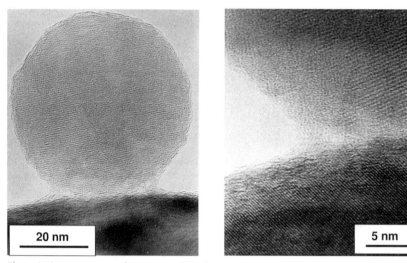

Figure 2.20 Two sintering alumina particles. The curvature-dependent vapor pressure causes material to evaporate at the positively curved surfaces of the particles, and to condense in the wedge (or neck) between the two particles, where the curvature is negative.

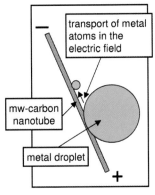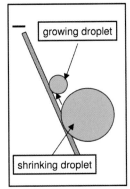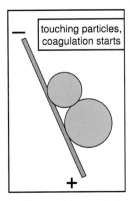

Figure 2.21 The basic concept of a nanomotor based on surface energy according to *Regan et al.* [18,19]. The motor consists of a multiwall carbon nanotube and two droplets of liquid metal. When an electric field is applied across the nanotube, metal atoms are transported from the larger droplet to the smaller, until they touch. At this moment the droplets coagulate and the process is re-started with material transport from the large drop to a point where a new drop can be formed.

in an electric field at the surface of carbon nanotubes [20]. This phenomenon, which is especially pronounced with indium, has led to the idea of a relaxation oscillator using the arrangement shown in Figure 2.21. On a carbon nanotube, connected to a direct current (DC) source, two droplets of indium are placed within a close distance of each other. Within the electric field, at the surface of the carbon nanotube, indium is transported by electromigration from one droplet to the next; hence, one droplet shrinks and the other grows, until the moment when the two droplets touch each other and coagulate. During the coagulation process, the material of both drops is concentrated into a larger drop, which increases its size again. After coagulation, continuing material transport leads to reformation of the smaller second droplet, which is nucleated at a discontinuity at the surface of the nanotube. Provided that the system is free of evaporation losses, this set-up will oscillate for as long as it is connected to a DC source.

This entire process can be visualized in an electron microscope, and a series of electron micrographs obtained is shown in Figure 2.22. In these micrographs, the carbon nanotube, the droplets and, most importantly, the growth of the smaller particle at the expense of the larger one, can be seen easily in images a to c. In Figure 2.22d, the arrangement after coagulation can be seen, just before the process is about to be repeated.

The relaxation time – that is, the time for complete coagulation – was estimated to be in the range of 200 ps. This proof of principle for the reversible influence of a mechanical dimension by electrical currents led to the idea of a technically applicable device where the process works with solid indium (see Figure 2.23). The device consists of two nanotubes connected with a small indium crystal, the extension bar, and additionally a metal reservoir. Again, the material is transported by electromigration. Indium crystallizes tetragonally (an anisotropic structure), and as

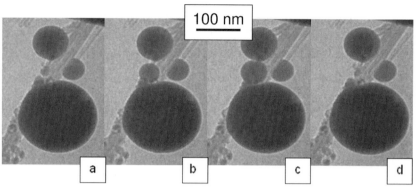

Figure 2.22 Electron micrographs showing the sequence of droplet growth (a–c) and after coagulation (d). The droplets consist of indium, the atoms of which are transported by electromigration at the surface of a carbon nanotube. Illustration reused with permission from Regan et al. [19]; Copyright 2005, American Institute of Physics.

the indium crystal connecting the nanotubes grows anisotropically and elongates, it causes the nanotubes to be moved. Changing the direction of the electric current leads to transport in the opposite direction, which means that the system is fully controllable.

A group of electron micrographs demonstrating such a system is shown in Figure 2.24. The two carbon nanotubes and the metal reservoir on the upper nanotube are visible in Figures 2.24 b and c, while the growing crystal can be seen between the nanotubes in Figure 2.24d. Interestingly, the material from the in-reservoir is not used uniformly from the surface, but it is in fact removed from the top of the reservoir. Again, this is a phenomenon of anisotropic surface energy. In the figure, an extension of the indium crystal of more than 100 nm is apparent, and this may act as ram.

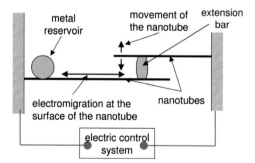

Figure 2.23 The basic concept of a nanomotor based on electromigration at the surface of a nanotube, and the anisotropy of surface energy of a noncubic metal, according to Regan et al. [20]. Metal atoms may be moved from and to the metal reservoir. The metal extension bar changes in length because, due to anisotropy of the surface energy, metal atoms are added or removed only at the planes directly adjacent to the carbon nanotubes.

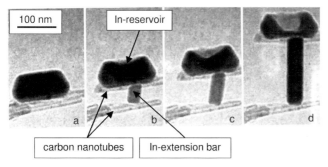

Figure 2.24 Electron micrographs showing the extension of an indium extension bar by adding additional material at the end surfaces. The material is transported by electromigration from the indium reservoir to the extension bar or, in case of opposite electric polarity, in the other direction. Note that the material from the reservoir is also removed in an anisotropic manner. Illustration taken with permission of the American Chemical Society from Regan *et al.* [18].

The extension of the indium crystal and the voltages that lead to the material transport are shown graphically, as a function of time, in Figure 2.25. In this case, the "nanomotor" is seen to cycle with the voltage, with the speed of approximately $1\,\text{nm}\,\text{s}^{-1}$ for extension and shrinking, respectively, being linear with time. Over a limited range the speed is proportional to the voltage, although at a lower voltage the thermally activated transport (and therefore the speed) are reduced significantly, as the electric losses (*Joule* heating) are reduced. The maximum speed is also limited by *Joule* heating, as the temperature of the system must not exceed the melting point of the metal.

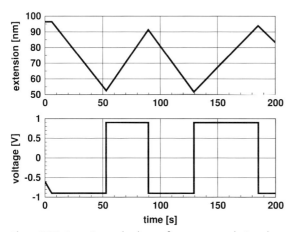

Figure 2.25 Extension and voltage of a nanomotor designed according to Figure 2.23 [20]. Within a limited range, the extension rate is controlled by the applied voltage.

References

1 Gleiter, H. (1992) *Nanostructured Mater,* **1,** 1–19.

2 Gurtin, M.E., Murdoch, A.I. and Continuum, A. (1975) *Arch. Rat. Mech. Anal.,* **57,** 291–323.

3 Fried, E. and Gurtin, M.E. (2004) *Advances in Applied Mechanics,* (eds H. Aref and E.V.D. Giessen), **40,** 1–177.

4 Fischer, F.D., Waitz, T., Vollath, D. and Simha, N.K. (2008) *Progr. Mater. Sci.,* **53,** 481–527.

5 Barnard, A.S. and Zapol, P. (2004) *J. Chem. Phys.,* **121,** 4276–4283.

6 Barnard, A.S. and Curtiss, L.A. (2005) *Nanoletters,* **5,** 1261–1266.

7 Nanophase Technologies Corporation, USA www.nanophase.com (2007).

8 Ascencio, J. (2007) Private communication.

9 Navrotsky, A. (2003) *Geochem. Trans.,* **4,** 34–37.

10 McHale, J.M., Auroux, A. and Navrotsky, A. (1997) *Science,* **277,** 788–791.

11 Mays, C.W., Vermaak, J.S. and Kuhlmann-Wilsdorf, D. (1968) *Surface Sci.,* **12,** 134–137.

12 Lamber, R., Wetjen, S. and Jaeger, I. (1995) *Phys. Rev. B.,* **51,** 10968–10971.

13 Qi, W.H. and Wang, M.P. (2005) *J. Nanoparticle Res.,* **7,** 51.

14 Ayyub, P., Multani, M., Barma, M., Palkar, V.R. and Vijayaraghavan, R. (1988) *J. Phys. C.,* **21,** 2229–2245.

15 Ayyub, P., Palkar, V.R., Chattopadhyay, S. and Multani, M. (1995) *Phys. Rev.,* **51,** 6135–6138.

16 Miedema, A.R. and Boom, R. (1978) *Z. Metallkde,* **69,** 183–190.

17 Miedema, A.R. (1978) *Z. Metallkde,* **69,** 287–292.

18 Regan, B.C., Aloni, S., Jensen, K., Ritchie, R.O. and Zettl, A. (2005) *Nanoletters,* **5,** 1730–1733.

19 Regan, B.C., Aloni, S., Jensen, K. and Zettl, A. (2005) *Appl. Phys. Lett.,* **86,** 1–3.

20 Regan, B.C., Aloni, S., Ritchie, R.O., Dahmen, U. and Zettl, A. (2004) *Nature,* **428,** 924–927.

3
Phase Transformations of Nanoparticles

3.1
Thermodynamics of Nanoparticles

Although thermodynamics may be treated on a variety of levels of complexity and precision, we present here an elementary introduction and therefore, in all cases, the simplest possible description is used, neglecting any influential factors required for an exact description of equilibria. The influence of the vapor phase is not considered in any of the cases. However, because of the large surface of nanoparticulate materials, energy stored as surface energy must always be taken into account when considering the thermodynamics of systems. It will be shown that, in many cases, the amount of energy stored at the surface is in the same range as the energy of phase transformations in the bulk. Accordingly, surface energy controls the stability of multiphase systems, and therefore the *Gibb*'s free enthalpy must be written as

$$G = U - TS + \gamma A \tag{3.1}$$

In this equation, G, U, S, and T have their usual meanings of free enthalpy, enthalpy, S entropy, and temperature, respectively, with each parameter always being related to one mole. Here, γ is the surface energy and A the surface area per mol of the system. It was shown in Section 2.2 that, in the case of small particles, the energy connected to the surface is in the range of the energy of formation for oxides. Therefore, a strong influence of particle size to phase transformations is expected. This is shown graphically in Figure 3.1 for the melting of aluminum, where the surface energy in the solid state $G_{surface-solid} = A_{solid}\gamma_{solid}$ and liquid state $G_{surface-liquid} = A_{liquid}\gamma_{liquid}$ and their differences are plotted. For comparison, the enthalpy of melting is also shown in the graph. It should be noted that the difference in surface energy between the solid and liquid states is in the range of the enthalpy of melting; therefore, it is clear that particle size (in this case aluminum particles) might have a major influence on melting. In more general terms, particle size will have a significant influence on phase transformations.

Nanomaterials: An Introduction to Synthesis, Properties and Application. Dieter Vollath
Copyright © 2008 WILEY-VCH Verlag GmbH & Co. KGaA, Weinheim
ISBN: 978-3-527-31531-4

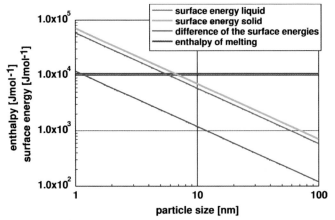

Figure 3.1 Surface energy of solid and liquid aluminum as a function of particle size, and enthalpy of melting. The difference in surface energy in the solid and liquid states is of the same order of magnitude as enthalpy of melting; thus, a significant influence of particle size on melting is expected.

3.2
Heat Capacity of Nanoparticles

The heat capacity c_V at constant volume is defined as $c_V - \left(\frac{\partial U}{\partial T}\right)_V = T\left(\frac{\partial S}{\partial T}\right)_V$. Although, heat capacity is one of the properties where a significant influence of particle size is expected and, in theory, is well understood, the experimental data obtained lead to different conclusions. The first approach is to describe a simplified model of heat capacity; this is created using a linear crystal, as this simplified model shows all the necessary features. A chain of atoms – a "linear crystal" – characterized by the number of atoms N, the distance of two points in the chain, the lattice constant a, is shown in Figure 3.2.

The vibrations of such a chain have nodes at the ends. Additionally, as the vibrations are quantized, vibration nodes are possible only at the position of an atom. Two parameters give the size of the crystal as $L = Na$. At temperatures above 0 K, the atoms begin to vibrate, although as they are connected within a crystal only a limited, well-defined, number of vibrations is possible. The finite number of lattice points defines the limited number of lattice vibrations. Clearly, the longest half-wave which fits into the lattice has the length L, and this leads to a wavelength of $\lambda_{max} = 2L = 2Na$. The shortest wavelength possible in such a lattice is $\lambda_{min} = 2a$.

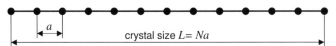

Figure 3.2 A linear "crystal" represented by a chain of N atoms with a distance, the lattice constant a, leading to a crystal size $L = Na$.

This is valid independently if one considers longitudinal or transversal vibrations, with the corresponding frequencies being $v_{max} = c/2a$ and $v_{min} = c/2Na$, where c is the speed of elastic waves in the material. Each one of these vibrations is connected to an energy hv, where h is *Planck's* constant. In order to derive the energy of a crystal, all energies of the lattice vibrations must be summed; hence, the thermal energy E due to lattice vibrations of a crystal is described by

$$E = \sum_i n_i v_i h \tag{3.2}$$

The number n_i of vibrations with frequency v_i is calculated using *Bose–Einstein* statistics. The possible frequencies v_i are a function of the particle size, and the following wavelengths are possible:

$$\lambda = \frac{2Na}{1}, \frac{2Na}{2}, \frac{2Na}{3}, \dots \frac{2Na}{i}, \dots \frac{2Na}{N} \tag{3.3a}$$

This leads to the allowed frequencies

$$v = \frac{c}{2Na}, \frac{2c}{2Na}, \frac{3c}{2Na}, \dots \frac{ic}{2Na}, \dots \frac{Nc}{2Na} \tag{3.3b}$$

In Equation (3.2) the only temperature-dependent term is the number of vibrations n_i of the frequency v_i. From Equations (3.3a) and (3.3b) it is clear that, by reducing the particle size, the energy of the vibrations with the longest wavelength $\lambda_{max} = 2Na = 2L$ (the one with the lowest energy) is increasing. As these vibrations are excited primarily at low temperatures, a reduction of the heat capacity at low temperatures may be expected.

This simple model does not take into account the increased degrees of freedom for vibrations of the atoms at the surface. In fact, surface atoms may make a significant contribution to the heat capacity, provided that the particles are sufficiently small. A precise and detailed theory of heat capacity as a function of the particle size is provided by Malinovskaya and Sachkov [1]. Although this theory leads (as expected) to a decrease in heat capacity with decreasing particle size, in the case of extremely small particles – when the particle virtually now consists only of surface – an increased heat capacity is predicted. The results of detailed calculations for the heat capacity at 298 K are shown in Figure 3.3, for In_2O_3, where there is a remarkable and sudden increase in c_V at particle sizes below ca. 1.2 nm. Clearly, for these sizes the calculations showed increased degrees of freedom for almost all atoms. When used as a simple model, it is possible to compare the degrees of freedom for the vibration of atoms at the surface with those of a liquid.

The results of this plausible model are not reproduced directly by experimental data. As an example, the heat capacity of nanocrystalline and coarse-grained copper and palladium is depicted in Figures 3.4a and b [2]. For both the metals, a larger heat capacity is found for the nanocrystalline material as compared to the coarse-grained counterpart. The material used for the measurements depicted in Figures 3.4a and b was sintered, and as polycristalline material is known to have a large volume fraction of grain boundaries such an increased heat capacity would not be too surprising.

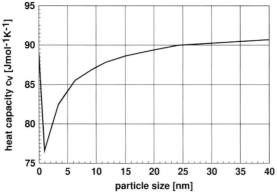

Figure 3.3 Heat capacity of In_2O_3 as a function of the particle size according to detailed theoretical treatment of Malinovskaya and Sachkov [1]. Note the dramatic increase in heat capacity for particle sizes <1.2 nm.

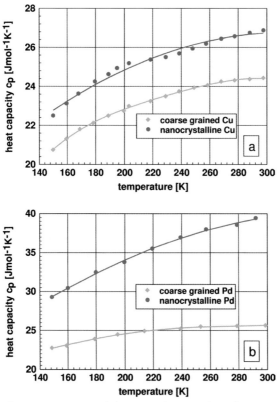

Figure 3.4 Comparative heat capacity for sintered metallic nano-crystalline materials and coarse-grained material [2]. (a) Copper materials; (b) palladium materials. In both cases, the heat capacity for nanocrystalline is greater than for coarse-grained material.

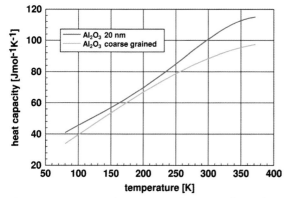

Figure 3.5 Comparative heat capacity of nanocrystalline and coarse-grained alumina. As for metals (see Figure 3.4), the heat capacity of the nanocrystalline material is greater [4].

Additionally, nanomaterials have an increased tendency to dissolve light-element impurities which, with their larger degrees of freedom for vibration, contribute significantly to the heat capacity; this was demonstrated by Tschöpe and Birringer for nanocrystalline platinum [3]. Although, in this chapter, we generally refer to heat capacity at a constant pressure, c_p, in a solid material the difference in heat capacity at constant pressure and constant volume, c_V, is negligible.

An increased heat capacity of nanocrystalline materials is not only found in metals, comparable phenomena having also been observed in ceramic materials. As an example, the heat capacity of sintered alumina with a grain size of 20 nm, compared to a coarse-grained material, is shown in Figure 3.5 [4]. In both the cases, the material consisted of α-phase material, while a small content (ca. 1%) of γ-phase in the nanocrystalline sample was assumed to have no influence. In Figure 3.5, an increased heat capacity can be seen at low temperature and, even more strikingly, at temperatures above 250 K. The authors explained this behavior by there being an increased freedom for vibration of the ions at the grain boundaries, and stressed the fact that the material had a reduced density of 89%, most likely due to the grain boundaries. Although the width of grain boundaries is normally assumed to be less than 1 nm, even for a width of 2 nm a reduced grain density must be assumed. This explains the increased number of vibration modes and, consequently, the higher heat capacity.

A reduced density – or more generally, a reduced degree of order – appears to be a general phenomenon that is associated with very small nanoparticles. Its connection with phase transformations of nanoparticles is discussed in the following section.

3.3
Phase Transformations of Nanoparticles

In general, phase transformations are connected with changes in physical properties, and in most cases it is the density of the material that is changing. In relation to particles, a changing density means a change in the surface; hence, if the energy connected to the surface of nanoparticles is large then these changes may have a significant influence on phase transformations. At the temperature where a phase transformation occurs, T_{trans}, the following equilibrium condition is valid:

$$U_{\text{old}} - T_{\text{trans}} S_{\text{old}} + \gamma_{\text{old}} A_{\text{old}} = U_{\text{new}} - T_{\text{trans}} S_{\text{new}} + \gamma_{\text{new}} A_{\text{new}} \tag{3.4}$$

In Equation (3.4), U is the enthalpy, S the entropy, γ the surface energy, and A the surface of the *old* respectively *new* phase (always related to one mole). Now, the question arises as to whether the transformation temperature is a function of the particle size, or not. This somewhat aged question was first posed in connection with the crystallization of organic phases at the end of the 19th century. When related to nanoparticles, this problem was found to be most important and generalized to all types of phase transformations. By using the differences $\Delta U_{\text{trans}} = H_{\text{new}} - H_{\text{old}}$ and $\Delta S_{\text{trans}} = S_{\text{new}} - S_{\text{old}}$, Equation (3.5) reduces to

$$\Delta G_{\text{trans-nano}} = \Delta U_{\text{trans}} - T_{\text{trans}} \Delta S_{\text{trans}} + \gamma_{\text{new}} A_{\text{new}} - \gamma_{\text{old}} A_{\text{old}} = 0 \tag{3.5}$$

This change of surface is related to particle size; hence, when assuming spherical particles one obtains for the surface per mol $A = 6M/\rho D$, where M is the molar weight, ρ the density, and D the particle diameter. To do this, the surface per mol must be calculated as a function of the particle size.

By using $\frac{D_{\text{new}}}{D_{\text{old}}} = \left(\frac{\rho_{\text{old}}}{\rho_{\text{new}}}\right)^{1/3}$, one obtains

$$\Delta G_{\text{trans-nano}} = \Delta U_{\text{trans}} - T_{\text{trans}} \Delta S_{\text{trans}} + \gamma_{\text{new}} \frac{6M}{\rho_{\text{new}} D_{\text{new}}}$$
$$- \gamma_{\text{old}} \frac{6M}{\rho_{\text{new}} D_{\text{new}}} \left(\frac{\rho_{\text{new}}}{\rho_{\text{old}}}\right)^{2/3} = 0 \tag{3.6}$$

From the equilibrium condition, one obtains for the temperature of transformation

$$T_{\text{trans}} = \frac{\Delta U_{\text{trans}}}{\Delta S_{\text{trans}}} - \frac{6M\gamma_{\text{new}}}{\rho_{\text{new}} D_{\text{new}} \Delta S_{\text{trans}}} \left[1 - \left(\frac{\gamma_{\text{old}}}{\gamma_{\text{new}}}\right)\left(\frac{\rho_{\text{new}}}{\rho_{\text{old}}}\right)^{2/3}\right] \tag{3.7}$$

Therefore, using the abbreviation $T_{\text{coarse}} = \frac{\Delta U_{\text{trans}}}{\Delta S_{\text{trans}}}$ for the transformation temperature of the coarse material, one finally obtains

$$\Delta T = T_{\text{coarse}} - T_{\text{trans}} = \frac{6MT_{\text{coarse}}\gamma_{\text{new}}}{D_{\text{new}} \Delta U_{\text{trans}} \rho_{\text{new}}} \left[1 - \left(\frac{\gamma_{\text{old}}}{\gamma_{\text{new}}}\right)\left(\frac{\rho_{\text{new}}}{\rho_{\text{old}}}\right)^{2/3}\right] \tag{3.8}$$

Finally, Equation (3.8) represents an inverse linear relationship between the reduction of the phase transformation temperature and the particle size. Assuming that the difference in particle size before and after transformation is small, this results in the well-known and important relationship

$$\Delta T_{\text{trans}} = \alpha \frac{\gamma T_{\text{coarse}}}{\Delta U_{\text{trans}} D} \tag{3.9}$$

This equation simply says that, in a first approximation, the temperature of phase transformation changes inversely to the particle size (this is also known as *Thomson's law* [5]). As in the case of melting processes,

$$\alpha = 1 - \left(\frac{\gamma_{\text{old}}}{\gamma_{\text{new}}}\right)\left(\frac{\rho_{\text{new}}}{\rho_{\text{old}}}\right)^{2/3} = 1 - \beta \tag{3.10}$$

is usually positive, there is a rule that melting temperatures will decrease with decreasing particle size, and this has severe consequences for phase diagrams of nanoparticulate materials. When considering only the materials' properties, it is clear that β now rules the change of temperature for phase transformation. When considering the inverse transformation, for β the inverse value must be used, and therefore α changes its sign. However, as ΔU_{trans} also changes sign, the sign of ΔT remains unchanged. Here, β is used to compare the behavior of different materials during phase transformation, and consequently Equation (3.8) is rewritten as

$$\Delta T_{\text{trans}} = \frac{T_{\text{coarse}}}{\Delta U_{\text{trans}}} \frac{6 M \gamma_{\text{old}}}{\rho_{\text{old}} D_{\text{old}}} (1 - \beta) \tag{3.11}$$

The considerations above do not take into account the thermal expansion and temperature dependence of the surface energy; therefore, strictly speaking, they are valid only under isothermal conditions. Although thermal expansion brings about a minor correction, the general laws are not changed. Castro *et al.* [6] extended this approach to the melting of nanoparticles by considering thermal expansion and temperature-dependent surface tension. The graph shown in Figure 3.6 provides a

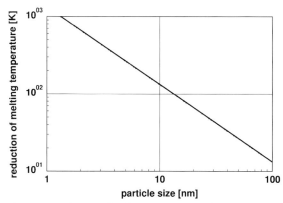

Figure 3.6 Reduction of melting temperature of aluminum as a function of particle size. Note that the surface term causes a significant reduction in the melting temperature.

general view of the change in melting temperature of aluminum as a function of particle size in a double logarithmic scale.

The data in Figure 3.6 demonstrate the possibility that a material which may crystallize well in coarse grain sizes in the nanometer range may not crystallize as nanoparticles, as the depression in melting point due to surface energy may be greater than the melting temperature. This phenomenon is often observed in the case of ceramic nanoparticles such as Al_2O_3 or Fe_2O_3. However, in order to estimate this in great detail, a significantly more precise theory must be applied. Additionally, it must also be borne in mind that the above description is insofar incomplete, as the elastic response of the two phases in consideration is not taken into account.

Within the context of phase transformations of small particles, the majority of extensive studies have been conducted with respect to the melting of metal particles. Previously, a number of extensive, theoretical studies of the melting of nanoparticles have been conducted by Pawlow [8], Reiss and Wilson [7], Hanszen [9], and Zhao *et al.* [10], with each presenting a unique theory tailored to interpret the experimental results obtained.

A first example – the decrease of the melting point of aluminum, a low-melting metal – is demonstrated in Figure 3.7, where experimental data on the melting point of aluminum as a function of particle size are displayed [11]. In Figure 3.7a the melting point is plotted against particle size, whereas in Figure 3.7b [in association with Equation (3.7)] the inverse particle size is selected as abscissa. Within the precision of the measured values, it is clear that the melting point of aluminum nanoparticles follow, at least in the size range from 10 to 40 nm, exactly the simplified theory. For comparison, the melting temperature of coarse-grained material is indicated in both graphs.

The simplified theory for the change in transformation temperature as a function of particle size explained above can be extended. Rearranging Equations (3.5) and (3.6) allows an estimation to be made of the enthalpy of transformation as a function of particle size.

$$\Delta U_{trans-nano} = \Delta U_{trans} + \gamma_{new}A_{new} - \gamma_{old}A_{old}$$
$$= \Delta U_{trans} - \frac{6M\gamma_{new}}{\rho_{new}D_{new}}\left[1 - \left(\frac{\gamma_{old}}{\gamma_{new}}\right)\left(\frac{\rho_{new}}{\rho_{old}}\right)^{2/3}\right] \qquad (3.12)$$

This formula can also be simplified as the experimentally well-proven relationship

$$\Delta H_{trans-nano} = \Delta H_{trans} - \kappa\frac{1}{D} \qquad (3.13)$$

which shows a decrease in enthalpy for transformation with decreasing particle size. In addition to determining the melting point of aluminum nanoparticles with calorimetric methods, Eckert *et al.* [11] also measured the enthalpy of melting as a function of particle size (see Figure 3.8). In this figure, the enthalpy of melting is plotted against the inverse particle size, thus confirming the linear relationship between melting enthalpy and inverse particle size as predicted by Equations (3.12) and (3.13).

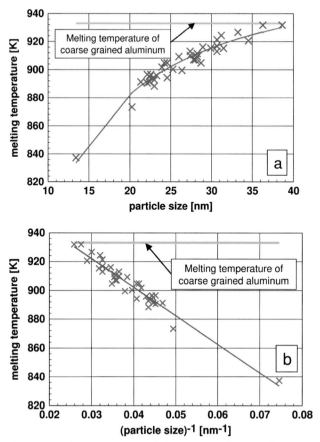

Figure 3.7 Melting temperature of aluminum as a function of grain size, according to Eckert *et al*. [11]. The melting temperature of the bulk material is indicated by the bold line. (a) Aluminum melting points plotted versus particle size; (b) aluminum melting points plotted versus inverse particle size. Note the inverse proportionality as described in Equation (3.7).

The "perfect fits" as shown in Figures 3.7b and 3.8 are rather rare cases, and most experimental data show more or less severe deviations. For example, Figures 3.9a and b show the melting point of lead nanoparticles over a size range of ca. 3 to 50 nm [12].

The data in Figure 3.9b show that the above-mentioned linear relationship between melting point and inverse particle size is valid only for the range of very small particles. There are many possible reasons for this deviation, including interaction with the surrounding gaseous atmosphere and the kinetic processes of melting. Coombes [12] suggested that melting started in a surface layer which was estimated to have a thickness of ca. 3 nm; therefore, it was not surprising that the linear

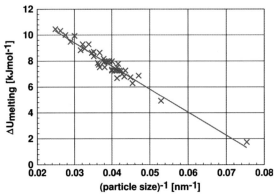

Figure 3.8 Enthalpy of melting of aluminum, according to Eckert et al. [11]. Note the inverse proportionality as described in Equation (3.13).

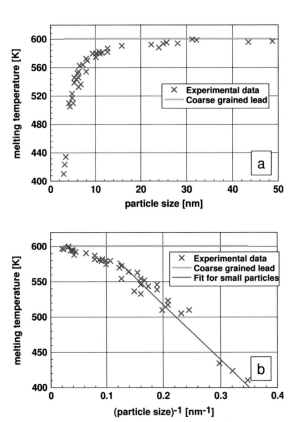

Figure 3.9 The melting point of lead as a function of particle size, according to Coombes [12]. (a) Melting point plotted versus particle size; (b) melting point plotted versus inverse particle size.

Table 3.1 Characteristic constants (β) [according to Equation (3.8)] responsible for changes in the liquid–solid transition temperature for metals, as derived from their materials data.

Metal	$\dfrac{\gamma_{liquid}}{\gamma_{solid}}$	$\left(\dfrac{\rho_{solid}}{\rho_{liquid}}\right)$	$\left(\dfrac{\rho_{solid}}{\rho_{liquid}}\right)^{2/3}$	$\dfrac{\gamma_{liquid}}{\gamma_{solid}}\left(\dfrac{\rho_{solid}}{\rho_{liquid}}\right)^{2/3}$
Copper	0.90	1.11	1.07	0.97
Gold	0.87	1.11	1.07	0.93
Silver	0.82	1.12	1.08	0.89

approximation fitted up to approximately $1/D = 0.145\ \text{nm}^{-1}$, corresponding a particle diameter of approximately 7 nm.

By replacing surface energy with the solid–liquid interface energy, the *Thomson* equation explains, in a simple manner, the supercooling of liquids without nuclei for crystallization, as in order to form the first crystal nuclei (homogenous nucleation) the temperature of the melt must be reduced to a level where the smallest nuclei are formed.

When considering the crystallization of metal nanoparticles, the phase transformation (where the largest pool of experimental data exists) for most examples produces β-values of less than 1 [see Equation (3.10)]. If β is less than 1, the freezing point decreases with decreasing particle size, and vice versa. Some typical values are listed in Table 3.1.

As might be expected, the β-values in Table 3.1 are generally less than 1, and therefore the melting point is seen to decrease with decreasing particle size. Bismuth might be an exception here, as it shows a volume expansion during crystallization (as does water); however, as the published data for bismuth are wide-ranging such an estimation would be meaningless. Because of even more unreliable data, similar estimations – as are displayed for some metals – are not shown for ceramic materials. This situation may be entirely different for small metal particles in another liquid metal, where an increase in the melting point with decreasing particle size is often expected and observed. The situation may be entirely different for the phase-transformation processes of nanoparticles in a solid matrix, as it must also be considered that the surrounding matrix, which confines the particles, would hinder any volume expansion.

A decrease in the melting temperature for nanoparticles is also assumed for ceramic particles. It is well known that some ceramic nanoparticles show a size limit for crystallization which, in the case of alumina (Al_2O_3) is 8 nm, and for iron oxide (Fe_2O_3) is 3 nm. In the case of zirconia (ZrO_2) this limit is well below 1 nm.

For ceramic materials, the monoclinic–tetragonal transformation of zirconia is often studied as a function of the particle size. At room temperature with particle sizes in the micrometer range, zirconia is monoclinic and transforms at 1475 K to the tetragonal phase, and later at 2650 K into the cubic phase. However, by adding metals with valencies of 2 or 3 it is possible to shift these transformations to lower temperatures. The most successful method for stabilizing the

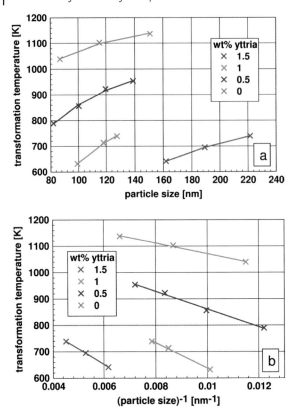

Figure 3.10 Temperature of monoclinic–tetragonal transformation of yttria-doped zirconia as a function of particle size. These data are taken from Suresh and Mayo [13] and Mayo *et al.* [14]. (a) Monoclinic–tetragonal transformation temperature of zirconia plotted versus particle size. The yttria content is used as a parameter for the curves. (b) Monoclinic–tetragonal transformation temperature of yttria-doped zirconia plotted versus inverse particle size. These data also verify the validity of Equation (3.8) for solid-state transformations.

tetragonal and cubic phases is the addition of Y_2O_3 and MgO. As zirconia dissolves yttria (Y_2O_3), and yttria stabilizes the tetragonal phase, it would be interesting to see how the particle size and yttria content might influence the monoclinic–tetragonal transformation. Such experimental findings are shown graphically in Figure 3.10 [14,15].

In Figure 3.10a, it is clear that the temperature of transformation increases as the particle size increases, and that an increasing yttria content has a similar influence to that of a reduced particle size – which is exactly the expected behavior. It is very surprising that, in Figure 3.10b, the transformation temperature plotted as a function of inverse particle size is strictly linear, especially as most of the particles are more than 100 nm in size. These results show that, despite a relatively large particle size,

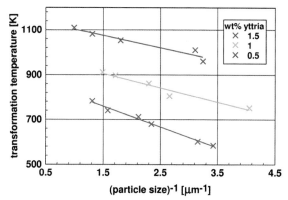

Figure 3.11 Monoclinic–tetragonal transformation of yttria-doped zirconia as a function of inverse grain size. Unlike Figure 3.10b, the material was sintered [14]. These data also verify the validity of Equation (3.8).

the simplified transformation law of Equation (3.8) is still valid, perhaps due to the high surface energy of the ceramic material. The strong influence of yttria additions on the transformation temperature is also clearly visible.

Although the reduction in transformation temperature is found not only in free particles but also in sintered bodies, experimental evidence to support this is much more difficult to obtain due to grain growth occurring during sintering. Mayo *et al.* [14] have demonstrated a decrease in the temperature of the tetragonal–monoclinic transformation of yttria-doped zirconia (see Figure 3.11). As noted in the case of free particles, yttria additions act like a reduced grain size, and consequently for yttria-free zirconia this phenomenon was not observed due to the large grain size. For yttria contents in the range of 0.5 to 1.5 wt.%, the linear decrease in transformation temperature with the inverse grain size, as derived from Equation (3.8), has been clearly demonstrated.

Suresh and Mayo [13] determined the enthalpy of the phase transformation by using calorimetric methods, some characteristic features of which are shown in Figure 3.12. Initially, a significant decrease in the absolute value for the enthalpy of transformation with increasing yttria content is apparent, as might be expected since the transformation temperature decreases with increasing yttria content. For all levels of yttria doping, and for decreasing particle sizes, the enthalpy of transformation remained constant until a particle size was reached where the influence of the surface sets in. However, such a particle size is larger than that where the $1/D$ relationship for the tetragonal–monoclinic transformation temperature is valid (see Figure 3.11). A further decrease in particle size greatly increases the enthalpy of transformation. Based on the data in Figure 3.12, it is clear that a complete theoretical description must also explain this sharp transition, but obviously the simple theoretical approach leading to Equation (3.8) is not sufficient.

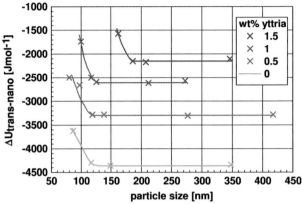

Figure 3.12 Enthalpy of the monoclinic–tetragonal transformation of yttria-doped zirconia as a function of yttria content and grain size [13]. Note that Equation (3.8) is fulfilled only for the smallest particles. The increase in enthalpy of transformation begins suddenly at a grain size which is dependent on the yttria content.

3.4
Phase Transformation and Coagulation

At this point it should be considered whether a phase transformation might be caused by the temperature flash that occurs during the coagulation of two particles (see Section 2.2). The situation for the coagulation of two aluminum particles of equal size at room temperature is shown in Figure 3.13, together with the melting point of the coagulated particle with the diameter $D_{coagulated} = 2^{1/3} D$.

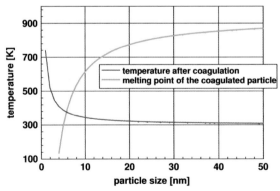

Figure 3.13 Temperature after coagulation of two aluminum particles of equal size. The melting temperature of aluminum nanoparticles is plotted as a function of particle size. Temperature flashing during coagulation may cause melting of the coagulated particle.

In Figure 3.13, the temperature of the coagulated particle is plotted, starting from room temperature (300 K). Clearly, below a limit of approximately 6 nm, the temperature of the melting point is exceeded during coagulation, and this explains why small metal nanoparticles are always found as perfect spheres. The melting point of aluminum falls below 300 K but, in the case of metals, amorphization is not observed. This represents just one point where the simplified theory applied is no longer valid, although for many metals it is often observed that small clusters may have structures that differ from those found in the bulk material. This situation is discussed in the following section.

3.5
Structures of Nanoparticles

The phenomena described above for zirconia are not restricted to compounds with comparatively "simple" structures. In the case of zirconia, the temperature of phase transformation was found to decrease with decreasing particle size, which led to the fact that the tetragonal phase – a high-temperature phase – is in fact found at room temperature. Additionally, in the case of particles with sizes less than 5 nm the cubic phase is quite often found. Ayyub *et al.* [15] framed a rule which stated that, with decreasing particle size nanoparticles prefer the phase with a higher symmetry. As, in general, the latter phase is the high-temperature phase (the phase with the highest entropy), this led to the proposition that nanoparticles would tend to crystallize in the high-temperature phase, provided that they were small enough. This concept is outlined in the following section, using a few ceramic materials as examples.

Ayyub *et al.* [15] reported the details of lattice constant determinations for Al_2O_3 and Fe_2O_3. These two oxides have common characteristics, with conventional grain size at room temperature; both crystallize in the hexagonal α-phase, and in both cases the cubic γ-phase is observed at high temperatures. Although alumina has a few more intermediate phases, these are not discussed in this context.

The sequence of phases is shown diagrammatically in Figure 3.14, where the normalized unit cell volumes are plotted as a function of the particle size. This normalization is necessary, as a valid comparison of unit cell volumes of different crystallographic structures is possible only on the basis of a constant number of formula units (=molecules). When considering the phases, in both cases the decreasing particle size has the same influence as increasing the temperature, which means that small nanoparticles crystallize in the high-temperature structure. Exactly the same phenomenon as shown here for Al_2O_3 and Fe_2O_3 is also found with zirconia. Figure 3.14 also demonstrates an additional influence of the decreasing particle size: in contrast to metallic nanoparticles (see Section 2.2), the lattice constant increases in the case of oxides, indicating that the structural changes are caused by the huge surface-to-volume ratio. This is clearly visible in the case of the γ-phases, and a minor effect is also visible in the other phases.

This phenomenon is not restricted to compounds with comparably simple structures; rather, it is also found in the case of more complex structures, with

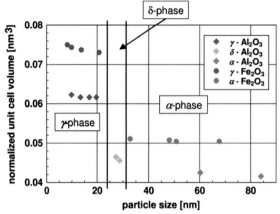

Figure 3.14 Normalized unit cell volume for different Al_2O_3 and Fe_2O_3 phases as a function of grain size [15]. Normalization provides a constant number of formula units per unit cell; otherwise, comparison is impossible. In the γ-phase, the unit cell volume is increased as the particle size decreases. A preference for high-temperature structures as the particle size decreases is clearly visible.

typical examples being ferroelectric or antiferroelectric compounds. At high temperature, these compounds are cubic; however, by reducing the temperature there occurs a transformation to the tetragonal perovskite structure, which is ferroelectric below the *Curie* point. This technologically extremely important class of compounds shows, as nanoparticles, an unusual pattern of phase transformation for the cubic–tetragonal transformation. Comparable with Al_2O_3 or Fe_2O_3, and in the case of zirconia, particle size plays a similar role as temperature. In all cases, the phase transformation is significantly influenced by the particle size.

In Figure 3.15, the lattice parameter of nanoparticulate $BaTiO_3$ (a ferroelectric material with a perovskite structure) is displayed as a function of the annealing temperature [16]. As for the cases discussed above, at small particle size $BaTiO_3$ crystallizes in the cubic, high-temperature structure. The lattice parameters are measured after the annealing process at room temperature. In Figure 3.15 the decrease in the lattice parameter with increasing annealing temperature can be clearly seen; this is equivalent to an increasing particle size, as the particle size was increasing during annealing. The starting grain size was less than 10 nm. Figure 3.15 is remarkable for two features: (i) with an increasing annealing temperature, equivalent to an increasing grain size, a decrease in the lattice constant of the cubic phase occurs; and (ii) there is a quasicontinuous transition from the cubic phase to the tetragonal phase. The grain sizes (35, 100 and 400 nm) were determined using electron microscopy.

Both findings – the decrease in the lattice constant with increasing particle size and the continuous or at least quasicontinuous transition from the cubic to the tetragonal distorted phase – are unusual. In conventional materials, phase transformations are

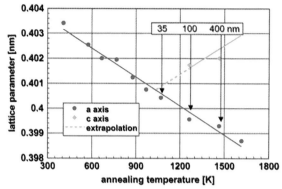

Figure 3.15 Lattice parameter of nanoparticulate BaTiO$_3$, a ferroelectric material with perovskite structure, as a function of the annealing temperature. During annealing, grain growth occurred [16]; hence, an increasing annealing temperature was equivalent to an increasing grain size. All lattice parameters were measured at room temperature, and particle sizes determined by electron microscopy. For details, see the text.

characterized by an abrupt change in structure, and the lattice constant is independent of the grain size. By evaluating the X-ray diffraction profiles of Frey and Payne [16] with respect to lattice constant and grain size, and combining this with the data for the lattice constant given in Figure 3.15, one obtains the dependency of the lattice constant as a function of the particle size for the cubic phase. This is displayed in Figure 3.16. The remarkable increase in the lattice constant observed for particle sizes below ca. 10 nm is explained by completing the chemical reaction and the release of residual reaction products. Frey and Payne showed this by chemical composition; however, the decrease in the lattice constant within the range of 10 to

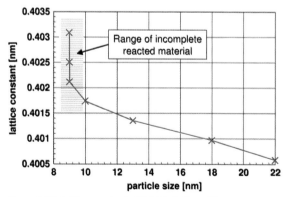

Figure 3.16 Lattice constant of BaTiO$_3$ in the cubic structure as a function of particle size [16]. The lattice constant decreases with increasing particle size. Values in the shaded area were deemed unreliable as the material was not completely reacted.

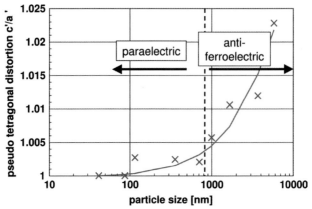

Figure 3.17 Pseudotetragonal distortion and the transition paraelectric–antiferroelectric of $PbZrO_4$ as a function of grain size [15]. A reduction in grain size had a similar effect as an increase in temperature. The paraelectric, cubic phase was the high-temperature phase; the antiferroelectric, tetragonal distorted phase was the low-temperature phase.

22 nm was clearly a function of the particle size. Essentially the same phenomenon is also observed with other ceramic nanoparticles.

The continuous transition from cubic to tetragonal phase of compounds with a perovskite structure as a function of the particle size is observed quite often. A further example is provided in Figure 3.17 for antiferroelectric $PbZrO_4$ [15], where the pseudotetragonal distortion, the a'/c' ratio, is plotted against the particle size. (Although $PbZrO_4$ is orthorhombic, a pseudotetragonal unit cell may be defined with the constants $a' = a/\sqrt{2} = b/2\sqrt{2}$ and $c' = c/2$.) The ratio a'/c' describes the deviation from cubic symmetry; hence, at $a'/c' = 1$ the material is cubic. As shown in Figure 3.15, the data in Figure 3.17 illustrate a continuous decrease in the a'/c' ratio in the direction of a cubic cell, which is achieved at particle sizes of approximately 90 nm. The paraelectric–antiferroelectric transition, which is observed at 500 K for materials with a conventional grain size, occurs with a particle size below approximately 800 nm at room temperature.

In a similar way, Huang *et al.* [17] showed the same phase transformation for $PbZr_{0.3}Ti_{0.7}O_3$ using the electric properties. Figures 3.18a and b depict the dielectric constant and *Curie* temperature, respectively, as a function of the particle radius; here, a lattice constant ($a = 0.8$ nm) was selected as the unit for the abscissa. The phase transformation was observed at a particle radius of 5.6 lattice constants ($= 4.5$ nm). In addition, a phase transformation was observed from cubic at small grain sizes to tetragonal with increasing particle size.

The continuous transition from one phase to another (distorted) phase, in combination with an increase in the lattice parameter with decreasing particle size appears to be general, and was even observed for high-temperature superconductors [15]. Whenever materials exhibiting phase transformations are analyzed, the

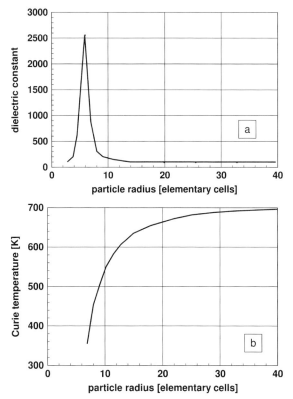

Figure 3.18 Dielectric constant and *Curie* temperature of PbZr$_{0.3}$Ti$_{0.7}$O [17]. For small particle sizes transition occurs to the paraelectric, high-temperature phase. A reduction in particle size and high temperature has similar effects on particle structure and properties. The lattice constant ($a = 0.8$ nm) was selected as the dimension for the ordinate. (a) Dielectric constant of PbZr$_{0.3}$Ti$_{0.7}$O$_3$. At small particle sizes (< 5.6 lattice constants; ca 4.5 nm) the material was in the paraelectric, high-temperature phase, whereas at larger particle sizes the material was ferroelectric. (b) *Curie* temperature of PbZr$_{0.3}$Ti$_{0.7}$O$_3$. Below a particle size of eight lattice constants, the *Curie* temperature was no longer defined. The material was paraelectric, independent of the temperature.

same observations are made: with decreasing particle size, a transition to the most symmetric (usually high-temperature) phase is observed. Additionally, it is remarkable that in many first-order phase transitions, the lattice changes are continuous and not, as has been observed for conventional materials, abrupt in nature. This behavior is quite strange, although in this context it must not be forgotten that the determination of phases and particle sizes is usually made from an evaluation of X-ray diffraction line profiles. There are, however, two points which cause these evaluations to be problematic: (i) the tetragonality of the structures is very small; and (ii) in the case of small particles the diffraction lines, which are broadened due to the small particle size, are not split. This makes it difficult to decide whether a diffraction line profile consists of a split line of the tetragonal phase, or whether it is a superposition of the diffraction lines of the tetragonal and cubic phases. In the latter case, the transition from one phase to the next would not be quasicontinuous but rather abrupt, possibly superimposed by fluctuation processes between the two phases.

3.6
A Closer Look at Nanoparticle Melting

When discussing the melting of nanoparticles, the point was mentioned that in the case of larger particles the melting process starts from a thin surface layer. In case of lead, a value of approximately 3 nm was found for the thickness of the surface layer. Hence, the question arises of how this behavior which is so fundamentally different compared to that of particles with conventional grain sizes, can be explained. Chang and Johnson [18] showed, on the basis of theoretical considerations using *Landau*'s parameter of ordering M, that small nanoparticles do not have the degree of ordering that is observed in bulk materials. In that case, the ordering parameter is defined in a way that the value is 1 for ideal crystals and 0 for the melted phase. Chang and Johnson [18] also provided a formula to calculate this order parameter M for tin as a function of the radius and particle sizes. In Figure 3.19, M is shown as a function of the radius for particles of different sizes. In the case of large particles (e.g., radius 10 nm), a perfect ordering is achieved in the interior of the particle, and a reduced ordering close to the surface. The thickness of this layer with reduced ordering is approximately 3 nm. When considering very small particles (e.g., those with a radius of 1 nm), a perfect ordering is not attained even in the center of the particle. In such a case the maximum ordering would be less than 0.3, which means that small-sized particles would act more like a melted material than a crystallized one. By applying

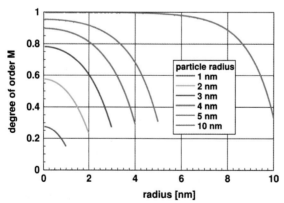

Figure 3.19 *Landau*'s order parameter M for nanoparticles of tin as a function of radius and particle size [18]. The degree of order decreased with decreasing particle radius, and also from the interior to the surface. For perfectly crystallized particles M = 1; for melted particles, M = 0.

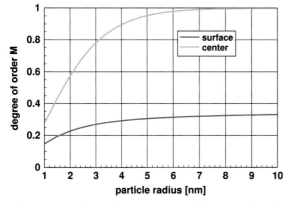

Figure 3.20 *Landau*'s order parameter M in the center and at the surface of nanoparticles [18]. Small nanoparticles with radii <5 nm never show perfect crystallization.

this theory, Chang and Johnson were able to fit the data for size-dependent melting point of tin quite well.

In order to demonstrate this phenomenon in greater detail, a graph showing the degree of ordering at the surface and in the center of particles with different radii is shown in Figure 3.20. The data in the figure make it clear that in this case, tin particles of less than 10 nm diameter never reach the ordering as it is observed in bulk materials. This also explains the experimental findings that larger nanoparticles begin to melt from a surface layer, whereas the smaller nanoparticles melt as a whole.

In addition, there appears to be a steady increase in the ordering parameter at the surface with increasing particle size; consequently, similar phenomena in materials with conventional grain sizes cannot be expected.

From the examples given above it seems clear that, as with many other phase transitions, the melting of nanoparticles is a gradual rather than discontinuous process. This is consistent with the experimental determination of the melting point of lead nanoparticles [12] and the transition from the cubic to tetragonal phase of $BaTiO_3$ [16]. However, despite this being an extremely successful way of describing and interpreting phase transitions, it is important to realize that the order parameter M is a mathematical construct that is dependent on an external parameter, the correlation length, which is chosen freely to fit the experimental data.

Although until now those considerations that have been made were purely thermodynamic in nature, nanoparticles are so small that thermal fluctuations are in fact observed. Typical examples of these phenomena are superparamagnetism and its analogue, superferroelectricity (see Section 5.1), and electron microscopy studies on the melting of small metal nanoparticles have provided information on similar phenomena. A series of electron micrographs of 2-nm gold particles, recorded over a period of 5 min, is shown in Figure 3.21, and illustrates the changing shape and structure of the particle [19]. It should be noted that the relatively poor quality of these images is due to the extremely short intervals (1/60 s) between frames; nonetheless, the quality of these images cannot be overestimated.

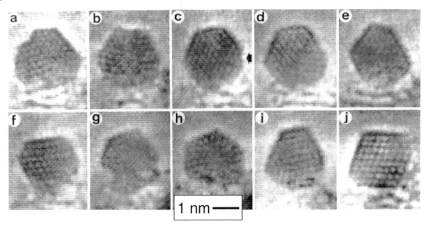

Figure 3.21 A series of electron micrographs of 2-nm gold particles [19], taken at intervals of 1/60 s. The images show spontaneous changes in particle habitus at a temperature of approximately 370 K, from single twins (a, d, and i) to multiple twinned icosahedral particles (b and h) and further to cuboctahedral shapes (e, f, and i). Copyright: American Physical Society 1986.

The particle shown in Figure 3.21 changes its shape from single twins (a, d, and i) to multiply twinned icosahedral particles (b and h) and to cuboctahedral particles (e, f, and i). In all cases, the lattice visible fringes correspond to the (1 1 1) lattice plane. The temperature of the gold particle was not significantly higher than 370 K, and the series of images suggests a thermal instability. Clearly, the difference in free energy of the different particle shapes is so small that they are energetically more or less equivalent.

Based on such a concept, Ajayan and Marks [20] developed a thermodynamic description of this phenomenon by calculating the free energy of the particles as a function of the shape and twinning. These calculations led to the proposal that a correlation exists between particle size and shape, and temperature. A phase diagram proposed by these authors showed clearly separated ranges for particles of different shape, single crystals, and multiply twinned crystals. In addition, a size range was indicated where the energy differences between these different possible shapes were so small that a fluctuation occurred between the different shapes. This phase diagram, for gold nanoparticles, is shown in Figure 3.22.

This new phase diagram suggests that, for sufficiently small particle sizes and at not too high a temperature, single crystals are never stable. Rather, at the lowest temperature, multiply twinned icosahedral particles are the most stable form. By increasing the temperature, one enters the region where the most stable particles are multiply twinned decahedrals, although prior to melting there is a large range of particle sizes and temperatures where the particle may change either its phase or habitus. Materials in this range are termed "quasimelted" [20].

The phenomenon of quasimelting is observed thermodynamically in the size and temperature range where crystallized material is expected. In fact, a related phenomenon might be expected on the side where thermodynamics anticipates a melted

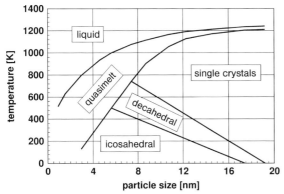

Figure 3.22 Phase diagram of gold nanoparticles, showing regions of different well-defined habitus. A region termed quasimelt is also indicated where the particles change their habitus spontaneously [20].

material, and indeed Oshima and Takayanagi [21] observed this phenomenon. Three series of electron micrographs of a 6 nm-diameter tin particle are shown in Figure 3.23, where panels a–f, g–l, and m–r have 1/60-s time differences, respectively, from picture to picture. Within each picture, the crystal embryos can be seen appearing and disappearing.

These tiny crystals are visible in frames (c) and (m); however, this is not to say that they are the only crystals and do not appear in the other frames. It is possible – and also highly probable – that their orientation is such that the lattice fringes are not observable. Based on many observations made, it was possible to design a phase diagram where the size and temperature range characteristic for the appearance of these crystal embryos is approximately localized. Such a phase diagram, for tin nanoparticles, is shown in Figure 3.24; an inset in this figure shows a schematic drawing of the particles according to Figures 3.21b or m, in the pseudocrystalline region.

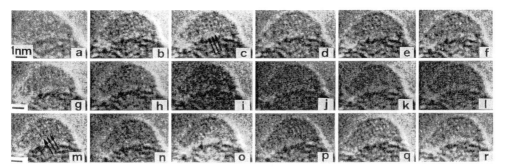

Figure 3.23 A series of electron micrographs of tin particles, taken at intervals of 1/60 s [21]. In these images the appearance and disappearance of small crystallized regions, called embryos, can be seen within the particles (frames c and m). The phenomenon of "pseudocrystalline particles" can be attributed to a well-defined particle size–temperature range in a phase diagram (Reprinted with permission from [21]; Copyright: Springer 1993.)

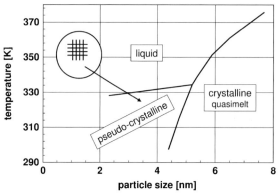

Figure 3.24 Temperature–particle size phase diagram for tin nanoparticles according to Oshima and Takayanagi [21]; conditions as noted for pseudocrystalline particle formation in Figure 3.23 were observed. A schematic drawing of the pseudocrystalline particles is also shown. In the region denominated as crystalline quasimelt, the particles fluctuate between the different possibilities of their habitus.

The region where the crystal embryos occur is referred to by the authors as "pseudocrystalline". Clearly, the range of the pseudocrystalline phase is beyond the line given by $T_{nano} = a - b/D$, the thermodynamic boundary of melting derived from Equation (3.8). Until now, it has not been clear whether this severe deviation from the elementary description is due to an insufficient experimental database, or if the simple theory of melting is not applicable. It is important to note here that, in the whole range denoted as crystalline quasimelt, fluctuations between different particle shapes occur. This is essentially the same observation which was described by Ajayan et al. [20] as quasimelted in the case of gold particles. It is important to realize here that Oshima and Takayanagi [21] showed that instabilities occurred on both sides of the thermodynamic limit, between the crystalline and the melted phases.

The phenomenon of pseudocrystallinity is not limited to metallic particles. The appearance of crystal embryos in amorphous WO_3 particles is shown in Figure 3.25, where the crystal embryos are indicated by arrows. It is remarkable that in the case of this oxide, the crystallized nuclei (embryos) are larger than the metallic nanoparticles. The appearance of small, crystallized regions in an amorphous matrix was also observed in other oxides such as Cr_2O_3. Clearly, this very general phenomenon is not restricted to metallic nanoparticles.

3.7
Structural Fluctuations

Based on the fluctuations described above, leading particle size-dependent phase diagrams may be explained by simple thermodynamic considerations. The simplest case of fluctuations – superparamagnetism (see Chapter 5) – is observed when the

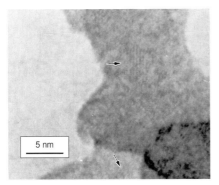

Figure 3.25 The occurrence of crystal embryos in WO_3 nanoparticles. This phenomenon is observed in both metallic nanoparticles and ceramic particles. The crystallized nuclei were seen to be larger than those of metallic particles.

thermal energy kT (k is the *Boltzmann* constant, T is temperature) is larger than the energy of magnetic anisotropy Kv (K is material-dependent constant of magnetic anisotropy; v is the volume of the magnetic particle). Fluctuations, in this case superparamagnetism, occur when the condition

$$Kv \leq kT \tag{3.14}$$

is fulfilled. It is clear that the conditions in Equation (3.14) are fulfilled only for sufficiently small particles.

In the case of structural fluctuations, it is necessary to examine more closely the thermodynamics in the vicinity of a phase transformation [22]. Such a situation is shown graphically in Figure 3.26, where the free enthalpy G_1 for phase 1 and G_2 for phase 2 are plotted against the temperature. The transformation temperature T_{trans} is assumed at the intersection of the two lines representing G_1 and G_2.

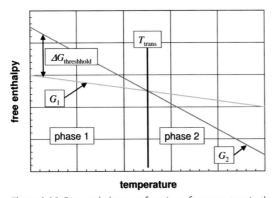

Figure 3.26 Free enthalpy as a function of temperature in the vicinity of a phase transformation. Two areas occur where phases 1 and 2 are stable. However, when $\Delta g_{threshhold} < kT$ is fulfilled, spontaneous fluctuations of the individual particles are possible.

In order to transform the phase at any temperature different from the transformation temperature, the threshold $\Delta G_{\text{threshhold}}$ must be overcome. As each particle fluctuates individually, then when considering one particle this energy threshold is given by

$$\Delta g_{\text{threshhold}} = \Delta G_{\text{threshhold}} \frac{m}{M} \tag{3.15}$$

In Equation (3.15), m is the mass of the particle and M the molar weight. In this context, variables related to one particle are written in lower-case letters, while those related to 1 mol are written in upper-case letters. As the condition for thermal instability, the onset of fluctuations, the following relationship is valid:

$$|\Delta g_{\text{threshhold}}| < kT \tag{3.16a}$$

In Equation (3.16a) it is necessary to take the absolute value of the difference of the free enthalpy because fluctuations in both directions must be considered. Furthermore, as G_1 and G_2 are linear functions of the temperature, the system is symmetric with respect to T_{trans}, leading to

$$|\Delta g_{\text{threshhold}}| \propto |T - T_{\text{trans}}| \tag{3.16b}$$

Neglecting any kinetic influences, this equation states that when approaching the transformation temperature from either side, the probability for fluctuations is equal, and is simply dependent on the distance to the transformation temperature.

In the following section, the consequences of these assumptions are explained; however, in a first approach, thermal expansion is neglected.

As an example, phase transformations (e.g., melting of particles) are used to explain the basic principles. For reasons of clarity, these considerations are restricted to the isothermal case, and to do this it is assumed that the particles are embedded in an infinite isothermal bath.

Initially, the transformation temperature of small particles is estimated, which led to Equations (3.8) and (3.9). In order to obtain the temperature limits of fluctuations between the two phases, it is necessary to expand Equation (3.16a).

$$\Delta g_{\text{trans-nano}} = \Delta U_{\text{trans}} \frac{\pi D_{\text{new}}^3 \rho_{\text{new}}}{6M} - T\Delta S_{\text{trans}} \frac{\pi D_{\text{new}}^3 \rho_{\text{new}}}{6M}$$
$$+ \gamma_{\text{new}} \pi D_{\text{new}}^2 - \gamma_{\text{old}} \pi D_{\text{old}}^2 = kT \tag{3.17}$$

From Equation (3.17) one obtains for the temperature of the onset of the fluctuations as a function of the particle size:

$$T_{\text{fluct}} = \frac{\Delta U_{\text{trans}} + \gamma_{\text{new}} \dfrac{6M}{D_{\text{new}} \rho_{\text{new}}} - \gamma_{\text{old}} \dfrac{6M}{\rho_{\text{new}} D_{\text{new}}} \left(\dfrac{\rho_{\text{new}}}{\rho_{\text{old}}}\right)^{2/3}}{\Delta S_{\text{trans}} + k \dfrac{6M}{\pi D_{\text{new}}^3 \rho_{\text{new}}}} \tag{3.18}$$

Except for the term $k \frac{6M}{\pi D_{new}^3 \rho_{new}} = \vartheta$, Equations (3.8) and (3.18) are identical in terms of the denominator. As the *Boltzmann* constant k is very small, this term is influential only at very small particle sizes. In the case of increasing particle size, the size-dependent term δ in the denominator approaches zero faster than the surface-related terms in the numerator. Therefore, with increasing particle size, the temperature of the onset of fluctuations T_{fluct} approaches asymptotically the temperature of transformation T_{trans}. In order to observe fluctuations, the two terms in the denominator must be of similar size; when the particle size approaches zero, the fluctuations characterize the system. The term δ depends only on the geometry of the particle; it is material-independent. Therefore, Equation (3.18) reflects the well-known fact that besides geometry, fluctuations are ruled only by the entropy of transformation ΔS_{trans}; δ also has the dimension of entropy.

A phase diagram designed by using Equations (3.9) and (3.12) resembles those calculated by Ajayan and Marks [20] for the phase limits of the quasimelt. However, when examining the experimental results reported by Oshima and Takayanagi [21], it is possible that these equations do not describe the experimental findings. One reason for this discrepancy might be that the denominator in Equation (3.18) is not constant, and indeed both experimental indications and theoretical results [18] have shown that the thermodynamic quantities ΔU_{trans} and ΔS_{trans} are particle size-dependent. This is necessary because it has been well documented experimentally – and backed up by theory – that small particles, notably those less than 5 nm, show an inherently high degree of disorder which increases with decreasing particle size.

In general, one never analyzes the behavior of a single particle, but rather that of a system of many particles, termed an *ensemble*. Here, the use of a central theorem of statistical thermodynamics – the ergodic theorem – simplifies the task. According to *Boltzmann* and *Gibbs*, in the ergodic theorem a time average can be replaced by an ensemble average, when phase transformation occurs from one level (old phase) to a second level (new phase). The probabilities of the relative time periods can then be measured either with one particle in two different phases, or with an ensemble of particles, the numbers of which can be counted in each phase.

In a two-level system, based on statistical thermodynamics, it is known that the probability of occupation for the two levels, which is in an ensemble of many particles equal to the fraction, is

$$p_1 = c_1 = \exp\left[\frac{-(G_2 - G_1)}{RT}\right] = \exp(-\Delta G_{trans}) \quad c_2 = 1 - c_1 \tag{3.19}$$

where $G_i = H_i - TS_i$, $i = 1$ or 2. Consequently, the occupation of the second level is c_2.

At this point the melting of gold nanoparticles is cited as an example. As great uncertainty exists regarding the material data of nanoparticles, experimental results on the melting of gold nanoparticles have been used to determine such data appropriate to nanoparticulate gold. Among many reports, that of Castro *et al.* [6] was selected, from which Figure 3.27 shows the melting temperature of gold as a function of the inverse particle size.

Except for three points, which related to particle sizes < 1 nm, the data followed a straight line when the melting temperature was plotted against the inverse particle

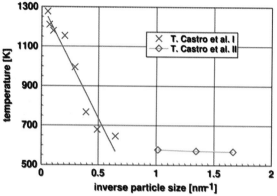

Figure 3.27 Melting temperature of gold nanoparticles. From Castro *et al.* [6].

size. The melting points for these small particles were at near-constant temperature (points shown as Castro *et al.* II in Figure 3.27) [6]. These outliers were omitted from an evaluation of the material data, after which Equation (3.8) was used to determine the enthalpy and entropy of transformation.

For the surface energy γ_{liquid} and γ_{solid}, the values were taken from the consistent data sets of Miedema and Boom [23,24] as $\gamma_{solid} = 1.55 - T \cdot 1.4 \times 10^{-4}$, $\gamma_{liquid} = 1.34 - T 1.6 \times 10^{-4}$, (with γ in J m^{-2} and T in K). For all the other thermodynamic constants, standard values were taken. The calculations resulted in an enthalpy of transformation (ΔH_{trans}) of 13 230 J mol^{-1}, and to an entropy of transformation (ΔS_{trans}) of 9.97 J K^{-1} mol^{-1}. The fractions of solid and liquid 1.4-nm gold nanoparticles, using Equation (3.19), are shown in Figure 3.28.

By determining the relative amounts of solid and melted particles as a function of the particle size, it is possible to calculate phase diagrams where two-phase regions caused by fluctuations are indicated.

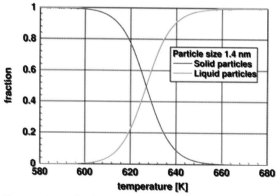

Figure 3.28 Number fractions of gold particles with a diameter of 1.4 nm. Both phases are stable over a broad temperature range [22].

References

1 Malinovskaya, T.D. and Sachkov, V.I. (2003) *Russ. Phys. J.*, **46**, 1280–1282.

2 Rupp, J. and Birringer, R. (1987) *Phys. Rev.*, **36**, 7888–7890.

3 Tschöpe, A. and Birringer, R. (1993) *Phil. Mag. B.*, **68**, 223–229.

4 Wang, L., Tan, Z., Meng, S., Liang, D. and Li, G. (2001) *J. Nanoparticle Res.*, **3**, 483–487.

5 Thomson, W. (1871) *Phil. Mag.*, **42**, 448–452.

6 Castro, T., Reifenberger, R., Choi, E. and Andres, R.P. (1990) *Phys. Rev.*, **42**, 8548–8556.

7 Reiss, H. and Wilson, I.W. (1948) *J. Colloid Sci.*, **3**, 551–561.

8 Pawlow, P. (1908) *Z. Phys. Chem.*, **65**, 1.

9 Hanszen, K.-J. (1960) *Z. Phys.*, **157**, 523–553.

10 Zhao, M., Zhou, X. H. and Jiang, Q. (2001) *J. Mater. Res.*, **16**, 3304–3308.

11 Eckert, J., Holzer, J.C., Ahn, C.C., Fu, Z. and Johnson, W.L. (1993) *Nanostruct. Mater.*, **2**, 407–413.

12 Coombes, C.J. (1972) *J. Phys. F.*, **2**, 441–448.

13 Suresh, A. and Mayo, M.J. (2003) *J. Mater. Sci.*, **18**, 2913–2921.

14 Mayo, M.J., Suresh, A. and Porter, W.D. (2003) *Rev. Adv. Mater. Sci.*, **5**, 100–109.

15 Ayyub, P., Palkar, V.R., Chattopadhyay, S. and Multani, M. (1995) *Phys. Rev.*, **51**, 6135–6138.

16 Frey, M.H. and Payne, D.A. (1996) *Phys. Rev.*, **54**, 3158–3168.

17 Huang, H., Sun, C.Q. and Hing, P. (2000) *J. Phys. Cond. Matter*, **12**, L127–L132.

18 Chang, J. and Johnson, E. (2005) *Phil. Mag.*, **85**, 3617–3627.

19 Iijima, S. and Ichihashi, T. (1986) *Phys. Rev. Lett.*, **56**, 616–619.

20 Ajayan, P.M. and Marks, L.D. (1988) *Phys. Rev. Lett.*, **60**, 585–587.

21 Oshima, Y. and Takayanagi, K. (1993) *Z. Phys. D.*, **27**, 287–294.

22 Vollath, D. and Fischer, F.D. (2008) *J. Nanopart. Res.*, DOI:10.1007/s11051-007-9326-3.

23 Miedema, A.R. and Boom, R. (1978) *Z. Metallkde*, **69**, 183–190.

24 Miedema, A.R. and Boom, R. (1978) *Z. Metallkde*, **69**, 87–292.

4
Gas-Phase Synthesis of Nanoparticles

4.1
Fundamental Considerations

Gas-phase processes were the methods first applied to synthesize nanoparticles. Previously, and significantly earlier, colloid chemistry had been used to obtain nanoparticles in suspensions (= colloids), although after being separated from the liquid the colloidal particles formed agglomerates. In gas-phase processes, a vapor of the material condenses to form small particles, the concentration of which in general is very low. The process of particle formation may be divided into four major steps:

- *Nucleation*: A nucleus is the smallest stable unit, and usually consists of two or, in most cases, three atoms or molecules. The nucleus acts as core for further

- *condensation* of atoms or molecules. Therefore, the nucleus grows and forms clusters and later a particle. Condensation is a stochastic process, which is ruled by the dynamics of gas species. Colliding clusters or small particles may coagulate. During

- *coagulation* by exchange of surface energy, a new particle is formed. Having reached a certain size, the difference in surface energy will be so small that further coagulation of particles is impossible. Now, the process of

- *agglomeration* may start. Agglomerates consist of two or more individual particles. Often, one distinguishes between soft and hard agglomerates. Soft agglomerates are bond by *van der Waals* forces, whereas hard agglomerates are sintered. Only soft agglomerates can be separated. High temperatures and small particle sizes lead to the formation of hard agglomerates; therefore, after synthesis and particle formation the temperature should be reduced as rapidly as possible. This quenching process reduces the probability of the formation of hard agglomerates.

As the processes of condensation and coagulation are random by nature, the distribution of particle sizes obtained is relatively broad. However, by quenching or charging the particle electrically, one has the possibility to bias these processes, to influence the particle size distribution of the product. Although, in the following sections, some of the most important gas-phase processes used to synthesize

Nanomaterials: An Introduction to Synthesis, Properties and Application. Dieter Vollath
Copyright © 2008 WILEY-VCH Verlag GmbH & Co. KGaA, Weinheim
ISBN: 978-3-527-31531-4

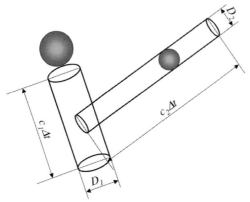

Figure 4.1 Model situation to estimate the probability of the collision of two particles of different sizes. The lines indicate the limitation of the cylinders circumscribing the trajectories of the particles.

nanoparticles will be described, it is first necessary to recognize how the processes of condensation and coagulation work.

In order to understand these processes, it is necessary to study the size-dependent probability of the condensation and coagulation processes [1]. First, one must estimate the probability of the collision of two particles with diameters D_1 and D_2; this situation is depicted in Figure 4.1.

To estimate the collision probability, one must calculate the volume V of the cylinder passed by a particle with velocity $\bar{c}$ in the short time interval Δt.

$$V = \frac{D^2 \bar{c} \Delta t \pi}{4} \tag{4.1}$$

Figure 4.1 and the following estimations assume that the particles are smaller than the mean free path of the particles in the gas. In the case of nanoparticles, this presumption is fulfilled in nearly all ranges of gas pressure applied for synthesis. The mean free path length λ in a gas is estimated by the formula

$$\lambda = \frac{4}{\sqrt{2}\pi N (D_G + D)^2} \tag{4.2}$$

where D is the diameter of the particle, D_G the diameter of the atoms or molecules of the gas species, and N the number of gas atoms or molecules per unit volume. As λ is indirectly proportional to the density of gas molecules per unit volume, the mean free path length decreases with increasing pressure in the system. Figure 4.2 gives the free path length in the gas; argon ($D_G = 0.29$ nm) was selected as example, as a function of the system pressure. Compared to the number of gas molecules, the concentration of particles is assumed to be negligible.

Figure 4.2 illustrates diagrammatically the mean free path for the gas pressures in the range from 10^2 to 10^5 Pa (= atmospheric pressure). Additionally, the separation line for particle sizes smaller or larger than the mean free path is shown. This line

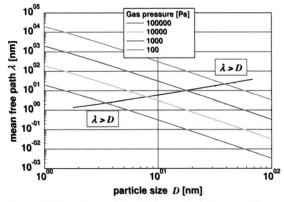

Figure 4.2 Mean free path length for a particle in argon. The parameter for the curves is gas pressure. Note the line separating the range, where the free path length is smaller or larger than the particle size.

limits the range, where the assumptions leading to Figure 4.1 and the following considerations are valid. For the processes of nucleation, condensation, and even coagulation, this assumption is valid even up to atmospheric pressure.

Kinetic theory of gases gives the value for the mean value of the velocity $\bar{c}$ of a particle, depending on the particle mass m, the temperature, T, and the *Boltzmann constant, k*:

$$\bar{c} = \left(\frac{2kT}{m}\right)^{0.5} \tag{4.3}$$

The probability of finding a certain particle during the time interval Δt in a well-defined volume element passed by the particle of a system with the volume V_{total} is

$$p = \frac{v}{V_{total}} = \frac{D^2 \bar{c} \, \Delta t \pi}{4 V_{total}} = \left(\frac{2kT}{m}\right)^{0.5} \frac{D^2 \pi}{4 V_{total}} = \left(\frac{2kT}{\rho}\right)^{0.5} \frac{D^{0.5} \pi}{4 V_{total}} = \kappa (DT)^{0.5} \tag{4.4}$$

where κ is a constant value that is independent of any geometry or temperature. From Equation (4.4) it can be seen that the probability of finding a certain particle on a point within or space with the volume V_{total} increases with the square root of the temperature and the particle diameter. This is quite a plausible result. As with increasing temperature, the velocity of the particles increase, the volume passed in a certain time interval increases, and therefore the probability of finding a particle increases. The same argument is valid for the particle size. However, as the velocity of the particles decreases with increasing particle mass, this increase is less pronounced than expected intuitively. When considering the probability of the collision of two particles with diameters D_1 and D_2, one first seeks the probability of finding these particles in the time interval Δt in the same volume element.

$$p_{1-2} = p_1 p_2 == \kappa^2 (D_1 T)^{0.5} (D_2 T)^{0.5} = \kappa_1 T (D_1 D_2)^{0.5} \tag{4.5}$$

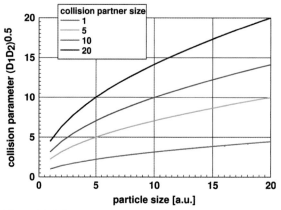

Figure 4.3 Collision parameter according to Equation (4.5) as a function of particle size. The parameter of the curves is the size of the collision partner. Note: The collision parameter increases with increasing particle size.

where κ_1 is a constant value. In the following section, constant values that are independent of essential parameters will be denoted consecutively as κ_i. In order to obtain the collision probability, one must multiply the probability defined with Equation (4.4) by the concentration of particles with diameters D_1 and D_2. Clearly, the term $(D_1 D_2)^{0.5}$ controls the collision of two particles with different diameters. In a simplified manner, when this parameter depends only on the geometry of the particles it is called the "collision parameter". Equation (4.5) shows that the collision probability increases linearly with temperature; therefore, in order to obtain nanoparticles – which means minimizing particle growth by coagulation – the temperature must be reduced as much as possible.

Calculated with Equation (4.5), Figure 4.3 depicts this collision parameter as the function of particles size with collision partners of different size.

As expected intuitively, for particles with increasing size, the probability of collision increases with increasing size of the collision partner. As larger particles are moving slower than smaller ones, with increasing size of the collision partner, those particles show a less pronounced increase of the collision parameter. This steeper increase for larger particles makes it impossible to obtain a product within a narrow range of particle sizes. This situation is similar to *Ostwald* ripening in colloid chemistry, where the larger particles consume the small ones, leading to a disappearance of the small particle fraction and an extension of the particle size distribution in the direction of the larger ones. As a consequence, particle size distribution functions of particles produced by a process of random collisions are always nonsymmetric functions. A typical particle size distribution for zirconia, as determined by the evaluation of electron micrographs, is shown in Figure 4.4 [2]. This material was produced using the inert gas condensation technique.

In Figure 4.4, the characteristic asymmetric particle size distribution that is usually fitted with a log-normal distribution, is clearly visible. Two points should

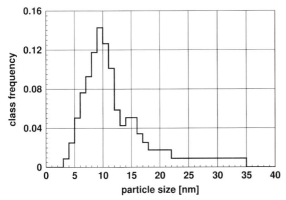

Figure 4.4 Size distribution for a zirconia powder synthesized using the inert gas condensation technique. The asymmetric particle size distribution has a long tail towards the large particle sizes, which is typical for products made by random processes. Such a size distribution is often fitted using the log-normal distribution function [2].

be mentioned in particular: (i) no particles with sizes below 3 nm were detected; and (ii) there was a long tail towards the larger particle sizes. Both features are characteristic for particles produced by a purely random process. In addition to this particle size distribution function, Figure 4.5 shows an electron micrograph of titania powder with a broad particle size distribution, wherein the particle sizes range from 5 to 50 nm.

In order to obtain a narrower particle size distribution, it is necessary to influence the process of particle formation in such a way that the simple collision law as described by Equation (4.4) is no longer ruling the system. One of these possibilities has already been mentioned, in that one can minimize coagulation by reducing the temperature, and stop the process from forming hard agglomerates by quenching with cold gas directly after particle formation. A further most elegant method of

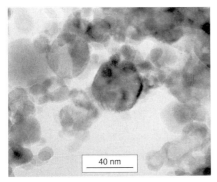

Figure 4.5 Electron micrograph of titania powder with broad particle size distribution, as obtained by purely random synthesis processes. The particle size ranges from 5 to 50 nm (Vollath, Sickafus [3]).

reducing the probability of coagulation is to load the particles with electrical charges of equal sign. In this case, the particles would be expected to repel each other.

In the following considerations on the behavior of electrically charged particles, the particles are treated as spherical capacitors. The capacitance C of a sphere is equal to its diameter, $C = D$; hence, at an electric potential V a capacitor carries the charge $Q = VC$. Then, assuming that all particles are charged to the same potential (i.e., $V = Q/D = constant$), then depending on the sphere's diameter the particles carry the charge

$$Q = VC = VD \tag{4.6}$$

From Equation (4.6) it can be derived that the electrical charge carried by a particle is proportional to the particle diameter; in other words, small particles carry less electrical charges than larger particles. This relationship is well known in aerosol physics [4], and the detailed analysis of charged aerosol particles by Zieman *et al.* [4] may be summarized in the following description of the electrical charges as a function of particle size distribution.

$$D \leq D_0 \Rightarrow Q = 1$$
$$D > D_0 \Rightarrow Q = 1 + \kappa_3(D - D_0) \tag{4.7}$$

In Equation (4.7), D_0 is a limiting diameter, and the electric charges are given in units of the elementary charge (= charge of one electron). Equation (4.7) is always correct because the smallest possible electrical charge is equal to the elementary charge. In this concept one expects that Q is a multiple of the elementary charge. Therefore, one may gain the impression that Equations (4.6) and (4.7) are incorrect, as they do not take care of the quantized character of the electrical charge. Although such an objection is correct for considerations directed towards a single particle, within this context we are considering only mean values over many particles, and therefore quantization of the electrical charge is nullified.

Assuming that the particles carry electrical charges of equal sign, they repel each other. Hence, the repelling force F between two particles with charges Q_1 and Q_2 that are related to particles with diameters D_1 and D_2 at a distance of r is

$$F = \frac{Q_1 Q_2}{r^2} = \frac{\kappa_2}{r^2} D_1 D_2 \tag{4.8}$$

This force results in an acceleration $Q_1 Q_2 / m r^2$, which reduces the speed and changes the direction of the path of the particles. As a result, the collision volume passed during time interval Δt is reduced, and consequently the collision probability is also reduced. An exact solution for this problem is extremely complex, as care must be taken to include all possible directions of the particle flight. However, for qualitative purposes, in a first approximation, the reduction of the collision volume can be described by the factor $1/Q_1 Q_2 \propto \frac{1}{D_1 D_2}$. Inserting this factor into Equation (4.5) leads to the following modified collision parameter:

$$p_{1-2} = p_1 p_2 = \kappa_3 T (D_1 D_2)^{0.5} \frac{1}{D_1 D_2} = \kappa_3 T (D_1 D_2)^{-0.5} \tag{4.9}$$

As a consequence of introducing a repelling term, Equation (4.9) now describes a reduced collision probability with increasing particle size. The particle growth by coagulation and agglomeration is limited. In analogy to Equation (4.5), the term $(D_1 D_2)^{-0.5}$ is now called the "collision parameter", while the temperature dependence remains unchanged linearly.

The consequence of particle charging on the "collision parameter" is visible in Figure 4.6a and b, where this modified "collision parameter" is plotted against particle size for different collision partners. In contrast to Figure 4.3, where the collision parameter increases with increasing particle size, in Figure 4.6a, calculated for $D_0 = 0$, a continuous decrease of the collision parameter with increasing size of the collision partner is also observed. When considering particles of 5 and more, in the case of charged particles the collision parameter is more than one order of magnitude smaller as compared to neutral particles. This is entirely different in Figure 4.6b, where $D_0 = 3$ was assumed. Here, a maximum of the collision parameter

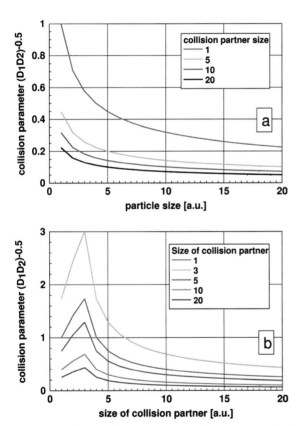

Figure 4.6 Influence of electrical charge on the collision parameter. The collision parameter is plotted against particle size (as in Figure 4.3). The parameter for the curves is the size of the collision partner. The decreasing collision parameter with increasing particle size limits particle growth during synthesis. (a) Collision parameter for electrically charged particles, where the charge Q is proportional to the particle diameter D, $Q \propto D$. (b) Collision parameter for particles, where the electrical charge Q follows the relationship: $D \leq D_0 \Rightarrow Q = 1$ and $D > D_0 \Rightarrow Q = 1 + \kappa_3 (D - D_0)$; $D_0 = 3$.

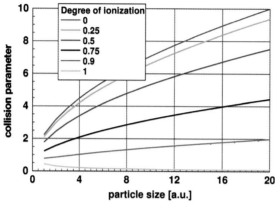

Figure 4.7 Influence of partial ionization on the collision parameter for the collision of two particles as a function of particle size. The size of the collision partner was assumed to be 5. The parameter for the curves is the degree of ionization. A reduction in collision parameter, limiting particle growth, is observed only near 100% ionization.

is realized for the particle size $D_1 = D_2 = D_0 = 3$. Clearly, it will be very difficult to obtain particles with sizes significantly larger than D_0. Both, Figure 4.6a and b show that charging the particles limits the size of the particles, and the long tail on the side of the large particles in the size distribution may be avoided. The experimental results have confirmed these considerations.

Nonetheless, the question still arising is how to create an experimental set-up in order to exploit the advantages of a synthesis process with charged particles. As the radius of nanoparticles is small and the processing temperature high, the first idea is to rely on thermal electron emission, even though all of the particles may not be ionized. The influence of partial ionization of the particles on the collision parameter is shown in Figure 4.7.

The collision parameter in Figure 4.7 was estimated by linear superposition of the collision parameter of charged and uncharged particles. At first it is clear that a significant size-limiting effect might be expected only above 90% ionization of the particles. However, this result excludes thermal electron emission as a mechanism to limit particle size, and therefore special arrangements must be designed to exploit this phenomenon. Size limitations by charging the particles are observed at the microwave plasma process and special variants of the flame synthesis process, both of which are described later in the chapter.

4.2
Inert Gas Condensation Process

Historically, the most important – and certainly the oldest – process for synthesizing nanoparticles in the gas phase is that of inert gas condensation [5]. This process applies thermal evaporation to a metal within a vacuum chamber filled with a small amount of

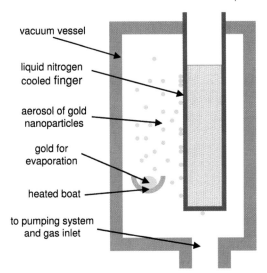

vacuum vessel

liquid nitrogen
cooled finger

aerosol of gold
nanoparticles

gold for
evaporation

heated boat

to pumping system
and gas inlet

Figure 4.8 Typical set-up for nanoparticle synthesis using the
inert gas condensation process. A metal (e.g., gold) is evaporated
in a vacuum vessel, filled at reduced pressure with an inert gas.
The metal vapor loses thermal energy by colliding with the inert
gas atoms, and forms nanoparticles. The product moves to a
liquid nitrogen-cooled finger (by the process of thermophoresis),
and is collected from the surface.

inert gas. For example, in order to produce gold nanoparticles, gold is evaporated in a
"boat" which is heated to a sufficiently high temperature. The atoms of gold vapor which
emanate from the boat collide with atoms of the inert gas, losing energy with each
collision. The gold vapor is then thermalized, and collisions of gold atoms with other
atoms of the same type lead to nucleation and subsequently to particle formation. The
particles formed in the gas phase travel by thermophoresis to a cold finger, where they are
collected. The general layout of the equipment used is shown schematically in Figure 4.8.

Following a production cycle, the metal particles are carefully scraped from the cold
finger; this design allows further processing of the product without breaking the
extremely pure vacuum conditions. In order to obtain oxides, small quantities of
oxygen are introduced into the system before scraping the powder from the cold
finger, such that the metal powder is oxidized very slowly. Care must be taken as rapid
oxidation leads to overheating and sintering of the product. As the formation of the
particles is a purely random process, the inert gas evaporation process typically leads
to a product with a broad particle size distribution.

4.3
Physical and Chemical Vapor Synthesis Processes

The basic principle of the inert gas condensation process leads to many variants,
as the systems employed differ in how the metal is introduced and subsequently

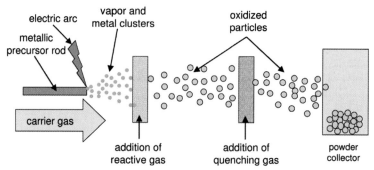

Figure 4.9 Typical set-up for physical vapor synthesis of oxide nanoparticles. A metallic precursor is evaporated using an electric arc or an electron beam. A stream of inert carrier gas transports the vapor into the reaction zone, where the reaction gas is added. To limit particle growth and agglomeration, the gas carrying the particles is quenched. Finally, the product is collected.

evaporated. One of the most interesting possibilities is heating with an electron beam [6]. The technical up-scaling of an inert gas condensation process may lead to the introduction of elements that limit particle size growth. However, two possible measures exist by which particle size and particle size distribution may be controlled: (i) a reduction of the residence time of the particles in the reaction zone; and (ii) rapid cooling of the particles after they have left the reaction zone. For both measures the original diffusion-controlled process as shown in Figure 4.1 is not applicable. Rather, it is necessary to replace transport via thermal diffusion with transport using a carrier gas. When used as a heat source for mass production, an electrical arc has many advantages and is utilized on a regular basis. The layout of such a system is shown schematically in Figure 4.9.

The system shown in Figure 4.9 utilizes an electric arc as a source of energy to evaporate the metallic precursor. This is a quite difficult process to control as the extremely high temperature of the arc may lead to high evaporation rates. Moreover, a high concentration of the evaporated precursor in the carrier gas usually results in large particles. In order to overcome these problems an extended knowledge of industrial processing is required, although the next two steps – addition of the reactive gas and quenching – are of similar difficulty.

The industrial product of Fe_2O_3 (see Figure 4.10) [7] typically shows a relatively broad particle size distribution, characteristic of purely stochastic processes. Furthermore, it is interesting to see "twins", characterized by changes in contrast (in this case, black lines) in many of these particles.

Although until now metal rods or powders have been used as precursors, this approach is not in all cases either economic or efficient. In fact, it is often more appropriate to use chemical compounds with a relatively high vapor pressure as the precursor. This variant of the synthesis process, which is referred to as "chemical vapor synthesis", utilizes a tubular furnace with temperatures up to 1500 K as source of heat. A carrier gas – in most cases argon or nitrogen – transports the evaporated

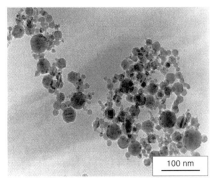

Figure 4.10 Fe₂O₃ powder produced by the physical vapor synthesis process. The broad particle size distribution (here 5 to 50 nm) is characteristic of this type of product [7]. The dark lines seen in larger particles indicate twin boundaries. (TEM micrograph reprinted with permission from Nanophase Technologies Corporation, Romeoville, IL, USA).

precursor through the heated reaction zone. As a precursor, chlorides, carbonyls or metal organic compounds are most often used, the ultimate compound selection depending on properties, availability, and price. It must not be forgotten that almost all precursors lead to typical reaction products in the off-gas, and may also leave some traces behind that might be dissolved in the matrix of the particles, or adsorbed at the particle's surface. This may cause severe disturbance; for example, chlorine reacts readily with organic materials applied to functionalize the surface of the particles. In order to obtain the metal oxide MeO$_y$, a typical reaction might be

$$MeCl_x + \frac{y}{2}O_2 \Rightarrow MeO_y + \frac{x}{2}Cl_2 \tag{4.10}$$

In most cases, this reaction requires a temperature in the range from 1200 to 1500 K, but adding water to the system leads to a significant reduction in the reaction temperature:

$$MeCl_x + \frac{x}{2}H_2O + \frac{2y-x}{2}O_2 \Rightarrow MeO_y + xHCl \tag{4.11}$$

However, in many cases, the advantage of a reduced reaction temperature is outweighed by the disadvantage of having highly corrosive hydrochloric acid as a byproduct in the system. If available, the use of a carbonyl is recommended; for example, in the synthesis of maghemite, γ-Fe₂O₃, would be:

$$2Fe(CO)_5 + \frac{13}{2}O_2 \Rightarrow Fe_2O_3 + 10CO_2 \tag{4.12}$$

For the available carbonyls, carbonyl chlorides or nitrosyl carbonyls, this process will operate successfully in most cases at temperatures below 600 or 700 K. In many cases, the advantage of a low temperature for evaporation of the precursor and for the reaction is more important than the high price and difficult handling. Handling is difficult, because most carbonyls are highly toxic and have limited stability in the open air. However, by selecting appropriate temperatures, the use of carbonyls will

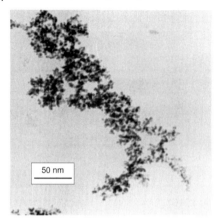

Figure 4.11 Fe_2O_3 powder prepared by decomposition of Fe
$(CO)_5$ and subsequent oxidation [8]. This product consists of
fractal agglomerates of amorphous 3-nm particles. The high
surface product has excellent catalytic properties. (TEM
micrograph reprinted with permission from Nanophase
Technologies Corporation, Romeoville, IL, USA.)

provide much more freedom in terms of the morphology of the intended powder.
Low temperatures lead to extremely fine and fluffy – in most cases glassy – powders,
whereas the products tend to be more crystalline when using a higher reaction
temperature. A typical example of such a fine powder, in this case glassy Fe_2O_3 [8], is
shown in Figure 4.11. These fine powders also tend to form extended agglomerates
(see Figure 4.11), whereas the primary particle size is quite uniform, in the range of
approximately 3 nm.

The other, most important groups of precursor compounds include metal alkoxides
and acetyl acetonates. In particular, for the metals of Groups II and IIa of the Periodic
Table, and for the rare earth elements, these are the most important precursors.

In addition to oxides, chemical vapor synthesis also allows the synthesis of carbides
and nitrides, although in this case great care must be taken when selecting the
precursor as the free enthalpy of formation of carbides and nitrides is significantly
smaller than for oxides. A typical example is the synthesis of SiC by Erhart and
Albe [9], via the pyrolysis of tetramethyl silane as precursor.

$$Si(CH_3)_4 \Rightarrow SiC + 3CH_4 \tag{4.13}$$

As the free enthalpy of formation is quite low, many side reactions, such as the
formation of Si_2C and elemental silicon and carbon, are possible. An additional,
often-applied reaction for the synthesis of SiC is

$$(CH_3)_3Cl_3Si \Rightarrow SiC + 2CH_4 + 3HCl_3 \tag{4.14}$$

However, even when the chemical reaction equation seems straightforward, this
reaction may be quite problematic, as an inappropriate selection of the reaction
temperature may lead to the formation of elemental Si particles and carbon (as soot).
In order to synthesize nitrides, nitrogen is used as the reaction and carrier gas.

However, as chlorides are more stable than nitrides it is necessary to add hydrogen to the system in order to shift the equilibrium towards nitride and HCl formation:

$$MeCl_n + \frac{m}{2}N_2 + \frac{n}{2}H_2 \Rightarrow MeN_m + nHCl \tag{4.15}$$

Apart from the problem of ammonium chloride (NH_4Cl) being formed as a byproduct, in most cases it is beneficial to use ammonia as the hydrogen carrier.

4.4
Laser Ablation Process

The situation is even more complex when applying a pulsed laser beam as a source of energy. However, this "laser ablation technique" has the advantage of allowing not only the use of metals but also oxides as precursors, which makes this process extremely versatile in its application. The general design of a production unit applying the laser ablation process is shown in Figure 4.12.

A system for powder production using the laser ablation process generally consists of two essential elements: the pulsed high-power laser, and with the optical focusing system and feeding device for the precursor. In order to produce larger quantities with this process, rotating targets and automatic wire feeding systems have been developed.

High-power laser pulses are focused onto the surface of the precursor target to evaporate the material. The target, whether metallic or nonmetallic (but even mixed targets are possible), is heated to temperatures of more than 1300 K. Due to the rapid evaporation in the high-power laser pulse, even the stoichiometry of a complex mixed target is preserved in the vapor phase. During the pulse, a supersonic jet of evaporated material (known as a *plume*) is ejected perpendicular to the target surface, and expands into the gas space above the target. Immediately after the laser pulse, the

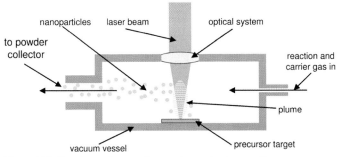

Figure 4.12 Experimental set-up for powder synthesis according to the laser ablation process. The laser beam is focused at the surface of a target (a metal or an oxide). In the high-intensity laser beam, a plume of evaporated material is ejected perpendicular to the target surface, expanding into the gas space above the target. The particles formed are transported with the carrier gas to the powder collector.

temperature in the plume reaches values of 3800 K and more [10]. During the adiabatic expansion of the plume, the temperature decreases, and the particles formed are transported by a continuous stream of carrier gas to the powder collector. The carrier gas may also contain reactive gas components; for example, to obtain oxides oxygen is added, for carbides CH_4, and for nitrides, NH_3.

Within the plume, there is a supersaturated vapor favoring the formation of particles. As the duration of the supersaturated conditions is limited by the adiabatic expansion of the plume, the gas pressure in the reaction vessel plays a crucial role in particle nucleation and growth. At a low gas pressure the plume expands very rapidly, and therefore the concentration of reactive species in the plume also decreases very rapidly, and this limits the particle growth. In contrast, at a higher gas pressure the supersaturation is higher. However, the higher the supersaturation, the smaller the size of the nucleus required for condensation, and this leads to a large number of nuclei and, consequently, again to smaller particle sizes. This simplified description of the complex processes in the plume is well supported by experimental results. As an example, Figure 4.13 shows, in graphical form, the dependency of Co_3O_4 particle size on gas pressure [11]; here, small particle sizes are apparent at lower and at higher gas pressures, whereas the particle size is largest over an intermediate pressure range.

With regards to laser type, either frequency-converted Nd-YAG or excimer lasers with pulse durations in the nanosecond range are usually applied. The interaction of these nanosecond pulses leads, especially at the surface of good thermal conductors, and prior to evaporation, to the formation of a pool of melted material. In the case of targets with a complex composition, this may lead to a powder composition, which differs from that intended. This problem may be avoided by applying picosecond lasers, as such a short, high-power laser pulse does not lead to melting at the surface; rather, the material evaporates instantaneously.

One general problem of the laser ablation process is that a high concentration of evaporated material gathers in the plume. In the case of an insufficiently rapid expansion, this may lead to the formation of agglomerates which, in most cases, are

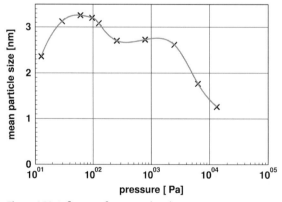

Figure 4.13 Influence of reaction chamber gas pressure on mean particle size of the product (in this case Co_3O_4) [11].

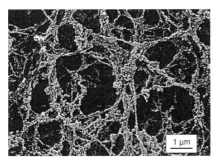

Figure 4.14 Products synthesized using laser ablation are typically highly agglomerated. This electron micrograph shows a highly agglomerated, web-like structure of silicon nanoparticles with a primary particle size of 10 nm. (Reprinted with permission from [14]; Copyright: Elsevier 1998.)

either fractal with a fractal dimension in the range from 1.7 to 1.9 [12], or web-like. A typical example of such an agglomerated powder is shown in Figure 4.14 [13], where silicon primary particles of about 10 nm are connected together and form a web-like structure, though the individual particles are not visible.

In most cases, the formation of agglomerates is not intended, and consequently many research investigations have been undertaken towards synthesizing individualized particles using this process. One typical – and successful – approach was reported by Wang *et al.* [13], who used an atypical YAG laser with a wavelength of 1064 nm and a laser pulse width of 0.3–20 ms. A wire of pure iron was used as precursor. The particle size of the product obtained (see Figure 4.15) was in the range from 5 to 55 nm, with only a minor proportion of material in the size range 50–90 nm.

The detailed size distribution spectrum (see Figure 4.16) shows a nonsymmetric particle size distribution typical of a random process of particle formation, and may

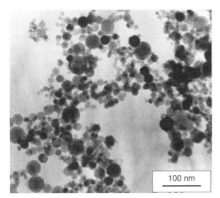

Figure 4.15 γ-Fe$_2$O$_3$ powder produced by a highly progressed laser ablation process. The starting material was metallic iron. (Reprinted with permission from [13]; Copyright: Elsevier 2006.) This product, with individual particles in the size range of 5 to 90 nm, contrasts with the product shown in Figure 4.14.

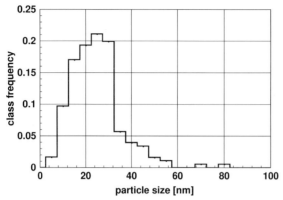

Figure 4.16 Particle size distribution of the γ-Fe$_2$O$_3$ powder product shown in Figure 4.15. The nonsymmetric particle size distribution is typical of synthesis methods based on purely random processes [14].

be fitted with a log-normal distribution function. Taken together, these data demonstrate the potential and importance of the laser ablation process for nonagglomerated materials, notably because it has no special precursor requirements.

4.5
The Microwave Plasma Process

The processes of chemical and physical vapor synthesis, as well as laser ablation, are purely random processes; hence, the only means by which particle size and size distribution can be influenced are the concentrations of active species in the gas, the temperature and, most importantly, the rapid cooling (quenching) of the gas after leaving the reaction zone. This situation is entirely different from that of the microwave plasma process, where the particles originating in the plasma zone carry electric charges. As a consequence, the probability for coagulation and agglomeration is significantly reduced, as the collision parameter decreases with increasing particle size. These effects are based on considerations leading to Equations (4.6)–(4.9) and Figure 4.6a and b. Vollath [15,16] developed a microwave plasma process for the synthesis of nanoparticles by exploiting the benefits of charged particles, as high production rates of unagglomerated particles and narrow particle size distribution are in direct contrast to the classical processes of gas-phase synthesis.

In order to understand the special properties of the microwave plasma process, it is first necessary to analyze the energy transfer in a microwave plasma. The energy U transferred to a particle with the electric charge Q in an oscillating electrical field is inversely proportional to the mass of the particle m and the squared frequency of the electrical field f.

$$U \propto \frac{Q}{mf^2} \tag{4.16}$$

As the mass of the electrons is a few thousand times smaller than that of the ions, a substantially larger amount of energy is transferred to the electrons, as compared to the energy transferred to the ions. Whilst Equation (4.16) is valid for one charged particle in an oscillating electrical field, in a plasma one finds free electrons, ions, dissociated gas, and precursor molecules in addition to neutral gas species. Therefore, collisions between charged and uncharged particles limit the mean free path of the charged particles accelerated in the electric field, ruling the energy transfer to the particles. Consequently, the collision frequency z of the gas species must be considered [17]:

$$U \propto \frac{Q}{m} \frac{z}{f^2 + z^2} \tag{4.17}$$

Equation (4.17) does not alter the mass relationship of the energy transfer, but it does show the reduction in energy transfer to the charged particles by collision with other neutral species. Equation (4.17) introduces a dependency of the collision frequency z in the plasma, depending essentially on the gas pressure. The free path length λ of the electrons now limits the maximum energy U_{max} to be transferred in an electric field with the field strength E:

$$U_{max} = Q\lambda E \tag{4.18}$$

As the strength of the electric field in a resonant microwave cavity (as applied to microwave plasma synthesis) is significantly above $10^3\,V\,cm^{-1}$, the energy transferred to an electron, depending on the gas pressure, may be in the range of kilo electron volts. The energy transferred to a charged particle as a function of the collision frequency z that is proportional to the gas pressure is shown in Figure 4.17. A detailed analysis of Equation (4.17) reveals a maximum energy transfer in the case

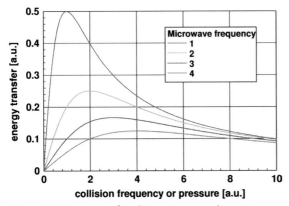

Figure 4.17 Energy transferred in a microwave plasma to an electrically charged particle as a function of the collision frequency and microwave frequency. The maximum energy is transferred when the collision frequency is equal to the microwave frequency. Stable operation is possible only at collision frequencies below the maximum of energy transfer.

$f=z$. For $z<f$, the energy transfer increases, whereas for $z>f$ the energy transfer decreases with increasing gas pressure (see Figure 4.17).

Practical experience has shown that stable operation is possible only in the range $z<f$ since, beyond the maximum of energy transfer the plasma is unstable and tends to quench. As the energy transfer in an electrical field is a function of the microwave power, gas pressure, and energy input, these provide powerful means for adjusting the conditions for synthesis, thereby obtaining the optimum conditions for the chemical reaction and the product.

In order to calculate the energy transferred to the free electrons, it is necessary to estimate the mean free path length of the free electrons; the results of such an estimation, assuming argon as the gas species, are shown in Figure 4.18.

Within the range of gas pressures usually applied for microwave plasma synthesis, the mean free path length for free electrons may be in the range of 10^{-3} to 10^{-2} m, leading to an energy of the electrons ranging from a few keV to more than 10 keV. The energy transferred to the ions is, at maximum, close to 100 meV, which is significantly above the thermal mean value.

Therefore, a microwave plasma is a nonequilibrium system, where the energy is deposited primarily to the electrons. The "temperature" of the electron is significantly higher than that of the ions or other charged species. Uncharged particles have the lowest temperature in the system, which is why the overall temperature of a microwave plasma is significantly lower than that in an AC or DC plasma, where temperatures of up to 10^4 K are attained. By using the appropriate conditions, the temperature in the microwave plasma process can be adjusted within the range from 400 to 1200 K.

The electrons gain sufficient energy to ionize the gas and precursor molecules, and to dissociate the latter. The electron energy is also sufficiently high to ionize those particles further which were already charged. The probability of the positively charged particles recombining with the high-energy electrons is small.

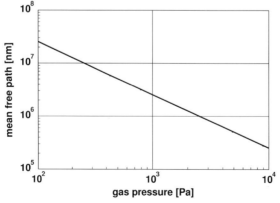

Figure 4.18 Estimated values for the mean free path length of free electrons as a function of the argon gas pressure.

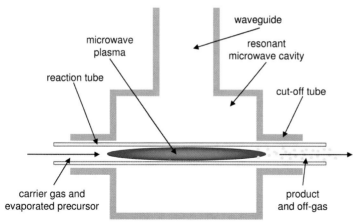

Figure 4.19 Set-up for particle synthesis using the microwave plasma process. The microwave plasma is ignited in a reaction tube which passes a resonant microwave cavity. A carrier gas containing the reaction gas transports an evaporated precursor into the plasma zone. The reaction product (nanoparticles) is collected after the reaction zone. Microwaves are coupled into the device with the waveguide.

The layout of a microwave plasma system for nanoparticle synthesis is shown in Figure 4.19. The central reaction tube is made from silica glass (pure silica must be used to avoid parasitic absorption of the incoming microwaves) and passes a resonant microwave cavity that is connected via a waveguide to the microwave generator and the tuning system [15]. At each end of the cavity there is a cut-off tube that attenuates the microwaves to avoid radiation leakage. The plasma is ignited at the intersection of the reaction tube and the microwave cavity; this is the zone where the reaction occurs. The evaporated precursor is introduced into the system via a stream of mixed carrier and reaction gases. Usually, the gas flow is selected so that the residence time of the particles in the reaction zone is less than 10 ms, while the gas pressure is selected in a range from 500 to 10^4 Pa, and the temperature is adjusted from 400 to 800 K. The so-called "kitchen frequency" of 2.45 GHz is most often used as a microwave frequency, although a frequency of 0.915 GHz may also be applied.

The mechanism of nanoparticle formation in the microwave plasma differs from that of a chemical vapor synthesis in a tubular furnace, as the reactants are ionized and dissociated. This also allows lower reaction temperatures.

In general, the chemical reactions are the same as in a conventional furnace, albeit at a lower temperature. However, care must be taken that the reaction or reaction products do not neutralize the electrical charges of the particles. A typical example of such an unfavorable reaction (e.g., for oxide synthesis) occurs after the addition of water. As shown in Equation (4.15), where a chloride is assumed as the precursor, in conventional terms the addition of water shifts the equilibrium more towards the oxide. However, in plasma the water dissociates and forms H^+ and OH^- radicals, so that two additional reaction routes are possible:

(i) $\quad MeCl_n + \dfrac{m}{2}O_2 + \dfrac{n}{2}H_2 \Rightarrow MeO_m + nHCl$

(ii) $\quad MeCl_n + (m+x)OH^- + (m+x)H^+ \Rightarrow$

$$(4.19)$$

$$MeO_m + xHClO + (n-x)HCl + \left(m+x-\dfrac{n}{2}\right)H_2 \qquad (4.20)$$

where x is assumed to be much greater than n.

The addition of water has two effects:

- the consequent formation of HCl (and possibly HClO) increases the reaction enthalpy, such that the temperature of the gas, determined after passing the reaction zone, is generally higher.
- the electrical charges of the particles are quenched, so that particle growth is inhibited.

In detail, the positively charged particles are neutralized by collision with OH^- ions, thereby blocking the mechanism that limits particle growth by the following process:

$$H_2O \Rightarrow H^+ + (OH)^-$$
$$particle^{n+} + n(OH)^- \Rightarrow (particle^{n+} + n(OH)^-)^{neutral} \qquad (4.21)$$

The neutralized particles carry a hydroxide layer at the surface, so the production of larger particles with a broad particle size distribution might also be expected. The near-perfect narrow size distribution of ZrO_2 powder produced without water addition, compared to the drastic effect of adding water to the reaction gas, under otherwise similar condition, on the synthesis product are shown in Figure 4.20a and b, respectively. The zirconia specimen produced without water addition showed a grain size approximately 8 nm, and most of the grains were of equal size. In contrast, the material produced with added water was characterized by a broad distribution of particle sizes, ranging from 10 to 50 nm. Such a dramatic difference between these two batches of the same material clearly demonstrated the validity of this simple model.

The narrow size distribution of particles shown in Figure 4.20a is characteristic of the microwave plasma process, and also occurs with smaller particles. An example of the particle size distribution of a ZrO_2 powder (mean particle size ca. 3 nm; distribution determined using particle mass spectrometry) is shown in Figure 4.21.

Interestingly, the particle size distribution in Figure 4.21 was extremely narrow (maximum 3.1 nm) when compared to the values shown in Figures 4.4 and 4.16. Despite the narrow size distribution, the distribution function remained asymmetric, there being few large particles in the system, which indicated that the particles had grown by a process of random collisions. Although a particle size of 3 nm is

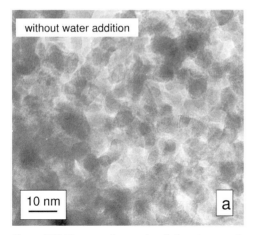
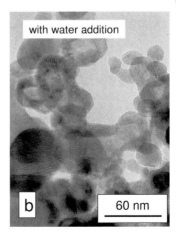

without water addition

with water addition

10 nm

a

b

60 nm

Figure 4.20 Electron micrographs of zirconia powders synthesized using the microwave plasma process from ZrCl$_4$ under different conditions [1]. (a) ZrO$_2$ ex ZrCl$_4$ product obtained in the absence of water; note the size-limiting effect of particle charging during synthesis. (b) ZrO$_2$ ex ZrCl$_4$ product obtained in the presence of water; the size-limiting effect of particle charging is lost, as the positively particles are neutralized by (OH)$^-$ radicals derived from the dissociated water. (Reproduced with permission from [1]; Copyright: Springer 2006).

extremely small, the smallest oxide particles produced in the microwave plasma process are approximately 2 nm.

Very similar results are obtained when synthesizing nitrides which, in general, are well crystallized and show a narrow particle size distribution. ZrN prepared using ZrCl$_4$ as the precursor and a mixture of N$_2$ and NH$_3$ as the carrier and reaction gases, respectively, is illustrated in Figure 4.22 [18].

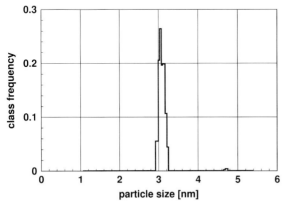

Figure 4.21 Particle size distribution of zirconia powder synthesized using the microwave plasma process. Data were measured online during synthesis using particle-sizing spectrometry (Roth [34]).

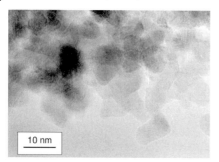

10 nm

Figure 4.22 Zirconium nitride (ZrN) particles (ex $ZrCl_4$) synthesized in a microwave plasma. A mixture of N_2 and NH_3 was applied as carrier and reaction gas, respectively. The lattice fringes visible within some particles indicate well-crystallized material. As with oxides, the particle size distribution is very narrow [18].

4.6
Flame Aerosol Process

Among all of the processes used to produce nanoparticulate powders, the flame aerosol process is the oldest. Additionally, it is the only one to be used for mass production in the kiloton range. Although this well-established industrial process has been used for many decades, the basic principles are still not well understood, not least because the processes of powder synthesis and particle formation take place at extremely high temperatures and over very short times.

Since prehistoric times, in China, carbon black (as a pigment for inks) has been produced by flame aerosol processes, and today this same technology – or one of its many variants – is still used to produce thousands of metric tons of carbon black, fumed silica, and titania (TiO_2) pigments. In addition, during recent years the application of this process has been expanded to incorporate many highly specialized products (for excellent reviews, see Refs. [19,20]). As the aerosol flame process has such a long history and broad application, many highly specific variants, leading to the production of particles with different morphology, size, and crystallinity, have been developed.

In the simplest case, a flame reactor set-up is as shown in Figure 4.23a and b. The flame reactor consists of a primary flame that is fueled with hydrogen, methane, or another hydrocarbon fuel. In most cases, the gaseous fuel is premixed with oxygen or air in the burner. In the case shown in Figure 4.23a, many small primary flames surround the secondary flame, where the reaction for particle formation occurs. Both, Figure 4.23a and b demonstrate the synthesis of silica, and therefore silane (SiH_4) or silicon tetrachloride ($SiCl_4$) were assumed as the precursor compounds. Reaction of the precursor with excess oxygen forms the secondary flame, while the particle size is adjusted by diluting the precursor with an inert gas such as argon or nitrogen. Instead of a gaseous or vaporized precursor, it is also possible to use liquid precursors, which may be sprayed via a two-phase nozzle into the primary flame. It is also possible to use a premixed gas for the primary flame or as the spraying gas in a two-phase nozzle [21,22].

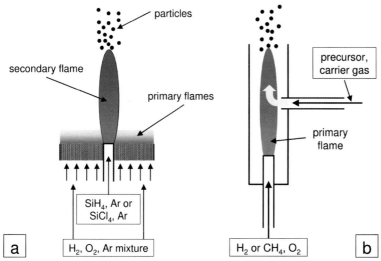

Figure 4.23 Typical arrangements for flame synthesis of
nanoparticles. (a) Primary flames surrounding a secondary flame,
in which the powder is produced, characterize this arrangement.
(b) In this design, the vaporized precursor is introduced into the
primary flame for reaction.

In order to obtain well reproducible results for this synthetics process, a stable
behavior of the flame is absolutely necessary. The appearance of a flame with different
additions of oxygen is shown in Figure 4.24; here, methane ($1.4\,L\,min^{-1}$) was used as
the fuel and hexamethyl disiloxane (HMDSO; $2.9\,L\,min^{-1}$) as the precursor to
produce 17 g of silica per hour. With regards to oxygen additions, two regimes are
observed: (i) at low flow rates, the flame is unstable and fluctuating; and (ii) with
increasing oxygen additions, the flame is more stable and hotter. The transition
between these two ranges is clearly pronounced, especially in terms of the particle
size produced [23].

As mentioned above, the transition from a fluctuating to a stable flame is reflected
significantly in the particle size obtained by the process. The data in Figure 4.25 show
clearly that there is a critical flow rate for oxygen (in this case $5.7\,L\,min^{-1}$), although
interestingly, the maximum particle size is observed at the transition. Over the range
characterized by a fluctuating flame it is difficult to obtain either stable or reproducible conditions, and consequently such a range is avoided.

The oxygen content in the flame not only influences the average particle size but
also has a major influence on the morphology of the powder. There are two reasons
for this:

- Under otherwise constant conditions, with increasing the addition of oxygen to the
  flame, the temperature of the flame increases.
- As the flow rate increases the flame becomes shorter (see Figure 4.24), and
  therefore the residence time of the powder particles in the flame is shortened.

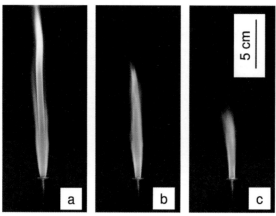

Figure 4.24 Influence of different oxygen additions on the appearance of a silica-producing flame. For the synthesis of 17 g silica h^{-1}, 2.9 L min^{-1} HMDSO was used as precursor and 1.4 L min^{-1} methane as fuel [23,35]. (a) O_2 flow 2.5 L min^{-1}; the flame was fluctuating; (b) O_2 flow 8.5 L min^{-1}; the flame was stable; (c) O_2 flow 24 L min^{-1}; this flame was the hottest.

It is for the latter reason that the particle size decreases with increasing oxygen addition, as is apparent in the micrographs shown in Figure 4.26.

As the flame temperatures are high and the product, silica, is not crystallized, the particles are spherical in shape, and this is typical for all products produced by flame synthesis. However, as the extent by which the flame shortens becomes correlated with a reduced residence time at high temperature, the probability of forming fractal agglomerates is increased. This can be clearly seen for the products shown in Figure 4.26a and b, which consist of spherical particles with a broad size distribution; this contrasts with the particles in Figure 4.26c, which are clearly agglomerated

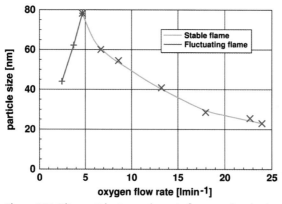

Figure 4.25 Silica particle size syntheses in flames and under the conditions shown in Figure 4.24. Stable and reproducible particle production is difficult or barely possible in the region marked as "fluctuating flame" [23].

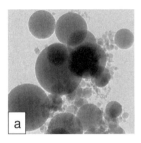
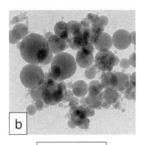
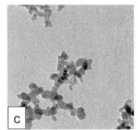

200 nm

Figure 4.26 Electron micrographs of silica powders produced with varying oxygen content in the flame under otherwise constant conditions. Conditions for synthesis were identical to those given in Figure 4.24 [35]. (a) O_2 addition 2.5 L min^{-1}; the product consisted of nonagglomerated particles, but with a broad distribution of particle size. (b) O_2 addition 8.5 L min^{-1}; similar to (a), the particles were not agglomerated. Although the distribution of particle sizes remained broad, the size of the largest particles was reduced. (c) O_2 addition 25 L min^{-1}; in contrast to the products depicted in (a) and (b), fractal agglomerates were formed; the particle size was quite uniform. (Reproduced with permission; Pratsinis [35].)

(typically, these are fractals). The change in synthesis conditions reflected in the transition from Figure 4.26a to b leads to a reduction in the maximal particle size, whereby conditions utilizing the largest oxygen supply will lead to small particles of a uniform size.

The flame temperature used for a typical flame synthesis can be influenced by using either air or pure oxygen as the oxidizing gas. The temperature distribution in the axis of a flame is shown in Figure 4.27 [24], and was obtained in an experiment to produce silica. The primary flame used a premixed methane : air mixture, while HMDSO was selected as the silicon-containing precursor. In Figure 4.27, temperature maxima in

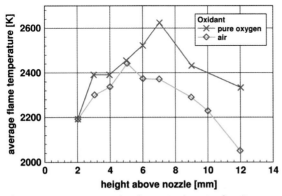

Figure 4.27 Temperature distribution in the axis of a silica-producing flame with pure oxygen or air as oxidant. As expected, the temperature in the flame using pure oxygen was significantly higher. HMDSO was used as silicon-containing precursor [24].

excess of 2400 K were achieved, at which many of the potential products had already melted, especially when a particle size-dependent reduction of the melting point is taken into consideration. In addition, the diffusion rate also increased exponentially with temperature, thus favoring the formation of spherical particles.

The high flame temperature has a significant influence on the morphology of the reaction product since, at high temperatures, the probability of forming agglomerates is very high. The formation of fractal or linear clusters was for a long time synonymous with the structure of powders synthesized in flames. However, as knowledge of the detailed processes occurring during particle formation by flame synthesis was acquired, it became possible to tailor the reaction in order to obtain the intended product [25]. A typical example of such tailored products is shown in Figure 4.28, where TiO_2 particles made from $TiCl_4$ in a methane diffusion flame reactor are depicted. Here, air or oxygen was used as the oxidant. The use of air as an oxidant results in lower maximum temperatures and shorter flame heights as compared to pure oxygen. The product obtained in the air flame (see Figure 4.28a) shows the characteristic agglomerated product, whereas the higher temperature of the oxygen flame resulted in nonagglomerated spherical titania particles (see Figure 4.28b). When comparing these products with those depicted in Figure 4.27 it can be concluded that, in this instance, the residence time in the hot zone of the flame is responsible for the different morphology of both products.

A series of systematic studies on the influence of the configuration for the addition of precursor gas, fuel, and air on the morphology of the reaction product using an experimental set-up (see Figure 4.29) led to important insights [26]. In the example of the synthesis of TiO_2 from $TiCl_4$, the basic configuration consisted of three concentric tubes that were each used in a different way to feed the system with precursor, methane, and air. Additionally, the precursor was diluted using argon. In all of these

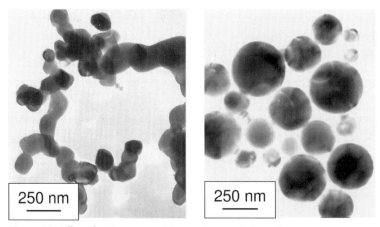

Figure 4.28 Effect of oxidant composition on the morphology of titania (TiO_2) powder. $TiCl_4$ was used as precursor [35]. (a) A flame with air as oxidant leads to the formation of aggregated particles. (b) Oxygen as oxidant leads to higher flame temperature with consequent nonagglomerated, spherical particles.

Figure 4.29 Experimental configuration to study the influence of different modes of gas supply on the morphology of the product [26].

experiments, two parameters were held constant: (i) the innermost tube was always used for the evaporated precursor; and (ii) the amount of each gas species was held constant. The different configurations and resultant mean particle sizes of the products are listed in Table 4.1.

The micrographs in Figure 4.30a–d show the most important characteristics of the products obtained in the four different configurations. Clearly, the smallest particle size (ca. 10 nm) was achieved in configuration 1, which was characterized by maximal dilution of the precursor gas with air. Mixing the precursor with fuel gas led to a maximal particle size of more than 100 nm. As the temperature level in the flame also controls particle growth, the highest temperature of the four configurations explained the most extensive particle growth (see Figure 4.30c and d).

Additionally, when comparing configurations 2 and 3 it is clear that simply by exchanging the gas connections between ports B and C, the mean particle size is increased from 25 to 80 nm. The experiments leading to the products depicted in Figure 4.30a–d indicate that, when setting up a flame synthesis plant, many

Table 4.1 Connection of different gases for the experimental configuration depicted in Figure 4.26, and mean particle sizes of the products [26].

| Configuration number | Connection of the different gases | | | Average particle size (nm) | Micrograph in Figure 4.30 |
	A	B	C		
1	$TiCl_4$, Ar, air	–	CH_4	11	(a)
2	$TiCl_4$, Ar	Air	CH_4	25	(b)
3	$TiCl_4$, Ar	CH_4	Air	80	(c)
4	$TiCl_4$, Ar, CH_4	–	Air	105	(d)

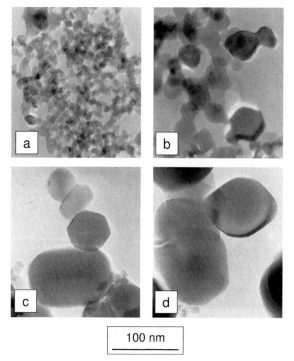

100 nm

Figure 4.30 Morphology of TiO$_2$ products obtained by flame synthesis in an experimental device according to Figure 4.29. Details of gas supply are listed in Table 4.1. The flow rates of the gases were equal in all configurations [26]. (a) Configuration 1: at gas connection 'A', TiCl$_4$ and air, diluted with Ar, were supplied. The fuel gas, CH$_4$, was connected to 'C'. (b) Configuration 2: the fuel, CH$_4$, was connected to 'C', and air to 'B'. TiCl$_4$ diluted with Ar, was supplied at connection 'A'. (c) Configuration 3: the connections of fuel and air were exchanged compared to configuration 2. (d) Configuration 4: the largest grain size was obtained by mixing the fuel into the gas stream of TiCl$_4$ diluted with Ar. (Reproduced with permission from [26]; Copyright: Elsevier, 2004).

investigations are required to identify the configuration that will deliver exactly the intended product.

As well as showing a tendency to deliver severely agglomerated products, one significant disadvantage of the flame synthesis process is the broad distribution of particle sizes, and many studies have been conducted in attempts to minimize this problem. One interesting approach has been to use the electrical charges of particles to avoid their coagulation and agglomeration, as is used in the microwave plasma process. Two such experimental set-ups are shown in Figure 4.31: in Figure 4.31a the electric field is set up between two plate electrodes, whereas in Figure 4.31b two needle electrodes are utilized, in the configuration shown. In both the cases, the electrical field strength was similar, at less than 2 kV cm^{-1}. Both arrangements led to a significant reduction in particle size with increasing strength of the electrical field, and the particle size distribution was also found to be narrower. The mechanisms acting in these two arrangements are different.

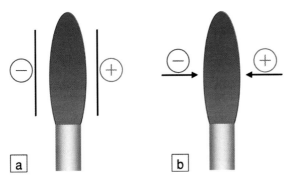

Figure 4.31 Arrangements for powder synthesis in a flame combined with a transverse electrical field to avoid grain growth by agglomeration. (a) The transverse electrical field is set up between two plate electrodes; this leads to a separation of particles with electrical charges of different sign. As the particles repel each other, agglomeration is reduced. (b) The corona discharge from the needle electrodes placed on each side of the flame leads to an emission of electrons that charge the particles in the flame negatively. As all particles now carry electrical charges of equal sign, they repel each other; this avoids agglomeration.

When analyzing the experimental set-ups shown in Figure 4.31 it is clear that the field strength in the case of the plate electrodes is by far insufficient to ionize the gas molecules or the particles. However, when considering the high temperatures in the flame (ca. 2500 K), it may be assumed that thermal ionization of the particles has occurred, or at least thermal ionization supported by the electrical field. In either case, the particles carry electrical charges and, as the energy of the free electrons is relatively small, an agglomeration at the uncharged particles is possible. Thus, it must be assumed that the particles are carrying electrical charges of both signs. Particles with electrical charges of different sign may be separated in the electrical field and move in the direction of the different electrodes. This reduces the probability of agglomeration, as particles with electrical charges of equal sign will repel each other. Certainly, an increasing influence of the electric field with increasing field strength would be expected, and these results have been confirmed experimentally.

A somewhat different mechanism, albeit with similarly good results, would be expected when using the experimental device shown in Figure 4.31b. Here, a corona discharge is observed from the tips of the electrodes, and the electrons emitted from the cathode cross the flame. While passing through the flame, the electrons donate a negative charge to the particles such that a large proportion of them carry negative electric charges which increases with increasing voltage between the two electrodes. As the charged particles repel each other, particle growth by agglomeration is thwarted, and consequently the particle size of the product decreases in line with the increasing field strength between the electrodes.

One good way to demonstrate the nature of the mechanism influencing the electric field is by the shape of the flame. The shape of a flame without a transverse electrical field and with an electrical field between the plate electrodes or needles is shown in Figure 4.32. The different shapes of the flames are obvious; in the case of the plate

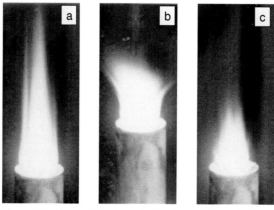

Figure 4.32 Shape of titania-producing flames with and without transversal electric field [35]. (a) The electrical field is switched off. (b) A transverse electrical field of 2 kV cm^{-1} is applied between the plate electrodes. Splitting of the flame is caused by the attraction of different electrically charged particles by the electrodes. (c) A transverse electric field of 2 kV cm^{-1} between the needle electrodes.

electrodes the flame is seen to be separated into two parts, reflecting the attraction of the electrodes to the charged particles. This is not apparent in the case of the needle electrodes.

When considering particle size as a function of the two electrode arrangements, the differences in the way that the two arrangements function become clear. The separation of differently charged particles begins at relatively low voltages, whereas the arrangement with needle electrodes becomes active only after the electrical field strength has reached a level where the cathode emits electrons into the system. Due to the relatively short extension of the electrical field in the length axis of the flame, the separation effects, as found between the plates, show no significant influence. Direct proof of this is provided by the way that the electrically charged particles are pulled out of the flame by the electrical field. The experimental arrangement detailed in Figure 4.31a is illustrated in Figure 4.33, from which it is clear that the flame is separated in the electric field while the electrodes become covered with particles.

When electrically charged particles are pulled from the flame by the electrical field between the electrodes (see Figure 4.33), the particles are deposited on the surface of the cold electrodes. The thickness of the particle furring on both electrodes is very similar, which indicates not only an equilibrium of electrically charged particles in the flame but also that thermal ionization processes are occurring.

As a further example of the success of this experimental strategy, the average grain size of powders obtained in a device using plate and needle electrodes was examined as a function of the electric field strength (see Figure 4.34). Here, the different phenomena acting in the two experimental set-ups are clearly visible. In the case of plate electrodes, the separation of the particles with opposite charges begins at

Figure 4.33 Flame synthesis of titania between plate electrodes. The flame is broadened and split up by the electrical field. The different electrically charged particles are then attracted by the plate electrodes and deposited [35].

relatively low electrical field strengths and increases significantly with increasing strength of the electrical field. This is different for the needle electrodes, where a significant effect begins at an electrical field strength of approximately $0.8 \, \text{kV cm}^{-1}$, where a corona discharge begins at the needle connected to the negative pole. At this voltage between the two needle electrodes, the corona discharge delivers free electrons. However, with increasing intensity of the corona discharge, the effect of the electrical field on the particle size increases, as does the probability of charging the

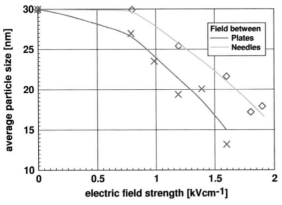

Figure 4.34 Synthesis of TiO$_2$ (ex TiCl$_4$) in a methane:oxygen flame with a transverse electrical field. The figure shows the dependency of particle size on electrical field strength between two plate or needle electrodes in an experimental arrangement according to Figures 4.31a and b. Between the plate electrodes, the separation of differently charged particles begins at a relatively low electric field strength. Between the needle electrodes, the intended effect starts at an electrical field strength where electron emission begins at the tip of the negative needle [27,28].

particles. Compared to the plate electrodes, the field effect for needle electrodes is less pronounced for two reasons: (i) in contrast to the plate electrodes the field influences only a small part of the flame; and (ii) field separation at lower voltages is not observed. Furthermore, from thermal ionization studies (as shown clearly in Figure 4.33), equal numbers of positively and negatively charged particles are formed. Hence, the positively charged particles are first neutralized and then negatively charged, which reduces the efficiency of this design.

Three micrographs of products synthesized in an experimental device with a transverse electrical field between two plate electrodes, according to Figure 4.31a, are shown in Figure 4.35. The powders, TiO_2 (ex $TiCl_4$) were produced in a premixed methane : oxygen flame with electric fields of 0, 1.4, and 1.6 kV cm^{-1}. As may be seen in Figure 4.35a–c, the size of the particles and the degree of agglomeration decreases with increasing strength of the electrical field.

The average particle sizes shown in Figure 4.34, together with the micrographs depicted in Figure 4.35a–c, show that the flame synthesis process which until now has been applied primarily to the production of huge quantities of inexpensive materials has – especially in combination with a transverse electrical field – the potential to produce highly sophisticated materials with small particle sizes and a low degree of agglomeration, and with a relatively narrow particle size distribution. Moreover, these products closely resemble those obtained using the microwave plasma process.

Flame synthesis has many more additional variants, one of which may be applied to burnable organic liquids instead of gaseous fuels. The precursor is dissolved in the

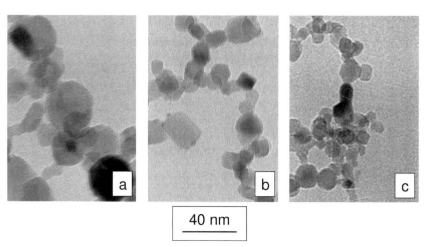

40 nm

Figure 4.35 Titania particles made from $TiCl_4$ in a premixed methane:oxygen flame assisted by an electrical field between plate electrodes across the flame according to Figure 4.28a [35]. (a) Reference material obtained without an electrical field. (b) Product obtained by applying an electrical field of 1.4 kV cm^{-1}, as compared to (a); a reduced particle size and a narrower size distribution were observed. (c) Product obtained by applying an electrical field of 1.6 kV cm^{-1}; compared to (a) and (b), the smallest particle size and least degree of agglomeration were observed. This product had the narrowest particle size distribution.

liquid fuel, with typical examples being solutions of water-free chlorides in acetonitrile (CH_3CN) or solutions of acetylacetonates (e.g., $C_5H_8O_2)_3Al$) in appropriate organic solvents. The selection of an organic solvent may be problematic, as the compounds selected should not have the propensity to produce soot. Therefore, the use of benzene should be avoided, although mixtures of benzene and ethanol have been applied with reasonable success.

4.7
Synthesis of Coated Particles

Although many applications of nanomaterials require the use of nanocomposites, the impossibility of obtaining well-distributed nanocomposites simply by blending processes led to the development of coated nanoparticles. The aims of coating may be manifold: in the simplest case, the coating serves simply as a second phase to dilute the nanoparticles; typical applications include the addition of ceramic nanoparticles to adjust the refractive index or improve the mechanical properties of a polymer. A directly related case, though much more sophisticated, is to use a coating as a distance holder to adjust particle interactions; this may be used in connection with magnetic nanoparticles. Finally, the most advanced case is to design coatings (or combinations of different coatings) to add additional properties to the particles; here, the most advanced examples include magnetic particles with luminescent coatings or similar constructs.

The processes used for particle coating must fulfill a series of requirements. The first and most important point is that the particles remain individualized and are not agglomerated. As explained above, this requires either extremely low particle concentrations in the gas atmosphere, or particles that carry electrical charges of equal sign. Additionally, the temperature in the coating step must be sufficiently low so as not to destroy the matter used for coating. This condition is of particular importance when the coating consists of organic compounds.

Pioneering studies on the synthesis of coated nanoparticles were conducted in connection with the microwave plasma process [29,30]. A set-up used to produce ceramic-coated nanoparticles using the microwave plasma process is shown in Figure 4.36. The system consists of two subsequently arranged microwave cavities and a reaction tube passing through both cavities. The reaction is carried out in the microwave plasma at the intersections between the reaction tube and the microwave cavities. As shown above (see Section 4.5), the particles leave the reaction zone with a positive electric charge.

As mentioned above, the successful coating of particles requires the latter to be electrically charged. However, after leaving the reaction zone, the free electrons rapidly lose their energy by collision with other particles. At each collision with a gas atom, a free electron loses between 10 and 15 eV; hence, assuming that the energy of a free electron is in the range of 10^4 eV in the plasma, an electron requires approximately 10^3 collisions until its energy is sufficiently small so as not to ionize the particles further, but to compensate the positive electrical charge. Unfortunately, as

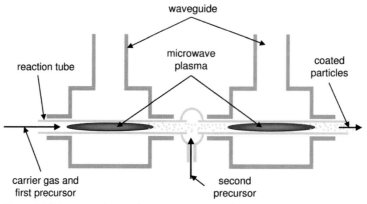

Figure 4.36 Set-up used to synthesize ceramic-coated ceramic nanoparticles in a microwave plasma. This was the first design to allow the production of significant quantities of coated nanoparticles [27,28].

the electrons move in all directions it is not a simple task to estimate the maximum distance where a significant number of particles will remain charged. In order to obtain a significant amount of nonagglomerated particles from the first reaction zone to the second zone, the distance between both reaction zones should so small as to have fewer than approximately 10^3 collisions of the electrons. Based on the data in Figure 4.18 it is clear that the free path length of an electron decreases significantly with increasing gas pressure; hence, the gas pressure in the system should be maintained as low as possible. Usually, a gas pressure below 2×10^3 Pa allows a distance between the two reaction zones of up to 50 cm.

A zirconia particle coated with alumina, and produced by the microwave plasma process, is shown in Figures 1.12 Figures 4.37. The particle which is visible in the

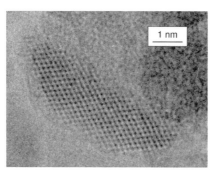

Figure 4.37 Zirconia particle coated with alumina; this material was produced using the microwave plasma process shown in Figure 4.36. The zirconia core is perfectly crystallized, as may be seen by the lattice fringes. The alumina coating is amorphous and therefore appears structureless in the electron micrograph. The facetted edges of the core particle are rounded to minimize surface energy.

center of Figure 1.12b is typically an agglomeration between a larger a smaller particle, while the other particles seem to be nonagglomerated. The thickness of the amorphous alumina coating is between 1.5 and 2 nm, and the zirconia core is crystallized. The nonspherical shape of the particle depicted in Figure 4.37 may also be caused by the agglomeration of two particles, and the thickness of the coating of this particle ranges between 0.2 and 0.5 nm. As can be seen by the lattice fringes, the zirconia core is perfectly crystallized, whereas the alumina coating is amorphous.

Instead of using two microwave cavities to generate a plasma, it is also possible to use two successive tubular furnaces, although by doing this one loses the advantage of nonagglomerated particles. As a consequence, such a system will coat clusters of particles rather than isolated particles. An example of such a product – a sintered agglomerate of zirconia that is coated with a few tenths of a nanometer of alumina – is shown in Figure 4.38. It is also remarkable that the coating is not of equal thickness around the particle, though this may have been caused by the high temperature leading to high mobility of the atoms at the surface during synthesis. As the particle is facetted, such a coating with different thickness, leading to rounded edges, is necessary in order to minimize the surface energy of the amorphous alumina. A similar, further-reaching, observation was provided in the Introduction.

The device displayed in Figure 4.36 is used not only to produce ceramic coatings on ceramic particles but also to obtain metallic coatings at the surface of ceramic particles. Unfortunately, however, if these layers are to be thin this is an impossible task, because the surface energy forces the metal layer to form small isolated clusters at the particle surface. Certainly, as the synthesis of oxide kernels requires an oxidizing atmosphere, this is possible only with noble metals, such as gold or platinum. A titania particle with a diameter of approximately 12 nm is shown in Figure 4.39; here, the dark dots present at the surface of the particle consist of platinum particles with diameters ranging between 2 and 3 nm. The decorated particles in Figure 4.39 were produced using the microwave plasma process, with

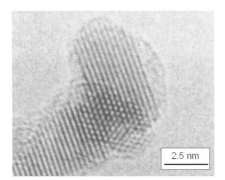

Figure 4.38 Electron micrograph of an agglomeration of zirconia particles coated with alumina. This material was produced in a conventional arrangement of tubular furnaces [31]. As in Figure 4.37, the kernel was crystallized and the coating amorphous. (Reproduced with permission from [31], Copyright: John Wiley & Sons. Ltd.)

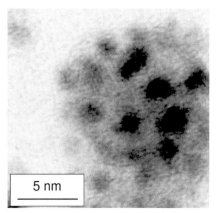

Figure 4.39 Titania particle decorated with platinum clusters; this material was synthesized using the microwave plasma process. Although, due to the high interface energy, platinum forms clusters rather than continuous coatings at the surfaces of oxide particles, it is also possible to produce thick coatings (Vollath, Szabó [32]).

platinum carbonyl chloride, $Pt(CO)Cl_2$, as the precursor. Materials such as transition metal oxide particles decorated with platinum or gold have been proven to serve as highly active catalysts which begin to function at relatively low temperatures.

In addition to coating oxide nanoparticles with an inorganic material, their coating with organic compounds is of major importance. Such coating may be performed with a polymer or a functional organic molecule, for example a luminescent compound. The coating of particles with organic matter requires temperatures that are sufficiently low so as not to destroy the molecules, and the oxidizing atmosphere when coating an oxide must also be taken into account. Therefore, except for very few exceptions, the *in-situ* coating of oxide particles is possible only by using the microwave plasma process. A set-up for the synthesis of organically coated nano-particles is shown in Figure 4.40.

In order to coat oxide particles with a polymer, the particles are first synthesized in the microwave plasma region, and the organic precursor is added after the cut-off tube. To obtain a polymer, the corresponding monomer or an easily evaporating oligomer is added. The temperature in the system is selected in such a way that the condensation of these organic precursor molecules at the surface of the nanoparticle is possible. To obtain the correct temperature and to avoid particle losses by thermophoresis, the condensation zone is heated with a tubular furnace or an alternative heating system. Then, provided that the compounds are selected correctly, under the influence of temperature and ultraviolet radiation from the microwave plasma, the monomer at the surface of the nanoparticles will begin to polymerize. In addition, in many cases the organic compounds react with the particle to form one huge molecule consisting of the particle and the polymer coating, an event which alters the properties of the particles and the polymer molecules dramatically. As these new giant molecules have new (e.g., optical) properties, the interaction between oxide

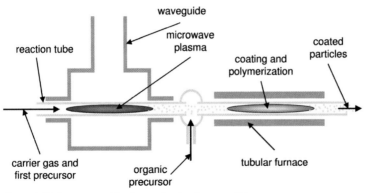

Figure 4.40 Arrangement for the synthesis of nanoparticles coated with organic molecules. It is necessary to maintain low temperatures; hence, the ceramic core should be produced using the microwave plasma process [33].

nanoparticles and polymethylmethacrylate (PMMA) at the surface is explained in Chapter 6.

An iron oxide particle coated with PMMA is shown in Figure 4.41, where the Fe_2O_3 particle has a diameter of 6–7 nm and the coating thickness is between 3 and 4 nm. This thickness was selected to obtain a useful contrast in the electron microscope. Thinner coatings, as are used for technical applications of superparamagnetic iron oxide particles, provide such a poor contrast for electron microscopy that the coating is more or less invisible. The device shown in Figure 4.40 may be set up with more than one stage for coating; for example, it may be possible to make a first coating of a luminescent material, and a second coating of a polymeric organic layer, for protection [1]. The development of particles on which two or more functional coatings may be combined is detailed in Chapter 6.

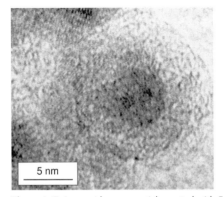

Figure 4.41 Iron oxide nanoparticle coated with PMMA. This composite was produced using a set-up as shown in Figure 4.40. The dark core is the iron oxide kernel.

References

1 Vollath, D. and Szabó, D.V. (2006) *J. Nanoparticle Res.*, **8**, 417–418.

2 Nitsche, R., Rodewald, M., Skandan, G., Fuessl, H., Hahn, H. (1996) *Nanostruct. Mater.*, **7**, 535–546.

3 Vollath, D. and Sickafus, K.E. (1992) unpublished results.

4 Ziemann, P.J., Kittelson, D.B., and McMurry, P.H. (1996) *J. Aerosol Sci.*, **27**, 587–606.

5 Kimoto, K. (1953) *J. Phys. Soc. Jap.*, **8**, 762.

6 Eastman, J.A., Thompson, L.J, and Marshall, D.J. (1993) *NanoStruct. Mater.*, **2**, 377–383.

7 Nanophase Technologies Corporation, 1319 Marquette Drive, Romeoville, IL 60446 http://www.nanophase.com.

8 MACH I, Inc., 340 East Church Road, King of Prussia, PA, 19406, USA, http://www.machichemicals.com.

9 Erhart, P. and Albe, K. (2005) *Adv. Eng. Mater.*, **7**, 937–945.

10 Puretzky, A.A., Geohegan, D.B., Fan, X, and Pennycook, S.J. (2000) *Appl. Phys.*, **A70**, 153–160.

11 Li, Q., Sasaki, T, and Koshizaki, N. (1999) *Appl. Phys.*, **A69**, 115–118.

12 Ullmann, M., Friedlander, S.K, and Schmidt-Ott, A. (2002) *J. Nanoparticle Res.*, **4**, 499–509.

13 Wang, Z., Liu, Y, and Zeng, X. (2006) *Powder Technol.*, **161**, 65–68.

14 Li, S. and El-Shall, M.S. (1998) *Appl. Surf. Sci.*, **127–129**, 330–338.

15 Vollath, D. and Sickafus, K.E. (1992) *NanoStruct. Mater.*, **1**, 427–438.

16 Vollath, D. (1994) *Mater. Res. Soc. Symp. Proc.*, **347**, 629.

17 MacDonald, A.D. (1966) *Microwave Breakdown in Gases*, Wiley, New York.

18 Vollath, D. and Sickafus, K.E. (1993) *NanoStruct. Mater.*, **2**, 451–456.

19 Pratsinis, S. (1998) *Prog. Energy Combust. Sci.*, **24**, 197–219.

20 Wooldrige, M.S. (1998) *Prog. Energy Combust. Sci.*, **24**, 63–87.

21 Vollath, D. (1989) *Euro-Ceramics*, **1**, 1.33–1.37.

22 Vollath, D. (1999) *J. Mater. Sci.*, **25**, 2227–2232.

23 Mueller, R., Kammler, H.K., Pratsinis, S.E., Vital, A., Beaucage, G, and Burtscher, P. (2004) *Powder Technol.*, **140**, 40–48.

24 Kamler, H.K., Mädler, L, and Pratsinis, S.E. (2001) *Chem. Eng. Technol.*, **24**, 583–596.

25 Zhu, W. and Pratsinis, S.E. (1996) in: (eds G.-M. Chow and K.E. Gonsalves), Nanotechnology, ACS Symposium Series, 622, 64–78.

26 Pratsinis, S.E., Zhu, W, and Vemury, S. (1996) *Powder Technol.*, **86**, 87–93.

27 Pratsinis, S.E. (1998) *Prog. Energy Combust. Sci.*, **24**, 197–219.

28 Kammler 2002 PhD Thesis, ETH #14622.

29 Vollath, D. and Szabó, D.V. (1994) *NanoStruct. Mater.*, **4**, 927–938.

30 Vollath, D. and Szabó, D.V. (1999) *J. Nanoparticle Res.*, **1**, 235–242.

31 Srdic, V.V., Winterer, M, and Hahn, H. (2001) *J. Am. Ceram. Soc.*, **84**, 2771–2776.

32 Vollath, D. and Szabó, D.V. (2000) unpublished results.

33 Vollath, D., Szabó, D.V, and Fuchs, J. (2004) *NanoStruct. Mater.*, **12**, 181–191.

34 Roth, P. (2001) University of Duisburg, Germany. Private communication.

35 Pratsinis, S.E. (2006) ETH Zurich, Switzerland. Private communication.

5
Magnetic Properties of Nanoparticles

5.1
Magnetic Materials

Materials are classified by their response to an external magnetic field as diamagnetic, paramagnetic, or ferromagnetic. Although, in general, all materials show inherently diamagnetic properties, only those materials not showing paramagnetic or ferromagnetic behavior in addition are known as *diamagnetic*.

The origin of diamagnetism is found in the orbital motion of electrons of the atoms acting like tiny electric current loops, producing magnetic fields. In an external magnetic field, these current loops align in a way so as to oppose the applied field, and therefore diamagnetic materials are exposed to a force pushing them out of the magnetic field.

Paramagnetism is, in most cases, significantly stronger than diamagnetism and produces magnetization in the direction of the applied field. In a paramagnetic material, the atoms act as tiny magnetic dipoles that may be oriented by an external magnetic field. This situation for a paramagnetic material in absence of an external magnetic field is shown in Figure 5.1a.

In *ferromagnetic* materials, the dipoles, represented by the unpaired electron spins of an atom, are interacting. This leads to a long-range ordering phenomenon causing to line up the dipoles parallel. This is depicted in Figure 5.1b.

For energetic reasons, the size of the ranges where this parallel orientation occurs in ferromagnetic materials is limited; these ranges – which are known as "magnetic domains" – are usually smaller than the grain size. Magnetic domains within a grain are separated by *Bloch* walls.

This situation is shown, in drastically simplified form, in Figure 5.2. Here, magnetic domains 1 and 2 with antiparallel orientation are separated by a 180° *Bloch* wall. In reality, the thickness of a *Bloch* wall is around 100 lattice constants, or even more. *Bloch* walls may connect magnetic domains with orientation differences of 90° or 180°, and the size of the domains and width of the *Bloch* walls are determined by thermodynamics. The direction of magnetization within a grain is changed by moving the *Bloch* walls. It is important to note that the existence of magnetic domains and *Bloch* walls makes it easier to change the direction of magnetization. For most ferromagnetic

Nanomaterials: An Introduction to Synthesis, Properties and Application. Dieter Vollath
Copyright © 2008 WILEY-VCH Verlag GmbH & Co. KGaA, Weinheim
ISBN: 978-3-527-31531-4

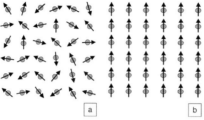

| a | b |

Figure 5.1 Distribution of molecular magnetic moments in different materials. (a) Paramagnetic material, the elementary magnetic moments are distributed arbitrarily. (b) Ferromagnetic material, the elementary magnetic dipoles are coupled and aligned in parallel.

materials, a specimen will remain magnetized to some extent after the removal of an external magnetic field; this effect is known as *remanence*. In general, the tendency for a material to remember its magnetic history is called *hysteresis*, and the magnetic field to compensate the remanence is called *coercitivity*. A schematic magnetization curve indicating the most characteristic points is shown in Figure 5.3. An additional important property of a magnetic material is the *energy product*, this being the product of coercitivity and remanence. The maximum temperature at which the ferromagnetic property exists is the *Curie* temperature, at which the thermal energy is larger than the energy coupling the atomic dipoles.

In case of ferrimagnetic compounds, the situation is somewhat more complicated. In substances such as MnO, FeO, and α-Fe_2O_3, an equal number of spins are arranged in two different sublattices exhibiting spontaneous antiparallel orientation. Therefore, the magnetic moment cancels out and these materials do not show net magnetic moments. This situation, which is depicted in Figure 5.3a, is referred to as *antiferromagnetic*.

It should be noted that antiferrimagnetic metals are also known to exist (typical examples are manganese and chromium), although this phenomenon is of greater importance in the case of compounds. As with ferromagnetic materials, a characteristic temperature – the *Néel* temperature – exists above which the material is

domain 1   Bloch wall   domain 2

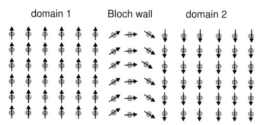

Figure 5.2 Separation of magnetic domains in ferromagnetic materials by *Bloch* walls. Here, the *Bloch* wall separates domains, with a difference in orientation of 180°. Generally, 90° *Bloch* walls are also possible.

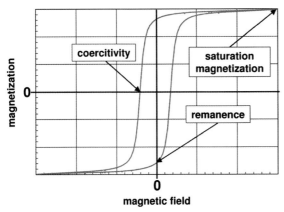

Figure 5.3 General appearance of a magnetization curve. Here, a magnetization curve exhibiting hysteresis is shown. The essential characteristics are saturation magnetization, coercitivity, and remanence.

paramagnetic, and below which it is antiferrimagnetic. In the situation where the magnetic moments of the two magnetic sublattices are (except for the antiparallel orientation) not equal, a magnetic net moment is observed. These materials, which are depicted in Figure 5.4b, are known as *ferrimagnets*. Ferrimagnetic materials consist either of two different metal ions or of metal ions of the same element with different valencies. An examples of a compound with different metal ions is $MnFe_2O_4 = MnO \cdot Fe_2O_3$, while an example of a compound where the metal ion appears in two different valencies is $Fe_3O_4 = Fe^{2+}O \cdot Fe_2^{3+}O_3$.

When considering at the dependency of coercitivity or remanence as a function of particle size, the behavior of one single isolated particle is as shown graphically in Figure 5.5. As mentioned above, large magnetic particles are subdivided by *Bloch* walls into magnetic domains, and the size of the *Bloch* walls and magnetic domains are controlled energetically. Thus, remanence and coercitivity are largely

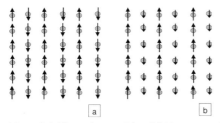

Figure 5.4 Magnetic materials exhibiting two antiparallel spin lattices. (a) Antiferromagnetic material. Here, the magnetic moments of the two antiparallel spin lattices cancel out such that the net magnetic moment of the material is nil. (b) Ferrimagnetic material. Here, the magnetic moments of the two antiparallel sublattices do not cancel out, such that a net magnetic moment remains.

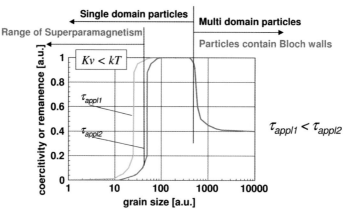

Figure 5.5 Dependency of coercitivity or remanence of magnetic materials as a function of grain size. While *Bloch* walls subdivide the grains, coercitivity and remanence are independent of the particle size. In one-domain particles without *Bloch* walls, coercitivity and remanence increase dramatically; this is the particle size range for magnetic data storage. At smaller particle sizes, coercitivity and remanence rapidly approach zero; this is the range of superparamagnetic materials. The grain size of this reduction depends on the time constant of the measurement.

independent of the particle size. Decreasing the particle size leads to a size range, where the particles consist of only a single magnetic domain. As the *Bloch* walls ease the change of the direction of magnetization, coercitivity and remanence will increase drastically. This range of particle sizes is employed in magnetic data storage. A further decrease of the particle size leads to a sudden decrease in remanence and coercitivity to zero.

The particle size at which this step occurs depends on the time constant of the measurement – the shorter the time constant τ_m of the measuring method, the more the step is shifted to smaller particle sizes. So, what is the reason for this phenomenon? The magnetic properties of an isolated single domain particle, or a group of such noninteracting particles as depicted in Figure 5.6a, are heavily influenced by the particle size.

The main reason for this strong influence is the fact that ferro- or ferrimagnetic materials retain the orientation of their magnetization by their magnetic anisotropy. The most dramatic effect is observed when the thermal energy of the particle is greater than the energy of anisotropy, at which point a thermal instability is observed. However, in the case of interacting superparamagnetic particles, an additional phenomenon may be observed. As in the case of paramagnetic materials, where interaction of the atomic dipoles leads to ferromagnetism, for superparamagnetic materials an interaction of the magnetic dipoles (represented by the particles) is also possible. This phenomenon, which was first observed by Morup [1], is known as *superferromagnetism*. A schematic diagram of a superferromagnetic material is shown in Figure 5.6b, where the one important essential for superferromagnetism is clearly apparent – that all the particles must be of very similar size.

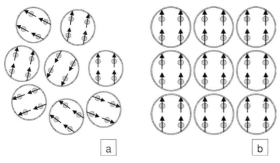

Figure 5.6 Superparamagnetic materials. (a) Within a superparamagnetic particle, the elementary magnetic dipoles are oriented in parallel. While the magnetic interaction of the particles is negligible, orientation of the magnetic moment of the particles is random. (b) Superferromagnetic material. Provided that the superparamagnetic particles are of equal size and their distance allows dipole–dipole interaction (as in ferromagnetic materials), magnetic coupling between the particles is possible.

5.2
Superparamagnetic Materials

As mentioned above, superparamagnetic materials excel in zero remanence and coercivity; moreover, superparamagnetism is limited to small nanoparticles.

In the case of a single isolated magnetic nanoparticle, the condition leading to superparamagnetism, a typical thermal instability, is

$$kT \geq Kv \tag{5.1}$$

where K is the constant of magnetic anisotropy (as shown in Table 5.1), v is the volume of the particle, Kv the energy of magnetic anisotropy, and kT the thermal energy (k is the Boltzmann constant, and T the temperature). If this condition is fulfilled, the material is superparamagnetic. The temperature T_B [see Equation (5.2)] is called the blocking temperature.

$$T_B = \frac{Kv}{k} \tag{5.2}$$

Clearly, although Equation (5.2) assumes monosized particles in a specimen, in most cases this assumption is not permitted. In this case, the volume v is replaced by the volume-weighted mean volume $\bar{v} = \sum_i p_i v_i$, where p_i is the probability for particles with the volume v_i. Superparamagnetism leads, as the vector of magnetization is fluctuating thermally, to a zero coercivity.

The explanation for this phenomenon is found in the magnetic crystal anisotropy. Magnetocrystalline anisotropy is an intrinsic property of any magnetic material, independent of grain size. The energy necessary to magnetize a ferro- or ferrimagnetic crystal depends on the direction of the magnetic field relative to the orientation of the crystal. Therefore, on distinguishing between magnetically "easy" and "hard" directions, these are the directions where the application of an external field leads easily to a high magnetization – the easy direction– or to a lower magnetization – the hard direction. In superparamagnetic materials, the vector of magnetization fluctuates between different easy magnetic directions, overcoming the hard directions.

In cubic materials, for any arbitrary direction, the energy of anisotropy can be reduced to two material constants K_1 and K_2. Assuming a direction with the angle α_1 to the [1 0 0], α_2 to the [0 1 0], and α_3 to the [0 0 1] direction, the energy of anisotropy is calculated from

$$K = K_0 + K_1(\cos^2\alpha_1 \cos^2\alpha_2 + \cos^2\alpha_2 \cos^2\alpha_3 + \cos^2\alpha_1 \cos^2\alpha_3) \\ + K_2 \cos^2\alpha_1 \cos^2\alpha_2 \cos^2\alpha_3 \tag{5.3}$$

Whilst the value of K_1 is well known for many magnetic substances, the value of K_2 is known only rarely, and K_0 is unknown for most materials. Values for K_1 for a range of magnetic oxides are listed in Table 5.1.

The data in Table 5.1 show that the constant of magnetic anisotropy K_1 may be greater or less than zero, while changing the sign of K_1 alters the crystallographic orientation of the easy and hard directions of magnetization. In order to demonstrate the dependency of the energy of anisotropy on crystallographic orientation, the situation in a (0 0 1) plane is shown. To rotate the vector of magnetization in the (0 0 1) plane, it is clear that α_3 is 90°, and therefore $\cos^2\alpha_3$ is zero. Furthermore, in this case Equation (5.3) may be simplified using $\alpha_2 = 90 - \alpha_1$, with the consequence that $\cos \alpha_2 = \sin \alpha_1$. This leads to the energy of anisotropy in the (0 0 1) plane

$$K = K_0 + K_1(\cos^2\alpha_1 \sin^2\alpha_1) \tag{5.4}$$

For cubic materials with a positive and a negative value of K_1, respectively, Figure 5.7a and b show polar plots of the energy of anisotropy as a function of the orientation. In these plots, the energy of anisotropy is plotted radially versus the orientation. For positive values of K_1, there is a clear minimum in the directions perpendicular to the face planes of the elementary cubes, whereas the maxima are in the directions of the diagonal of the faces of the cube. In case of negative values of K_1 the easy direction is rotated for 45°.

In reality, although Equation (5.3) describes a quite complex three-dimensional body, the basic phenomenon is clarified in Figure 5.7a and b. In the case of magnetite, Fe_3O_4, which is a cubic material, the easy and hard directions deviate fundamentally; these were found as follows:

$\langle 1 1 1 \rangle$ is the easy direction of magnetization,
$\langle 1 1 0 \rangle$ is the intermediate direction of magnetization,
$\langle 1 0 0 \rangle$ is the hard direction of magnetization.

Therefore, a sphere of magnetite will magnetize in one of the four $\langle 1 1 1 \rangle$ space diagonals of a cube, provided that the specimen is allowed to rotate freely.

In Figure 5.8, the same data are plotted as shown in Figure 5.7a: here, $K_1 > 0$, in *Cartesian* coordinates, but in the reduced range from the orientations $[1 \bar{1} 0]$ to $[1 1 0]$.

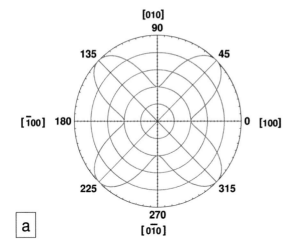

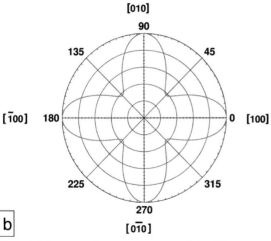

Figure 5.7 Constant of magnetic anisotropy K_1 for different directions in the (1 0 0) plane of a cubic material. (a) $K_1 > 0$. The hard directions are along the diagonals of the cube; the soft directions are along the edges. (b) $K_1 < 0$. The hard directions are along the edges of the cube; the soft directions are along the diagonals.

Table 5.1 Constant of magnetic anisotropy K_1 for different ferrimagnetic materials (ferrites).

Ferrite	Constant of anisotropy ($J\,m^{-3}$)
Fe_3O_4	-11×10^3
$MnFe_2O_4$	-2.8×10^3
$CoFe_2O_4$	90×10^3
$NiFe_2O_4$	-6.2×10^3
$MgFe_2O_4$	-2.5×10^3

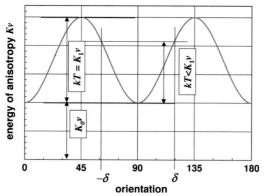

Figure 5.8 Energy of anisotropy for $K_1 > 0$ plotted as a function of the orientation in *Cartesian* coordinates. Energies $K_0 v$ and $K_1 v$ are indicated. For $kT < K_1 v$, the vector of magnetization fluctuates between $-\delta$ and $+\delta$; for $kT > K_1 v$, the vector of magnetization may change to another easy direction of magnetization.

In this plot, the orientation is used as abscissa and the energy of anisotropy as ordinate. Now, the energy $K_1 v$ necessary to rotate the orientation of the vector of magnetization from one easy direction to the next is indicated. It is also clear what happens when the thermal energy kT is smaller than the energy of anisotropy, in that the vector of magnetization oscillates around the minimum of 90° within the orientation range of $-\delta$ and δ.

The situation is different for noncubic materials. In hexagonal crystals, the soft directions usually lie in the hexagonal basal plane, whereas the c-axis, perpendicular to the basal plane is, magnetically, the hard direction.

The denotation "superparamagnetism" is made because the magnetization M of these materials follows mathematically the same law as found for paramagnetic materials. Therefore, for noninteracting particles, *Langevin*'s formula, describing paramagnetic materials, is valid.

$$M = nm\left[\coth\left(\frac{mH}{kT}\right) - \frac{kT}{mH}\right] \tag{5.5}$$

In Equation (5.5), m is the magnetic moment of one particle which is, in a first approximation, proportional to the volume of the particle, n is the number of particles, and H the magnetic field. In the case of paramagnetic materials, the magnetic moment of one molecule is – in most cases – equal to $1\,\mu_B$, the *Bohr* magneton. For superparamagnetic materials, the number of *Bohr* magnetons per particle may be in the range of 100 and more. The saturation moment of the specimen is nm.

The magnetization curves represent the behavior as a function of the magnetic field and temperature. Some typical examples of magnetization curves of polymer-coated γ-Fe_2O_3 particles as a function of the temperature are shown in Figure 5.9. At temperatures of 300, 200, and 100 K, superparamagnetic properties are apparent, these being characterized by a vanishing hysteresis, but at 50 K the magnetization

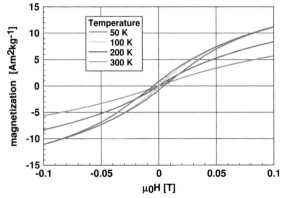

Figure 5.9 Magnetization curves of a superparamagnetic nanocomposite consisting of polymer-coated γ-Fe$_2$O$_3$ particles. At temperatures of 300, 200, and 100 K, this material is superparamagnetic, characterized by a vanishing hysteresis. At 50 K, the magnetization curve clearly shows hysteresis.

curve clearly shows hysteresis. Hence, for this specimen, the blocking temperature is clearly in the range between 50 and 100 K. It is important to realize that, in the superparamagnetic region, the magnetization increases with decreasing temperature. Additionally, in contrast to conventional magnetic materials, one characteristic feature is apparent from Figure 5.9 in that, as a consequence of the *Langevin* function describing superparamagnetism, there is no linear portion of the magnetization curve at low fields.

When considering Equation (5.5), the magnetic moment of superparamagnetic materials can be seen as a temperature-independent function of H/T. Therefore, plotting magnetization curves measured at different temperatures over H/T provides clear indications regarding the validity of *Langevin*'s function, the existence of superparamagnetism. Occasionally, the properties of superparamagnetic materials are given as a function of a temperature-independent magnetic quantity $\alpha = mH/kT$, called "reduced magnetic field". The magnetization curves plotted versus $\mu_0 H/T$ for temperatures of 200 and 300 K for a γ-Fe$_2$O$_3$ specimen consisting of pressed PMMA-coated particles is shown in Figure 5.10. Although both magnetization curves are almost identical, the slight difference may be explained by there being a minor interaction of the particles. Presumably, this was due to the coating being insufficiently thick as to thwart any particle interaction.

At a given field, less than that leading to saturation, the magnetization increases with decreasing temperature as long as the temperature is above blocking temperature. Below blocking temperature, the situation becomes more complex; this is depicted in Figure 5.11 [2], which shows the magnetization of γ-Fe$_2$O$_3$ at 0.05 T measured at increasing temperature. In the case of "zero field cooled," the specimen was cooled to minimum temperature at the magnetic field nil, and subsequently the magnetic field was increased to a level of 0.05 T. At low temperatures, there will be a quite low magnetization, while the spins are frozen and

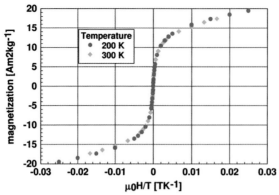

Figure 5.10 Temperature-compensated magnetization curves for pressed PMMA-coated γ-Fe$_2$O$_3$ specimen particles (experimental data from Figure 5.9). As expected from *Langevin*'s formula, the temperature-compensated magnetization curves determined at 200 and 300 K are identical. This serves as a highly sensitive proof of superparamagnetism.

cannot be rotated in the direction of the field. With increasing temperature, a larger number of spins are able to rotate in the direction of the field, such that the magnetization increases. The situation is different in the "field cooled" case, where the magnetic field was initially increased at room temperature and, as a second step, the specimen was cooled down. In this case, a few more spins in the specimen can be seen to turn in the direction of the magnetic field. The temperature at which the bifurcation occurs is the blocking temperature with respect to the time constant of the applied measurement method.

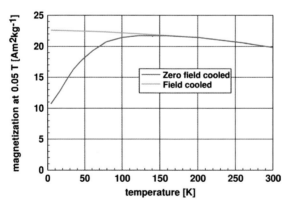

Figure 5.11 Magnetization of γ-Fe$_2$O$_3$ at 0.05 T measured at increasing temperature. "Zero field cooled" indicates that the specimen was cooled at the magnetic field nil; the magnetic field was subsequently increased to 0.05 T. "Field cooled" indicates that the specimen was cooled down in the magnetic field of 0.05 T. The blocking temperature with respect to the time constant of the applied measurement method is found at the temperature of bifurcation [2].

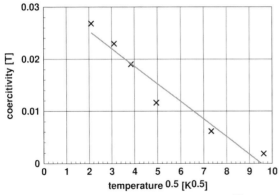

Figure 5.12 Coercitivity of γ-Fe$_2$O$_3$ as a function of $T^{0.5}$ below blocking temperature [2]. According to Equation (5.6), this plot allows blocking temperature to be calculated. Here, evaluation of the experimental data led to $T_B = 119.5$ K; $H_{C0} = 0.0337$ T.

Below blocking temperature, hysteresis appears. According to the theory of superparamagnetism [3], coercitivity is given by

$$\frac{H_C}{H_{C0}} = 1 - \left(\frac{T}{T_B}\right)^{0.5} \tag{5.6}$$

In Equation (5.6), H_C is the coercitivity at temperature T, H_{C0} is the coercitivity at 0 K, and T_B the blocking temperature. A graph showing coercitivity as a function of $T^{0.5}$ is shown in Figure 5.12 [2].

As expected from Equation (5.6), the relationship is essentially linear, and the equation is well suited to the determination of blocking temperature and coercitivity at 0 K. By using a linear fit, based on the experimental data of Figure 5.12, T_B was shown to be 119.5 K, while H_{C0} was 0.0337 T.

The magnetization curves depicted in Figure 5.9 are related only to the time constant of the measuring device which, in most cases, is $\sim$100 s. In order to define superparamagnetism the time constants are crucial; thus, to prove superparamagnetism with regards to shorter time constants, other more sophisticated methods are necessary.

When comparing the magnetization curves plotted in Figure 5.9 with those of materials with grain sizes in the micrometer region, significantly lower values of saturation magnetization are obtained. For γ-Fe$_2$O$_3$, in theory, a saturation magnetization of $\sim$75 A m^2 kg^{-1} is expected. The reduction of saturation magnetization observed in nanoparticles is a surface phenomenon. At the surface of magnetic materials spin canting phenomena can be observed, and this leads to a surface layer with very small saturation magnetization. Simply speaking, at the surface of magnetic particles the spins are not as well ordered as are observed in the interior [4]. As the surface:volume ratios of nanoparticles are larger by a few orders of magnitude than those of conventional materials, the contribution of spin canting at the surface to

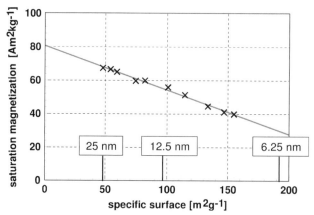

Figure 5.13 Linear decrease in saturation magnetization of MnFe$_2$O$_4$ nanoparticles in relation to specific surface area of the nanoparticulate powder [5].

the magnetization is of increasing importance as it causes a significant reduction in saturation magnetization.

Experimental evidence for these findings stems, for example, from the studies of Tang et al. [5], who showed that the saturation magnetization of MnFe$_2$O$_4$ nanoparticles decreased linearly with the increasing specific surface of the particles (see Figure 5.13).

A linear decrease with the specific particle surface is equivalent to a linear decrease with the inverse particle diameter. The experimental data in a range of ca. 50 to 150 m^2 g^{-1} shown in Figure 5.13 is equivalent to a range of mean particle diameters from 7 to 25 nm. Fitting of the experimental data and extrapolation to a specific surface area zero leads to a thickness of 0.7 nm of the nonmagnetic surface layer; extrapolation of the linear relationship to zero surface, equivalent to bulk materials, leads to a saturation magnetization of 81 A m^2 kg^{-1}.

Following the experimental findings that the saturation magnetization shows a linear decrease with the inverse surface area [5,6], in a first approximation it may be assumed that ferrite nanoparticles with diameter D and a nonmagnetic surface layer of thickness δ result in a magnetic active core with a diameter $(D - 2\delta)$ [7]. This approximation leads to a reduced saturation magnetization of

$$M_{\text{nanoparticle}} = \frac{(D - 2\delta)^3}{D^3} M_{\text{macroscopic}} \tag{5.7}$$

where $M_{\text{nanoparticle}}$ is the saturation magnetization of the specimen made of nanoparticles, and $M_{\text{macroscopic}}$ the value expected theoretically for macroscopic particles. The dependency of saturation magnetization as a function of particle size according to Han et al. [6], together with a fit based on Equation (5.7), is shown in Figure 5.14.

In these cases, the fits lead to a thickness of the nonmagnetic surface layer of 0.8 nm in the case of γ-Fe$_2$O$_3$, and of 1.0 nm for CoFe$_2$O$_4$. This is important when considering a particle size of 5 nm and a nonmagnetic surface layer thickness of

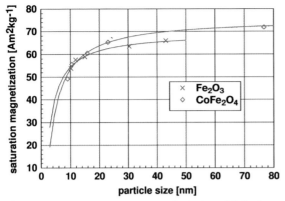

Figure 5.14 Saturation magnetization of γ-Fe$_2$O$_3$ and CoFe$_2$O$_4$ as a function of particle size. The fit was made according to Equation (5.7). Experimental data taken from Ref. [6].

1.0 nm, as only 20% of the saturation magnetization of bulk materials can be expected. This problem leads to a reduced use of nanoparticulate ferrites for applications demanding high magnetization.

Interestingly, in all experimentally verified cases, the nonmagnetic surface layer has a thickness of approximately one lattice constant; clearly, the magnetic structure of the first crystalline layer is disturbed by surface phenomena.

In bulk materials, dipole–dipole interactions of the particles cannot be excluded completely. Hence, for the successful application of superparamagnetic materials, an optimum must be found between the superparamagnetic properties and the density of the magnetically active particles in the matrix. Interaction of the particles in a matrix leads to magnetically larger particles that are, in some cases, no longer superparamagnetic. Such an interaction can be described by a fictitious particle size. Caizer *et al.* [8] described such an apparent increase in the magnetic particle size of γ-Fe$_2$O$_3$ in an amorphous silica matrix by a reduction of the thickness of the nonmagnetic layer. These authors showed that even a concentration as low as 0.68 vol.% did not eliminate particle interaction completely. In the case discussed here, a virtual increase in particle size, from 10.1 nm at room temperature to 11.7 nm at 70 K, was observed (see Figure 5.15). It is of interest to note that, with increasing temperature the thermal energy increasingly surmounts the dipole–dipole interaction, and this results in an apparent decrease in the magnetic particle size.

Although antiferrimagnetic materials have a magnetic moment that has been proven, both theoretically and experimentally, to be close to zero, the surface layer (as described above) leads to a remarkable magnetization, despite the particle core having no resulting magnetic moment. A typical example of this class of materials, chromia (Cr$_2$O$_3$), is antiferrimagnetic [9], and the magnetization curves of nanoparticulate chromia are shown in Figure 5.16. It is clear that this material has a resulting magnetic moment that stems from the disordered spins at the surface.

The data in Figure 5.16 also demonstrate a strong temperature-dependence of the magnetization. Down to a temperature of 40 K, the magnetization curves do not show

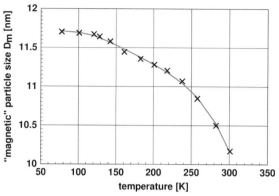

Figure 5.15 Magnetic determined particle size of γ-Fe_2O_3 in an amorphous silica matrix. By dipole–dipole interaction between the particles, the "magnetic particle size" is larger than the actual size. These results stem from a composite with 0.68 vol.% γ-Fe_2O_3 [8].

any hysteresis, though a minor hysteresis is seen at 10 K. When considering technical applications for these materials, the residual magnetic moment stemming from incomplete compensated spins at the surface may be of some value, as superparamagnetism may be demonstrated down to extremely low temperatures (see Figure 5.16). One typical application might be for magnetic cooling, and other compounds that may be of interest in this context include FeO and MnO.

In the case of a composite of noninteracting particles, the magnetic moment of the material can be calculated by linear superposition of the magnetic moment of each individual particle [7]. This may be used to retrieve the particle size distribution of a

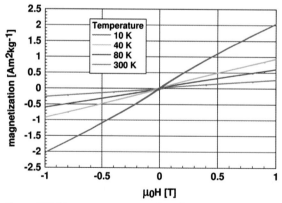

Figure 5.16 Magnetization curves of antiferromagnetic Cr_2O_3 at different temperatures. The measured magnetic moment stems primarily from the incompletely compensated spins in the surface layer. The material is superparamagnetic down to 40 K [9].

specimen from the magnetization curve. For each particle, the magnetization follows *Langevine's* formula:

$$M = m\left[\operatorname{cth}\left(\frac{mH}{kT}\right) - \frac{kT}{mH}\right] = mL(m, H, T) \tag{5.8}$$

In an arrangement of I classes of particles, each consisting of n_i particles with a magnetic moment m_i, the total magnetic moment is expressed by

$$M = \sum_{i=1}^{I} n_i m_i L(m_i, H, T) \tag{5.9}$$

Equation (5.9) is a linear relationship that allows calculation of the frequency n_i of particles with magnetic moment m_i. The magnetization curve of a nanoparticulate ferrite, together with the magnetic particle size distribution calculated using Equation (5.9), are shown in Figure 5.17a and b.

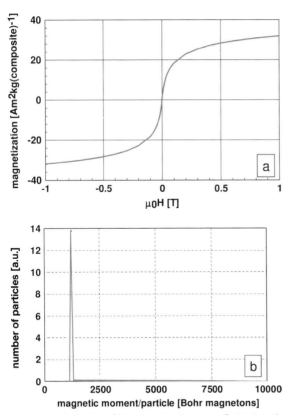

Figure 5.17 Evaluation of the magnetization curve of γ-Fe_2O_3 with respect to particle size [7]. (a) Magnetization curve. (b) Distribution of magnetic moments calculated according to Equations (5.8) and (5.9).

Fitting with Equation (5.8) leads to the magnetic moments in *Bohr* magnetons for different particle size classes. Using the tabulated specific magnetization of magnetic materials, the distribution of magnetic moments can be used to calculate particle sizes. To do this, one uses the fact that the magnetization is proportional to the particle volume, and this may in turn be used to calculate the "magnetic" particle size. The geometric particle size is best obtained by calibrating the calculations with the results of electron microscopy or X-ray diffraction studies. In any case, it is necessary to add the thickness of the nonmagnetic surface layer to the magnetic particle size. In particular, it must not be forgotten that this is a summand and not a factor enlarging the particle size. Particle sizes calculated from Figure 5.17a and b are shown in Figure 5.18.

As mentioned above, superparamagnetism is a property of isolated noninteracting particles. In a macroscopic material consisting of many particles, dipole–dipole interaction of the particles leads to magnetically large particles that are no longer superparamagnetic. Embedding the nanoparticles in a second, nonmagnetic, phase causes the particles to be spaced further apart, such that the interaction is reduced. This led to the production of nanocomposites. In order to ensure that a technical material is superparamagnetic, the individual particles should not touch each other. As mentioned previously (see Sections 1.1 and 4.7), the only nanocomposites where the distance between directly adjacent particles is clearly defined are produced as coated nanoparticles. The sample used to measure the magnetization curve shown in Figure 5.17a consisted of γ-Fe$_2$O$_3$ particles coated with PMMA and pressed into a pellet. Hence, by using coated nanoparticles it is possible to maintain superparamagnetism in technical components. An electron micrograph of such a specimen is shown in Figure 5.19a; this was prepared by cutting the pressed and sintered body into 20-nm slices using an ultramicrotome. During the cutting process, the "slip-stick" phenomena caused the material to show chatter marks (these are clearly visible

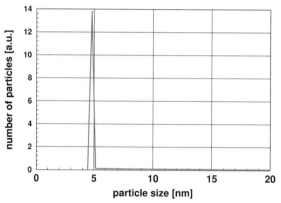

Figure 5.18 Distribution of magnetic particle sizes calculated from the results plotted in Figure 5.17a and b. Note the remarkably narrow distribution of particle sizes, and the very small proportion of particles larger than that at the distribution maximum.

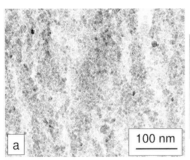
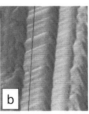

100 nm

a

b

Figure 5.19 Electron micrograph (a) and scanning force micrograph (b) of specimen material used to measure the magnetization curve in Figure 5.17a. Specimens were prepared by cutting the pressed and sintered body with an ultramicrotome into 20-nm slices. The visible striations are chatter marks stemming from slip-stick phenomena during cutting. The irregular horizontal striation in (b) was caused by notches in the diamond cutting knife (von Blankenhagen, reproduced with kind permission [24]).

in Figure 5.19b). (The irregular horizontal striations also seen were caused by notches in the diamond cutting knife.)

5.3
Susceptibility and Related Phenomena in Superparamagnets

The susceptibility of superparamagnetic materials can be calculated using Equation (5.5). Likewise, after developing *Langevine*'s formula in a *Taylor* series expansion for small values of H and using the first term, the magnetization M can be obtained:

$$M = nm \left[\frac{kT}{mH} + \frac{mH}{3kT} - \frac{kT}{mH} \right] = \frac{nm^2 H}{3kT}$$

and therefore, for the susceptibility μ:

$$\mu = \frac{\partial M}{\partial H} = \frac{nm^2}{3kT} \propto \frac{nv^2}{kT} = \frac{V^2_{specimen}}{nkT} \tag{5.10}$$

Equation (5.10) employs the fact that magnetic moment of a particle, m, is proportional to the particle volume, v. This equation suggests that susceptibility increases quadratically with the magnetic moment of the particles; hence, if two specimens have the same saturation moment nm, the one with the larger moment m of the particles has the higher susceptibility. In the case that a volume $V_{specimen}$ is subdivided into n particles $V_{specimen} = nv$, a proportionality is observed between μ and $1/n$. Or, put simply: the susceptibility of superparamagnetic particles increases with particle size. In general, nanoparticles of magnetic materials have a comparatively small susceptibility.

The thermal fluctuation of the magnetization is a random process, and therefore a frequency of fluctuation cannot be given; rather, the mean value of the time between two fluctuations – the relaxation time, τ – is utilized. The relaxation time τ according to Neél [10] is estimated by

$$\tau = \tau_0 \exp\left(\frac{Kv}{kT}\right) \tag{5.11}$$

As τ_0 is a material-dependent constant factor in the range between 10^{-9} and 10^{-13} s, the frequency of thermal relaxation may be well beyond 1 GHz at room temperature, provided that the composition of the ferrite is selected properly. The basis of this selection is the Aharoni [11] relationship for τ_0:

$$\tau_0 \propto \frac{m}{K} \propto \frac{v}{K} \tag{5.12}$$

From Equation (5.12) it is clear that a small value of τ_0 is obtained only with materials which show large values of the magnetic anisotropy K. Furthermore, τ_0 increases with increasing magnetic moment m of the particles. Although the variability in the magnetization m is limited as m is proportional to the particle volume, a small volume of the particles reduces susceptibility and magnetic moment. In this respect, the composition of the particle has a significant influence (see Table 5.1).

For technical applications, it is preferable to have small values of relaxation time that limit the frequency range for application, combined with a large susceptibility, and in this respect Equations (5.10) and (5.11) are contradictory. Therefore, super-paramagnetic materials may be optimized for either a high susceptibility or a short relaxation time; this in turn leads to a high-frequency limit of the applications.

In technical reality, the high-frequency application of superparamagnetic parts is hampered not only by particle size and the constant of magnetic anisotropy, but also by magnetic dipole–dipole interaction of the particles.

The magnetic susceptibility of different ferrite nanoparticles as a function of the frequency is shown in Figure 5.20.

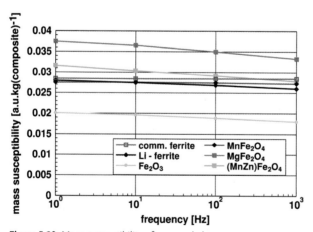

Figure 5.20 Mass susceptibility of nanoscaled versus conventional ferrites. The susceptibility of superparamagnetic ferrites decreased with increasing frequency; this was due to magnetic interaction of the individual particles.

In Figure 5.20, where the susceptibility of a conventional ferrite is plotted for comparison, two points are remarkable:

- The magnetic susceptibility of the nanocomposites is in the same range as is found for conventional ferrites; however
- In contrast to conventional ferrites, the susceptibility of nanoparticulate ferrites decreases with increasing frequency.

The reason for the decreasing susceptibility is found in the interaction and energy distribution of different particles. Those particles and interacting particles on the low energy side of the *Boltzmann* distribution where the fluctuation frequency is smaller than the inverse relaxation time are unable to follow the change in the direction of the external field, and consequently the susceptibility necessarily decreases with increasing frequency. This phenomenon may be reduced with a smaller particle size and a larger distance between the magnetic particles, but this will lead to reduced susceptibility since, in a given volume, the content of active material is reduced.

When considering superparamagnetism, distinction must be made between two different definitions: (i) an older definition, which stems from Elmore [12] and is based on the *Brown*ian motion of magnetic particles in a liquid; and (ii) a younger definition, as reported by *Néel* [10]. In the *Brown*ian case, the particles are rotating in a liquid when the direction of the magnetic field is changed. Based on these mechanisms, the *Brown*ian case is often referred to as "extrinsic" superparamagnetism, in contrast to the "intrinsic" superparamagnetism in *Néel*'s case. The relaxation time (the time needed by the particle to follow a change in the direction of the magnetic field) for the *Brown*ian superparamagnetism is $\tau_B = \frac{3v\eta}{kT}$, where η represents for the viscosity of the liquid carrier (the definitions of the other letters are outlined above). In the case of *Néel*'s superparamagnetism, the relaxation time is calculated using Equation (5.11) [13]. The *Néel* case will be discussed in detail later in this chapter. Both cases have in common that each particle consists of a single domain and, provided that the measuring time constant is adequate, hysteresis is not observed. Additionally, the magnetization curves follow *Langevin*'s law. When comparing the relaxation times for both cases, for 10-nm particles, in the *Brown*ian case one finds values in the region of microseconds, whereas in *Néel*'s case the relaxation time is about 1 ns. For magnetic particle with sizes of approximately 1 µm, suspended in a liquid, the *Brown*ian relaxation times are close to 1 s. The relaxation time rules the ability of the particles to follow a change of an external magnetic field with frequency *f*. In the case of $\tau < 1/f$, the particle is able to follow the external frequency, and therefore remanence and coercitivity are zero. In the case of $\tau > 1/f$, hysteresis is observed.

In spatially fixed nanoparticles, where the direction of the magnetization vector in the particle fluctuates, *Néel*'s superparamagnetism [10] is observed, characterized by a change in the *Mössbauer* spectrum, provided that the frequency of the thermal fluctuations is sufficiently high. The *Mössbauer* spectrum is a resonant γ-absorption spectrum which involves the emission and absorption of γ-rays from the excited states of a nucleus. (The Mössbauer effect is very complex, an exact description is possible only by quantum mechanics; hence, the following rather plausible explanation must be used with extreme care.) When an excited nucleus emits a γ-quantum with energy in the range from a few KeV to 100 keV, it must recoil to conserve momentum because the γ-photon has a nonzero mass. Therefore, the energy of the emitted γ-photon is reduced by the recoil energy in the range of a few 10^{-3} eV. Inversely, the same is occurring during the absorption of a γ-quantum, and this leads to a broadening of the energy distribution of the emission and absorption line; this in turn reduces the probability for resonance absorption, as the line width of the emission or absorption levels are significantly less than 10^{-5} eV. However, by placing the emitting and absorbing nuclei in a crystal, the crystal lattice is used for recoil. This reduces the recoil energy loss, to a value that emission and absorption lines are extremely narrow so that resonance absorption is possible. This phenomenon is used to detect extremely small energy shifts by moving either source or absorber with velocities on the order of millimeters per second. As an element, iron is extremely useful for *Mössbauer* studies since, as an emitter it is possible to use ^{57}Fe that is originated from the decay of ^{57}Co, in turn produced by neutron irradiation of natural cobalt. The decay of ^{57}Co leads to ^{57}Fe in an exited state that emits a γ-photon with energy of 14.4 keV. The natural line width of this emission is 10^{-8} eV.

The energy levels of the iron nucleus split up in the electric quadrupole field gradient and, additionally, in the magnetic field of the crystal are shown in Figure 5.21a and b, respectively. The magnetic splitting ΔE is proportional to the product of the magnetic moment µ of the nucleus in the ground state or in the excited state, and the magnetic crystal field B. It is important to realize that, in this case, the crystal field is the mean value of the field that is seen by the nucleus during one revolution. In superparamagnetic materials, however, one observes only the quadrupole splitting in the case that the inverse relaxation time ("fluctuation frequency") of the magnetization is larger than the *Lamor* frequency (= rotation frequency) of the iron nucleus.

The transition from the excited state to the ground state leads to a γ-emission. According to the selection rules for radiating transitions, only those with a difference in the quantum numbers between the ground state and the excited state of $\Delta M = 0, \pm 1$ are allowed. (For example, the quantum number in the excited state

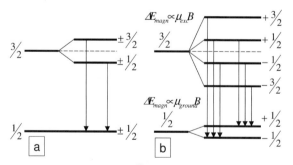

Figure 5.21 Energy levels in a ^{57}Fe nucleus with and without a magnetic field. (a) Energy levels without external magnetic field. In the excited state, electric quadrupole splitting is observed. According to the selection rules for radiative transitions, two emission lines are possible. (b) Energy levels in the ^{57}Fe nucleus exposed to an external magnetic field. The quadrupole levels show an additional magnetic splitting. The six allowed transitions are indicated.

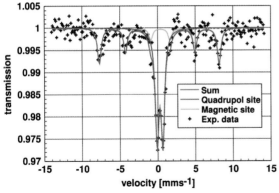

Figure 5.22 *Mössbauer* spectrum of a partly superparamagnetic specimen. The experimental data are fitted with two magnetic sites: (1) a quadrupole site stemming from the superparamagnetic fraction of the specimen with the small particle size, represented by the central doublet; and (2) a conventional magnetic site, attributed to the part of the specimen with the larger particle size, leading to the sextet.

$+3/2$ has an allowed transition to the ground state $+1/2$, whereas the transition to $-1/2$ is forbidden.) In the case of nonmagnetic, iron-containing materials, the *Mössbauer* spectrum consists of a doublet, split up in the electric field gradient of the crystal (Figure 5.21a). In the magnetic crystal field of a ferromagnetic crystal, this doublet is split up into a sextet (Figure 5.21b).

Finally, the *Mössbauer* spectrum provides the ultimate proof of *Néel*'s superparamagnetism, and is the only definition currently accepted. The *Mössbauer* spectrum of a partly superparamagnetic specimen is shown in Figure 5.22, whereby the superposition of a doublet with a sextet can be seen. The doublet characterizes the superparamagnetic fraction of the specimen, whereas the sextet is the "fingerprint" of the conventional magnetic fraction of the specimen. Clearly, the specimen consisted of a material with a relatively broad particle size distribution, where a certain fraction with the smaller particle size was superparamagnetic. The remainder of the specimen shows the spectrum of conventional magnetic materials.

The situation is different in the case of the spectra shown in Figures 5.23 and 5.24, which depict *Mössbauer* spectra of two different superparamagnetic ferrites. The spectrum displayed in Figure 5.23 stems from γ-Fe_2O_3 specimen, measured at 300 K, where the particles were coated with ZrO_2 as a distance holder. The material is purely superparamagnetic, and there is no trace of the sextet visible. The transition from the sextet to a doublet connected with the transition from normal ferrimagnetism to superparamagnetism is not restricted to pure γ-Fe_2O_3, but is also found at other ferrites, for example of the type $Me^{2+}Fe_2O_4$. This type of ferrite is derived from Fe_3O_4, where iron with valency 2 (Fe^{2+}) is replaced by another metal ion of the same valency. Most common are the additions of Mn or Mg, which reduce, or of Co, which increase, the constant of magnetic anisotropy.

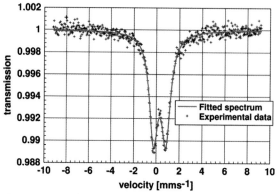

Figure 5.23 *Mössbauer* spectrum of ZrO$_2$-coated γ-Fe$_2$O$_3$ particles determined at 300 K. The spectrum consists only of the doublet representing superparamagnetic material.

Figure 5.24 shows a *Mössbauer* spectrum determined on superparamagnetic MnFe$_2$O$_4$. Here, it is clear that, as mentioned above, superparamagnetism is a property not only of pure γ-Fe$_2$O$_3$ but also of many other ferrites, provided that the condition in Equation (5.1) is fulfilled. In both spectra (Figures 5.23 and 5.24) the sextet is not visible, which indicates that specimens with a narrow particle size distribution exhibit pure superparamagnetism. When examining in detail the fit of the *Mössbauer* data in Figures 5.23 and 5.24, it is clear that there is an imperfect fit at the minima and maxima of the doublet. This occurs because these figures represent *Mössbauer* spectra of materials that consist of two components: the first component is the superparamagnetic ferrite core of the particle, while the second component represents the surface of the particles. As mentioned above, because of spin-canting phenomena this surface is magnetically less ordered than the core. In addition to the

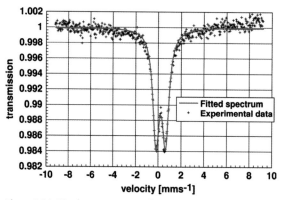

Figure 5.24 *Mössbauer* spectrum of MnFe$_2$O$_4$ measured at 300 K. As in the case shown in Figure 5.23, the entire material is characterized by the doublet representing superparamagnetic material.

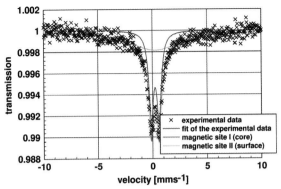

Figure 5.25 Complete fit of a *Mössbauer* spectrum for a γ-Fe₂O₃/ polymer nanocomposite. The experimental data are fitted with the doublet representing the superparamagnetic core and a broad unstructured feature characteristic for the magnetic disordered surface.

spectrum of the core (which is represented by the doublet, as described above), the surface layer is characterized by a broad unstructured minimum. An example showing a complete fit that takes the nonmagnetic surface layer into account is provided in Figure 5.25.

The spectrum of the superparamagnetic core (colored blue), representing the doublet (magnetic site 1), must be added to the spectrum the surface (colored green) (magnetic site 2) to provide the perfect fit (colored red). In this case, the surface layer represents 60% of the particle volume, whereas the magnetic core represents only 40%. Hence, the saturation magnetization expected will be only 40% of the theoretically possible value found with coarse-grained materials. This result should be examined in relation to the particle size dependency of saturation magnetization, as described by Equation (5.7). Assuming a 0.8-nm nonmagnetic surface layer, the particle size was ∼6 nm – a value which fitted well with electron microscopy measurements and the saturation of magnetization as described by Equation (5.7).

The energy levels of the iron nucleus under the influence of the magnetic crystal field are shown schematically in Figure 5.21b. As this splitting is directly proportional to the magnetic field, a detailed analysis of the *Mössbauer* spectrum will provide information concerning the magnetic crystal field. However, in superparamagnetic materials, great care must be taken when interpreting this result, as the splitting is proportional to the mean value of the crystal field that is seen by the nucleus during one revolution. Above the blocking temperature, this mean value is zero, but below the blocking temperature the measured field increases with decreasing temperature, as long as the fluctuation frequency is lower than the *Lamor* frequency of the nucleus. (Careful – this is a plausible explanation of a very complex phenomenon exactly described by quantum mechanics only!)

This situation is depicted in Figure 5.26, where the magnetic crystal field, determined from the splitting of the sextet in the *Mössbauer* spectrum, is plotted as a function of the temperature. A zero field is apparent above the blocking temperature

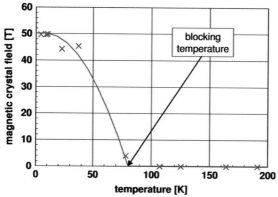

Figure 5.26 Magnetic crystal field of a ZrO_2-coated γ-Fe_2O_3 specimen. The crystal field was calculated using the splitting in the *Mössbauer* spectrum. A crystal field zero in the *Mössbauer* spectrum characterizes superparamagnetic material.

of 80 K, whereas below $\sim$10 K a constant value of 50 T is reached, this being the true value of the magnetic crystal field.

It should be mentioned here that this is only one possible definition of the blocking temperature, and is well suited to materials with a narrow particle size distribution. For materials with a broader particle size distribution, the temperature where 50% of the material is found in the doublet is normally used as blocking temperature.

The field discussed above is the magnetic crystal field of γ-Fe_2O_3 acting at the site of the iron nucleus. It may be of interest to mention that, in the case of an external magnetic field, the magnetization vectors of the particles are aligned in field direction. The vector of magnetization is no longer fluctuating, and therefore in a sufficiently high external magnetic field the *Mössbauer* spectra of superparamagnetic materials exhibit the sextet and not the doublet.

5.4
Applications of Superparamagnetic Materials

The majority of successful applications of magnetic nanomaterials use particulate composites, with superparamagnetism being necessary for the application of magnetic particles, for two reasons:

- Superparamagnetic particles avoid magnetic clustering.
- Superparamagnetic particles may be either attracted or released by switching the magnetic field.

From an economic viewpoint, the most interesting applications of superparamagnetic nanoparticles are related to medicine and biology. For this type of application, it is necessary to attach proteins at the particle surface that are specific to the application in mind. Some possible designs for particles connected with applications in

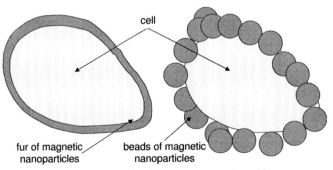

Figure 5.27 A living cell covered with superparamagnetic particles for magnetic cell separation. At the cell surface, particles may be attached either individually or clustered as a "fur" or beads, and functionalized with cell-specific proteins.

biotechnology are depicted in Figure 5.27. These comprise either: (i) a layer of particles at the surface, forming a type of fur of nanoparticles; or (ii) beads which consist of many superparamagnetic particles and are attached at the surface. However, it is not a simple task to produce beads while maintaining superparamagnetic properties.

The superparamagnetic particles or beads consist of a magnetic core, a coupling layer, and the proteins in question anchored at the surface (see Section 1.1 and Figure 1.13). As mentioned previously, the proteins attached at the surface are cell-type specific, from either cancer cells or well-defined organs.

In general, both designs of particle may be applied to magnetic cell separation techniques. When suspended in water, the magnetic particles attach specifically at the surface of one type of cell, as determined by the surface proteins present. When an external magnetic field is applied, one type of cell attached to the nanoparticles is removed from the suspension. To date, and based on practical experience, the design using superparamagnetic beads has proved to be the more successful.

A further successful application is related to medical diagnostics. Superparamagnetic nanoparticles are used to enhance the contrast in nuclear magnetic resonance (NMR) imaging [14] (see Figure 5.28a and b). NMR imaging functions by measuring differences of the concentration of protons which, in living structures, are quite low. However, the introduction of superparamagnetic particles into the region of interest changes the magnetic field locally, and this leads to local variations in the conditions for magnetic resonance; the result is a major improvement in imaging contrast. This phenomenon is shown in Figure 5.28, where the image is seen with and without enhanced contrast; the significant improvement in contrast is clearly visible. It is envisaged that, in time, superparamagnetic nanoparticles may replace the gadolinium salts currently used in this technique (an additional example is provided in Section 9.3).

One further potentially important application of superparamagnetic nanoparticles is that of magnetic refrigeration where, instead of using ozone-depleting refrigerants and energy-consuming compressors, nanocomposites moving in a magnetic field

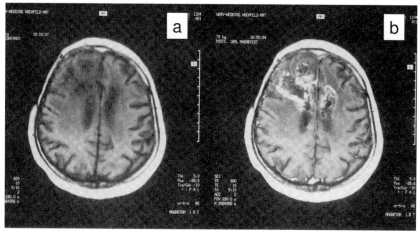

Figure 5.28 Contrast enhancement in NMR imaging of a brain tumor (primary benign brain tumor, astrocytoma) using superparamagnetic γ-Fe$_2$O$_3$ particles [14]. (a) Image recorded without contrast enhancement. (b) Image contrast improved by injecting water-suspended superparamagnetic γ-Fe$_2$O$_3$ particles into the patient's blood. (Reprinted with permission from Deutsche Apotheker Zeitung [14]).

might be employed. The concept behind magnetic cooling dates back several decades, having been applied in low-temperature physics.

Magnetic refrigeration is based on the fact that magnetic dipoles in a magnetic field are oriented in parallel; hence, the entropy of paramagnetic particles inside a magnetic field is reduced as compared to outside. Additionally, magnetic dipoles arranged in a row will attract each other, thereby reducing the enthalpy of the system. Hence, the removal of paramagnetic particles from the field will lead to a reduction in the temperature. Conventionally, this process – which is known as "adiabatic demagnetization" – uses paramagnetic salts of rare earth elements. Especially within the range of higher temperatures, the entropy change increases as the number of aligned spins in a particle increases; moreover, the efficiency of this process is significantly increased when using superparamagnetic particles rather than paramagnetic molecules. According to Shull [15], the temperature difference, ΔT, caused by a change in the magnetic field, ΔH, is in an ensemble with the total number of spins, N, distributed into n noninteracting particles, proportional to

$$\Delta T \propto n\left(\frac{N}{n}\right)^2 \frac{H}{T}\Delta H = Nn_s \frac{H}{T}\Delta H \qquad (5.13)$$

where $\left(\frac{N}{n}\right) = n_s$ is the number of spins per particle. Equation (5.13) states that ΔT is directly proportional to the number of spins per particle and the total number of spins in the system, assuming that the latter value is constant. The major advantage provided by using superparamagnetic particles rather than other paramagnetic

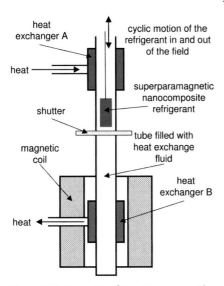

heat
exchanger A

cyclic motion of the
refrigerant in and out
of the field

heat

superparamagnetic
nanocomposite
refrigerant

shutter

magnetic
coil

tube filled with
heat exchange
fluid

heat
exchanger B

heat

Figure 5.29 Magnetic refrigeration system. The cooling process comprises two cycles: Cycle 1: The superparamagnetic material is exposed to the magnetic field; the heat exchanger cools the material. Cycle 2: The magnetic material is moved from the magnetic field to the second heat exchanger, which results in a temperature reduction of the material. In the second heat exchanger, the cooled material absorbs heat, which raises the temperature.

compounds means that a highly efficient magnetic refrigeration system can be developed by using permanent rather than superconducting magnets. Equation (5.13) makes it clear that magnetic cooling is most efficient at low temperatures, although for superparamagnetic particles it must be emphasized that the lowest temperature of application is the blocking temperature. By selecting appropriate materials, the use of magnetic refrigeration may be extended from close to absolute zero almost to room temperature.

The basic principle of a system for magnetic cooling is shown in Figure 5.29. In the simplest case, superparamagnetic material is moved between two heat exchangers, one of which is placed in a magnetic field. Initially, the material is exposed to the magnetic field, where it is cooled by the heat exchanger. The material is then removed from the magnetic field and this, according to Equation (5.11), will lead to a reduction in its temperature. In the second heat exchanger, the cooled material absorbs heat so that its temperature is increased; the warmed material is then moved again to the first (magnetized) heat exchanger, and the cycle is restarted. In a technical realization of the system the linear cyclic movement would be replaced with a rotational movement.

A further fascinating application of superparamagnetic particles, and one which is already widely used in technical products, is that of ferrofluids; these are stable suspensions of superparamagnetic particles in a liquid. In order to avoid the particles coagulating magnetically, they are coated with a second distance-holder phase. Ferrofluids are discussed in greater detail in Section 9.3.

5.5
Exchange-Coupled Magnetic Nanomaterials

Based on the details of nanoparticle interaction described in previous chapters, it might be concluded that this phenomenon is persistently disadvantageous. However, it has been shown by Kneller that a wise combination of different magnetic nanoparticles can provide significant improvements in the properties of magnetic materials [16]. These new materials consist of a mixture of hard and soft magnetic nanoparticles. The soft magnetic particles, which usually are super-paramagnetic, are located directly adjacent to a hard magnetic particle, the dipole moments of which force magnetization of the soft magnetic particles in the same direction. Consequently, the whole arrangement around the hard magnetic particles becomes oriented magnetically in the same direction (see Figure 5.30).

In the situation shown in Figure 5.30, the randomly oriented hard and soft magnetic nanoparticles are "exchange-coupled." As these particles are in contact with each other, any dipolar interactions need not be considered; hence, the assembly becomes one of exchange-coupled nanocrystals. In contrast to dipolar interaction, exchange coupling leads to a parallel orientation of the moments over the whole sample. Magnetic crystal anisotropy leads to a preferential alignment of the magnetic moments in the "easy" direction of magnetization for each particle. Now, theory suggests that the exchange energy connected to this process is proportional to the surface, whereas the energy of anisotropy is proportional to the volume of the particles. This leads to a size-dependent interplay of two energetic terms, and results in a "correlation volume" filled with parallel-aligned magnetic moments [17]. It has been shown by Herzer that the size of the correlation volume is given by

$$V_{corr} = \left(\frac{A}{K}\right)^6 \frac{1}{v^3} \quad \text{or} \quad N_{corr} = \frac{V_{corr}}{V_{particle}} = \left(\frac{A}{K}\right)^6 \frac{1}{v^4} \tag{5.14}$$

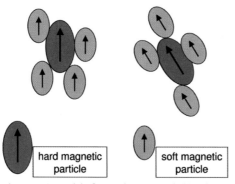

Figure 5.30 Model of an exchange-coupled hard magnetic material according to Kneller *et al.* [16]. The hard magnetic particles (red ovals) are coupled to soft magnetic particles (gray ovals). The exchange-coupled particles behave as though they have new properties.

where V_{corr} is the correlation volume, N_{corr} is the number of particles within the correlation volume, K the constant of magnetic anisotropy, and v the volume of each particle, assuming that all particles are of equal size. A, which is the exchange constant of nanoparticles, characterizes the process of exchange and is usually in the range of $10^{-12}\,\mathrm{J\,m^{-1}}$. Equation (5.14) states that the correlation volume increases with decreasing size of the particles and decreasing constant of anisotropy. In other words, the exchange volume of hard magnetic materials is smaller than that of soft magnetic materials. However, the correlation volume or equivalent number of exchanging particles N_{corr} is limited as the smallest size of the magnetic particles is also limited. Assuming magnetic particles with a diameter of 5 nm, in most cases the diameter of the correlation volume would be in the range of 100 nm. A further concept of description employs the term, correlation length

$$L = \left(\frac{A}{K}\right)^{0.5} \tag{5.15}$$

in relation to the particle size D in order to describe exchange-coupled materials [18].

The hard magnetic material proposed by Kneller consists of two different types of magnetic particle, namely a hard magnetic phase, and a soft magnetic phase with a constant of anisotropy that is at least two orders of magnitude smaller than that for the hard magnetic phase. Provided that the particles are sufficiently small, the effective constant of anisotropy of such a system is close to that of the hard magnetic phase, even when both magnetic phases have equal volume fractions. The advantage of such an exchange-coupled hard magnetic material is clear: normally, hard magnetic materials have high coercivity and low saturation magnetization and remanence, whereas soft magnetic materials are characterized by a very small coercivity and a high saturation magnetization. Exchange-coupled composites combine the advantages of both systems, as shown graphically and in simplified manner in Figure 5.31.

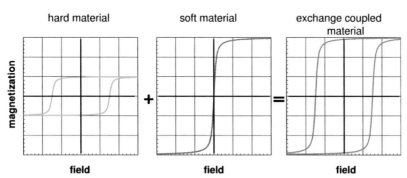

Figure 5.31 Hard exchange-coupled materials are composites of hard and soft magnetic materials; this combines the advantages of both groups. Hard magnetic materials have high coercivity but low saturation magnetization. Soft magnetic materials show high saturation magnetization but low coercivity. An exchange-coupled composite combines high saturation magnetization with high coercivity.

It can be seen in Figure 5.31 that the combination of hard and soft magnetic materials produces a new hard magnetic material with a significantly higher remanence, but the coercitivity is essentially unchanged. This design allows the production of permanent magnets with higher energy products (energy product = coercitivity × remanence) than are possible using conventional designs. It must also be noted that, in general, hard magnetic materials are significantly more expensive than soft magnetic materials, and therefore exchange-coupled hard magnetic materials are not only more effective but also more economic – a rare combination!

Generally, it is possible to use either a hard or a soft magnetic material with the appropriate inclusion of the other type of magnetic material. However, certain experimentally well-supported theoretical reasons exist as to why the application of hard magnetic islands in a soft magnetic matrix leads to the best possible results. This type of composite magnet is illustrated schematically in Figure 5.32.

Figure 5.32 also illustrates the difference between a magnet with random-oriented hard magnetic particles and one with an optimized structure. Normally, such an optimized structure would be obtained by using self-organization or precipitation processes where, in most cases α-Fe would be used for the soft magnetic phase, and an NdFeB alloy for the hard phase. The advantage of exchange-coupled permanent magnets therefore becomes obvious as a superior product is obtained despite using a reduced amount of the expensive hard magnetic material. The very best results are obtained when Fe_3Pt is used as the soft magnetic phase and FePt as the hard magnetic phase [19]. When producing this composite, Zeng *et al.* [19] started with a mixture of Fe_3O_4 and FePt, formed a body, and annealed this mixture at 650 °C in a reducing atmosphere. The starting mixture and the final product at high magnification are shown in Figure 5.33a and b, respectively. Clearly, the well-ordered starting mixture formed a mixture of FePt and Fe_3Pt grains. A high-resolution electron micrograph

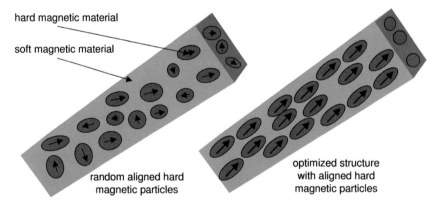

Figure 5.32 Two different types of exchange-coupled hard magnetic material. In both cases the magnetic hard phase is embedded in a soft magnetic matrix. The varieties differ in the arrangement of the hard magnetic phase. Left: the particles with broad size distribution are oriented in random. Right: the material exhibits an optimized structure, where the hard magnetic particles with uniform particle size are oriented parallel close to the axis of the work piece.

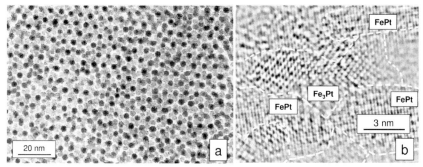

Figure 5.33 Starting mixture and final structure of an exchange-coupled hard magnetic composite based on Fe_3Pt as soft magnetic phase and $FePt$ as hard magnetic phase [19]. (a) Starting composite consisting of Fe_3O_4 and FePt. (b) High-resolution electron micrograph of the resultant Fe_3Pt/FePt hard magnetic composite (Reprinted with permission from Nature [19]; Copyright: Nature Publishing Group 2002).

of the final reaction product shows the perfect distribution of the magnetic phases after the reaction step (see Figure 5.33b).

In order to obtain a nanosized regular structure in the final product, with perfect distribution of the two different magnetic phases, a well-ordered regular structure (as shown in Figure 5.33a) is necessary within the material before the annealing stage is started. The energy product of this composite magnetic material exceeds the theoretical possible maximal value for FePt by more than 50%.

In a further attempt, Zeng et al. [20] produced an even more regular structure by coating $Fe_{58}Pt_{42}$ with Fe_3O_4 of different thickness. Such a structure, obtained with a particle size of 4 nm and a coating thickness of 0.5 nm, is shown in Figure 5.34, where

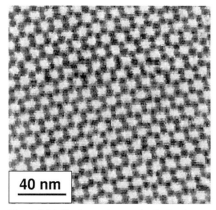

Figure 5.34 Electron micrograph of a hard magnetic composite consisting of $Fe_{58}Pt_{42}$ as hard magnetic phase and Fe_3O_4 as soft magnetic phase. The darker region represents the metallic particles; the lighter ring indicates the oxide coating (reprinted with permission from [20], Copyright: American Chemical Society 2007).

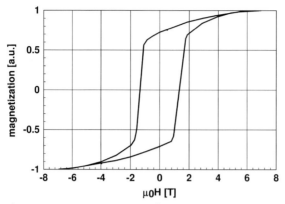

Figure 5.35 Magnetization curve of an exchange-coupled hard magnetic nanocomposite consisting of $Fe_{58}Pt_{42}$ as hard magnetic phase and Fe_3O_4 as soft magnetic phase. The oxide was applied as 1 nm-thick coating on the surface of the $Fe_{58}Pt_{42}$ particles [20].

the darker region represents the $Fe_{58}Pt_{42}$ particles and the lighter ring the Fe_3O_4 coating.

The magnetization curve of such a product with a coating thickness of 1 nm is shown in Figure 5.35. This magnetization curve displays a remarkable saturation magnetization and high coercitivity. The energy product of this composite is approximately 38% higher than the highest value that is theoretically possible for the pure PtFe bulk material. Similar interesting results and high coercitivities are found in Ni/NiO composites [21].

Until now, exchange-coupled magnetic materials have been discussed from the viewpoint of hard magnetic materials. However, some extremely interesting applications of this class of materials involve soft magnetic materials. When the constant of magnetic anisotropy becomes extremely small, it is clear from Equation (5.12) that the correlation volume may be huge for an "ultrasoft" magnetic material. The constant of anisotropy of these materials, which in most cases are metallic, is more than an order of magnitude smaller than that for soft ferrites. For materials such as FeSiBNbCu or FeZrB, the constant of anisotropy is well below $10^3 \, J \, m^{-3}$, and in this case the estimated correlation volume is of macroscopic size. As the exchange-coupled particles act as one particle, the susceptibility as estimated according to Equation (5.10) leads to extreme high values, because the magnetic moment of each particle becomes quadratic in the calculation of the susceptibility. However, in the case of metallic particles, it is an important prerequisite for successful application that each of the particles is coated with a thin insulating layer. For technical applications, this layer is produced either by oxidizing the surface of the particles or by coating with silica. This design leads to soft magnetic materials with unmatched high susceptibilities for electronic applications.

Composite materials based on this design are already available commercially, with typical examples being $(Ni,Fe)/SiO_2$, Co/SiO_2, $Fe,Co/SiO_2$, $NiFe_2O_4/SiO_2$,

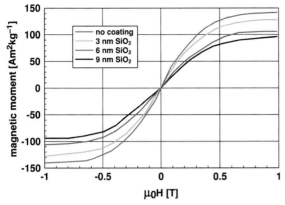

Figure 5.36 Magnetization curves of soft magnetic FeNi coated with SiO$_2$ of different thickness. As expected, the mass-related magnetization is reduced with increasing thickness of the nonmagnetic coating [23].

or $(Ni,Zn)Fe_2O_4/SiO_2$ [22]. Currently, research in this area is generally related to processes for synthesis.

Nonetheless, recent developments have been heading in the direction of increasing the frequency of application. Zhao *et al.* [23] have reported the details of such a material which demonstrates typical room temperature magnetization curves, and where the magnetically active material was FeNi coated with silica (see Figure 5.36). Here, the thickness of the coating is used as a parameter, and it is important to realize that in these magnetization curves, hysteresis is not visible. The size distribution of the particles was quite broad, and ranged from 10 to 150 nm. As expected, with increasing thickness of the coating, the saturation magnetization was seen to decrease, while the weight fraction of the magnetic active fraction decreased (Figure 5.36). Nonetheless, the saturation magnetization was remarkably high.

Perhaps the most important characteristic of this type of material is the high-frequency behavior. The major part of the susceptibility μ' and the quality factor of this type of material compared to competing conventional $(Co,Ni)Fe_2O_4$ ferrites are shown graphically in Figures 5.37 and 5.38.

The data in Figure 5.37 show that, although the susceptibility of the new exchange-coupled material is less than that of a commercial cobalt/nickel ferrite, in contrast to the ferrite the maximum frequency of application goes beyond 10^7 Hz, in this case at least up to 10^8 Hz. (The frequency of application for this ferrite is limited by *Bloch* wall resonance, as indicated in Figure 5.37, by an increase of susceptibility. During resonance, the magnetization losses are increased dramatically.) However, when considering the quality factor Q (the inverse of the energy loss, and an important figure of merit for high-frequency composite materials) the situation is quite different (see Figure 5.38).

The commercial cobalt/nickel ferrite showing *Bloch* wall resonance in a frequency range beginning close to 10^7 Hz showed a dramatic decrease in Q in this frequency

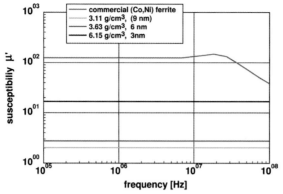

Figure 5.37 Real part of the complex susceptibility of soft magnetic composite consisting of FeNi kernels coated with SiO_2 of different thickness compared to a commercially available (Co, Ni)Fe_2O_4 ferrite. The susceptibility of the commercially available material was higher; however, due to *Bloch* wall resonances the application was limited to frequencies significantly below 10^7 Hz [23].

range. It is clear that, above 10^7 Hz, the exchange-coupled nanocomposites function better than the conventional materials. Furthermore, the Q-value increases with increasing coating thickness although, as shown in Figure 5.38, this is correlated with a significant loss of susceptibility. Such an effect is also apparent in the magnetization curves. Hence, when considering technical applications the materials must be selected by making the best compromise between frequency range, susceptibility, and losses.

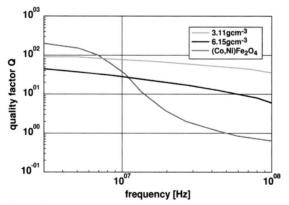

Figure 5.38 Quality factor (Q) of a soft magnetic composite consisting of FeNi kernels coated with SiO_2 of different thickness compared to a commercially available (Co, Ni)Fe_2O_4 ferrite. Whilst, above a few 10^7 Hz the conventional ferrite is no longer applicable, the nanocomposite can be used up to frequencies $>10^8$ Hz, even when Q is reduced [23].

References

1 Morup, S. and Christiansen, G. (1993) *J. Appl. Phys.*, **73**, 6955–6957.

2 Jeong, J.-R., Lee, S.-L. and Shin, S.-C. (2004) *Phys. Stat. Sol.*, **241**, 1593–1596.

3 Cullity, B.D. (1972) *Introduction to Magnetic Materials*, Addison-Wesley, Reading, USA, p. 410.

4 Kodama, R.H., Berkowitz, A.E., McNiff, E.J. and Foner, S. (1997) *J. Appl. Phys.*, **81**, 5552–5557.

5 Tang, Z.X., Sorensen, C.M., Klabunde, K.J. and Hadjipanayis, G.C. (1991) *Phys. Rev. Lett.*, **67**, 3602–3605.

6 Han, D.H., Wang, J.P. and Luo, H.L. (1994) *J. Magn. Magn. Mater.*, **136**, 176–182.

7 Vollath, D. and Szabó, D.V. (2002) in *Innovative Processing of Films and Nanocrystalline Powders* (ed. K.-L Choy), Imperial College Press, London, UK, pp. 219–251.

8 Caizer, C. and Hrianca, I. (2003) *Eur. Phys. J.*, **31**, 391–400.

9 Vollath, D., Szabó, D.V. and Willis, J.O. (1996) *Mater. Lett.*, **29**, 271–279.

10 Néel, L. (1949) *C. R. Acad. Sci. Paris*, **228**, 664–666.

11 Aharoni, A. (1964) *Phys. Rev. A*, **132**, 447–440.

12 Elmore, W.C. (1938) *Phys. Rev.*, **54**, 1092–1095.

13 Shliomis, M.J. (1974) *Sov. Phys.-Usp.*, **17**, 153–169.

14 Kresse, M., Pfefferer, D. and Lowaczeck, R. (1994) *Deutsch. Apotheker. Ztg.*, **134**, 3078–3089.

15 McMichael, R.D., Shull, R.D., Swartzendruber, L.J., Bennett, L.H. and Watson, R.E. (1992) *J. Magn. Magn. Mat.*, **111**, 29–33.

16 Kneller, E.F. and Hawig, R. (1991) *IEEE Trans. Magn.*, **27**, 3588–3560.

17 Herzer, G. (1997) in *Handbook of Magnetic Materials*, (ed. K.H.J. Buschow), Volume 10, Chapter 3, Elsevier Science, Amsterdam, p. 415.

18 Arcas, J., Hernando, A., Barandiaran, J.M., Prados, C., Vazquez, M., Marin, P. and Neuweiler, A. (1998) *Phys. Rev.*, **58**, 5193–5196.

19 Zeng, H., Jing, L., Liu, J.P., Wang, Z.L. and Shouheng, S. (2002) *Nature*, **420**, 395–398.

20 Zeng, H., Jing, L., Wang, Z.L., Liu, J.P. and Shouheng, S. (2004) *Nano Lett.*, **4**, 187–190.

21 Yi, J.B., Ding, J., Zaho, Z.L. and Liu, B.H. (2005) *Appl. Phys.*, **97**, 10K306-1-3.

22 http://www.inframat.com/magnetic.htm

23 Zhao, Y.W., Ni, C.Y., Kruczynski, D., Zhang, X.K. and Xiao, J.Q. (2004) *J. Phys. Chem.*, **108**, 3691–3693.

24 von Blankenhagen, P. (2002) Forschungszentrum Karlsruhe, Germany. Private communication.

6
Optical Properties of Nanoparticles

6.1
General Remarks

In view of technical applications, the optical properties of nanoparticles and nanocomposites are of major interest. Besides their economic importance, the scientific background of these properties is of fundamental importance in order to understand the behavior of nanomaterials.

When examining these optical properties, it is important to distinguish between several major cases:

- adjustment of the index of refraction by adding nanoparticles to polymers
- design of ultraviolet (UV) absorber transparent for visible light
- nanoparticulate pigments in transparent matrices
- luminescent materials
- photo- and electrochromic materials

As the information relating to each of these examples is vast, only the most basic facts will be explained at this point, although other optical properties will be considered under one of the following topics.

6.2
Adjustment of the Index of Refraction

In many applications, it is necessary to adjust the index of refraction of a polymer precisely to a given value. A typical example of this is the glue used to fix or connect optical glass fibers for information transmission. One way to do this is to add nanoparticles with an index of refraction which differs from that of the polymer. At low nanoparticle concentrations, the index of refraction of a composite $n_{composite}$ consisting of a matrix with indices of refraction n_{matrix} for the matrix and $n_{particle}$ for the nanoparticles, may be estimated by

$$n_{composite} = (1-c)n_{matrix} + cn_{particle} \qquad (6.1)$$

Nanomaterials: An Introduction to Synthesis, Properties and Application. Dieter Vollath
Copyright © 2008 WILEY-VCH Verlag GmbH & Co. KGaA, Weinheim
ISBN: 978-3-527-31531-4

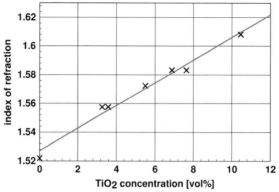

Figure 6.1 Index of refraction of TiO$_2$–PVA (polyvinyl alcohol) composites as a function of TiO$_2$ particle concentration. According to Nussbaumer *et al.* [1], the linear correlation given in Equation (6.1) is fulfilled perfectly within experimental accuracy, even at concentrations of more than 10 vol.% that are no longer "small".

This linear relationship of the index of refraction with particle volume fraction c given in Equation (6.1) has been experimentally very well verified; an example of the index of refraction of a TiO$_2$–polyvinyl alcohol (PVA) composite for different TiO$_2$ particle fractions is shown in Figure 6.1 [1]. Within experimental accuracy, the linear correlation given in Equation (6.1) is fulfilled exactly, even at concentrations in excess of 10 vol.%, which are no longer "low".

Particles that scatter light in a matrix reduce the transparency of the composite. Hence, in order to obtain a transparent material it is necessary to minimize light scattering at the nanoparticles in the composite. For spherical particles (i.e., smaller than the wavelength of the scattered light), the total power of the scattered light $P_{scatter}$ in such a composite is, according to *Rayleigh*, given by

$$P_{scatter} = \kappa P_0 c \frac{n_{particle} - n_{matrix}}{n_{matrix}^2} \frac{D^6}{\lambda^4} \tag{6.2}$$

In Equation (6.2), κ is a constant factor, D is the particle size, λ is the wavelength of the scattered light in the matrix with refractive index n_{matrix}, $\lambda = \lambda_0/n_{matrix}$ where λ_0 is the vacuum wavelength of the incident light, and P_0 is the intensity of the incident light. From Equation (6.2), it is clear that the particle size, D, is crucial as it has the power of 6. To minimize light scattering, the particle size must be maintained as small as possible, and in general perfectly transparent composites may be achieved if the largest particles are less than 10% of the wavelength under consideration. Thus, as the shortest wavelength visible to the human eye is 400 nm, the largest particle should not exceed 40 nm in diameter if a material is to be transparent over the whole range of visible light. It is important to note here that this is the *maximum* particle size, and not the *average* value; therefore a very narrow particle size distribution of small particles is indispensable as even a few larger particles will contribute more than proportionally to the scattering power. When D is the optically active particle size, it is important to note that this is the size of the nanoparticle clusters (if present). As clusters of particles cause a dramatic increase in the light scattering, Equation (6.2) does not allow for the presence of any clusters of particles in applications requiring transparent dispersions.

Equation (6.1) indicates that a certain index of refraction may be adjusted either by a smaller fraction c of nanoparticles with a higher index of refraction, or by a larger fraction of particles with a lower index of refraction. Equations (6.1) and (6.2) give no clue as to which of these two possibilities is better, but one possible answer may be obtained by considering the concentration. For statistical reasons, it is unavoidable that – even under the assumption of perfect blending – clusters consisting of two or more particles will occur.

Assuming a matrix with the volume fraction c of particles and perfect blending of the particles, the fraction c_i for cluster consisting of i particles is in a first approximation is given by

$$c_i = c^i \qquad (6.3)$$

Therefore, only small fractions of particles with a high refractive index can reduce the number of scattering clusters, and therewith the amount of scattered light. However, this is in contrast to the essence of Equation (6.2), which states that the total scattered power increases with the difference $n_{particle} - n_{matrix}$. This proposes the addition to the polymer of nanoparticles with a small difference in the index of refraction. However, the application of coated particles (in this case, ceramic particles individually coated with the matrix polymer) overcomes the problem of clustering completely.

When applying any ceramic nanoparticle to a polymer, care must be taken with regards to the catalytic interaction, because most oxides with a high index of refraction (e.g., TiO_2, ZrO_2) are strong photocatalysts. In the case of high UV intensities, this may lead to a system with self-destroying properties. To avoid this, the particles should be coated with a further catalytically inactive oxide; typical examples include alumina- or silica-coated particles, as neither alumina nor silica show any catalytic activity. These measures allow optimization of the system using particles with either large or small indices of refraction.

Changes in the index of refraction of polymethyl methacrylate (PMMA) filled with different ceramic nanoparticles are shown in Figure 6.2 [2], while the indices of refraction of the different materials used in Figure 6.2 are listed in Table 6.1.

The data provided in Figure 6.2 and Table 6.1 show clearly that adding particles with a refractive index which is higher than the matrix (e.g., ZrO_2, Al_2O_3) leads to an increase in the refractive index, and vice versa (as shown with the silica additions). It should be noted that, as the data in Figure 6.2 provide weight rather than volume content, a significant nonlinearity which differs from that in Equation (6.1) is observed.

Essentially the same considerations are necessary when designing transparent UV-absorbing materials, such as paints or lacquers, the only difference being a requisite strong absorption in the UV region. Typical absorbers, which are colorless to the

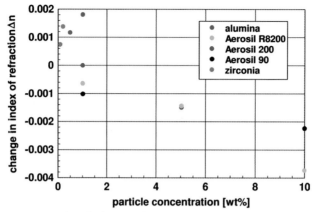

Figure 6.2 Index of refraction of PMMA filled with different
ceramic nanoparticles according to Böhm *et al.* [2]. As the data
relate to particle concentration in wt.% (not vol.%), the correlation
in Equation (6.1) is not easily visible. The indices of refraction for
the materials used are listed in Table 6.1.

human eye, include TiO_2, ZnO, and ZrO_2. Great care must be taken when selecting
these materials, and two important points must be considered:

- All oxides effectively used as UV absorber are photocatalytic active materials.
- The onset wavelength of the absorption is particle size-dependent.

The photocatalytic activity of the nanoparticles must not lead to a self-destruction of
the composite system, and it is essential to check this point before fixing a
combination of polymer matrix and nanoparticles. As mentioned above, in doubt,
the active UV-absorbing nanoparticle must be coated with a second, catalytically
inactive, ceramic material such as alumina or silica.

The particle size-dependency of UV absorption by TiO_2, over a wavelength range
from 200 to 600 nm, is shown in Figure 6.3 [3]. Here, the pigment material has a
mean particle size of 145 nm, while the ultrafine material size is about 45 nm. The
clearly visible blue shift of the absorption maximum with decreasing particle size is
typical for the optical properties of nanoparticles (see below). For both types of

Table 6.1 Indices of refraction for different materials.

Material	Index of refraction
PMMA used as matrix	1.49
ZrO_2	2.20
Aerosil[a] 90	1.45
Aerosil[a] 200	1.45
Aerosil[a] R8200	1.45
Al_2O_3-C	1.76

[a]Aerosil is a tradename for SiO_2 powders produced by DEGUSSA.

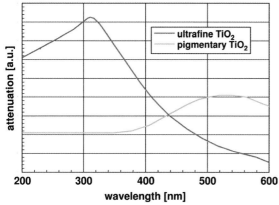

Figure 6.3 UV absorption of two different types of industrially produced titania over a wavelength range from 200 to 600 nm [3]. The mean particle sizes of the pigmentary and ultrafine materials were 145 and ca. 45 nm, respectively. The blue shift of the absorption maximum with decreasing particle size is evident; the flat onset of the absorption indicates a very broad particle size distribution.

material, the onset of absorption is quite flat, indicating a relatively broad particle size distribution. This is significantly different in the spectrum shown in Figure 6.4 [4], which shows the absorption of a nanocomposite of 0.037 wt.% TiO_2 in PVA (the near-negligible background absorption stemming from the PVA is subtracted). In this case the particle size was also significantly smaller, as might be realized by the remarkable blue shift of the onset of absorption at ~320 nm.

Furthermore, as a consequence of the narrow particle size distribution, the absorption maximum is, in contrast to the products shown in Figure 6.3, very narrow. Provided that the particles are isolated, a nanocomposite made of particles as shown in Figure 6.4 is perfectly transparent in the visible region. This contraposition of the optical absorbance of powders with broad and narrow particle size distribution makes it clear that, for each application, an optimal powder characteristic must be selected.

6.3
Optical Properties Related to Quantum Confinement

As noted in Section 6.2, the onset and the maximum of photon absorption exhibits a blue shift with decreasing particle size. The same phenomenon is observed for the most important optical property, light emission, and in this context a series of phenomena characteristic of nanoparticles may be identified. One of the most important groups of properties is connected with *quantum confinement*, which is observed in the interaction of very small nanoparticles with light, when free electrons and holes are created. The hole and the electron form a hydrogen-like complex, called an *exciton*. Depending on the properties of the particle, the radius of the exciton –

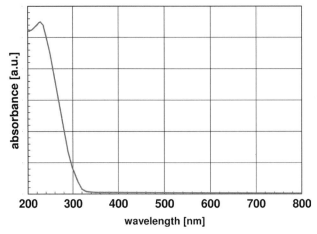

Figure 6.4 Optical absorption of a 0.037 wt.% TiO_2 in PVA nanocomposite in the wavelength range from 200 to 800 nm, according to Nussbaumer *et al.* [4]. Compared to Figure 6.3, the particle size was significantly smaller, leading to the remarkable blue shift of the onset of absorption. As a consequence of a narrow particle size distribution, the onset of the absorption is very steep and the maximum very narrow. Provided that the particles are individualized, a nanocomposite using a material like this is perfectly transparent in the visible region.

known as the "exciton *Bohr* radius" – may range from a tenth of a nanometer to a few nanometers. Quantum confinement occurs when one or more of the dimensions of a nanocrystal is/are smaller than the diameter of the exciton. In this case, the absorption and emission of light are strongly particle size-dependent. Interestingly, this is one of the very few cases, where a quantum phenomenon can be described in good approximation without solving *Schrödinger*'s equation.

In the simplest case – quantum confinement systems – the electrons in nanoparticles are described as a "particle in a box with infinitely high walls". The assumption of infinite walls is justified in the case of insulating ceramic materials and, in most cases with good approximation, also for semiconductors. This is an idealized description of a free electron in a nanoparticle with the diameter D. Generally, this problem of quantum confinement in a particle is solved using the *Schrödinger* equation; however, a "particle in a box" can be described with a simplified approach using the basic laws of quantum mechanics. In a one-dimensional system, the condition for a standing wave consisting of n half-waves of the wavelength λ in a box with the size L is

$$\frac{n\lambda}{2} = L \tag{6.4}$$

When substituted into the *DeBroglie* relationship, this leads to

$$p = mv = \frac{h}{\lambda} = \frac{nh}{2L} \tag{6.5}$$

In Equation (6.5), p is the momentum, m the mass, v the velocity of the electron, and h is *Planck*'s constant. Now one can calculate the energy E_n, obtaining

$$E_n = \frac{mv^2}{2} = \frac{p^2}{2m} = \frac{n^2h^2}{8\,mL^2} = \kappa\frac{n^2}{L^2} \tag{6.6}$$

Equation (6.6) states the main characteristic of quantum confinement systems, as it shows that the energy of the electron is inversely quadratic to the particle size. This describes the blue shift of the absorption edges with decreasing particle size. The energy difference ΔE between two quantum levels n and $n+1$ describes the energy of an emitted photon.

$$\Delta E = \frac{(2n+1)h^2}{8\,mL^2} \quad \text{leading to} \quad \lambda = \frac{8\,mcL^2}{(2n+1)h} \tag{6.7}$$

Equation (6.7) clearly shows the blue shift with decreasing size L of the box, which is equivalent to the particle size D, or the wavelength of the emitted photon increases quadratically with the particle size D. However, it must be kept in mind that the blue shift is not only observed in the quantum confinement case.

The solution of the *Schrödinger* equation leads to deeper insights. In order to solve the "particle in a box" problem, potential walls limiting the particle must be assumed. In the case of an insulating particle these walls are infinitely high, and therefore it is impossible for electrons to tunnel to the space outside the particle. With decreasing height of the potential wall, the probability of finding electrons outside the particle is increasing, as the electrons are tunneling into the space outside the box. On the left-hand side of Figure 6.5 the electron density distribution

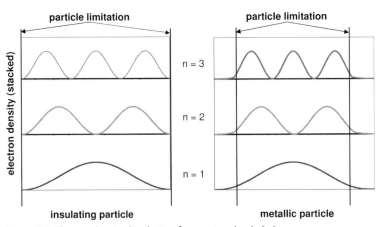

Figure 6.5 Electron density distribution for quantum levels 1, 2, and 3 in an insulating and a metallic nanoparticle obtained by solving *Schrödinger*'s equation for a particle in a box. In the insulating particle, independent of the quantum level, there is a zero probability of finding electrons outside the particle, whereas a metallic nanoparticle is surrounded by electrons. Electrons are tunneling into the space outside the particle.

in an insulator is shown for quantum levels 1, 2, and 3. For a metallic particle, the same data are plotted at the right-hand side of the figure. Independent of the quantum level, there is a nonzero probability of finding electrons outside the particle. However, when considering transition from the insulator to the semiconductor, and finally to a metal, theory predicts an increasing probability of finding electrons outside the particle. Finally, a metallic nanoparticle is surrounded by a cloud of electrons (see Figure 6.5).

In addition, *Pauli*'s principle enforces a blue shift with decreasing particle size. This is illustrated in Figure 6.6, which elucidates the transition from the energy levels of one atom to the energy bands in an insulating crystal. In an atom, each electron occupies one distinct energy level, and each energy level can be populated by only one electron. (Strictly speaking, taking the two possible spins into account, these are two electrons, but for the consideration here this is not essential.) By bringing two atoms together to form a molecule, a splitting of each energy level is observed, and this splitting continues with each atom added. For each additional atom, one new energy level is created. Finally, in a crystal, the energy levels form energy bands within which the energy differences are so small that they may be considered as "quasi-continuous". The energy difference E_g between two energy bands decreases with increasing size of the crystal (see Figure 6.6). Figure 6.6 illustrates the positions of the electron bands in general.

However, when analyzing experimental data the situation is not easy, as indicated by the simplified model leading to Equations (6.6) and (6.7). In general, the energy of the emitted photons as a function of particle size D is fitted by an equation such as:

$$\Delta E = E_0 + \frac{K}{D^\alpha} \tag{6.8}$$

where E_0 and K are constant values, and α is, in most cases, a noninteger number. One reason for the noninteger exponents is simply that, in most cases, there is not a single well-defined particle size but rather a particle size distribution.

When considering materials in the "real world", distinction must be made between metals, semiconductors, and insulators. In metals, in most cases, the highest occupied band is not completely filled, and therefore the movement of electrons (which is

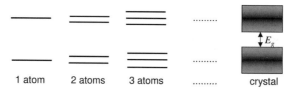

1 atom   2 atoms   3 atoms   ....   crystal

Figure 6.6 Energy levels for one atom and the transition from a molecule consisting of two or more atoms to the energy bands of an insulating crystal. Enforced by *Pauli*'s principle, in an atom, each electron occupies one distinct energy level, populated by one electron only. Bringing two or more atoms together causes a splitting of each energy level to occur, which continues with each further atom added. Finally, in a crystal, the energy levels form energy bands, where the energy differences are so small that the energy bands may be seen as "quasi-continuous". The energy difference E_g between two energy bands decreases with increasing number of atoms in the crystal.

equivalent to electric conduction) is possible. These are termed "free" electrons. In contrast, in semiconductors and insulators the last occupied band is filled completely and, as free electrons are not available, then electric conduction is impossible. In semiconductors, however, the energy gap is so narrow that some of the electrons have sufficient energy to jump, when thermally activated, into the next band. This creates electron "holes" in the originally filled band, and free electrons in the next (initially empty) band; as a consequence, the conduction of electricity becomes possible (see Figure 6.7). (In doped semiconductors, additional, isolated energy levels are created in this gap, and this also leads to electrical conductivity.)

In Figure 6.7, the level of the *Fermi* energy is also indicated. At absolute zero temperature, the *Fermi* energy E_F is the energy level of the least tightly bond electron within a solid. Together with the energy of the gap, E_g, the *Fermi* energy level is an important parameter when characterizing a solid. With nanoparticulate semiconductors it is clear that, in the case of decreasing particle size, the band gap widens, and this may have the consequence that semiconductors will increasingly acquire the properties of insulators.

The electrons in the incompletely filled conduction band of a metal are called "free electrons"; in crude terms, such electrons behave like water in a pot, in that they move collectively in discrete waves called "plasmons" (this point is discussed in detail in Section 6.5).

An increase of E_g with decreasing particle size leads necessarily to a blue shift of any absorption, or to the emission of photons. A typical example of the particle size-dependent band gap energy E_g is shown in Figure 6.8 for silicon and germanium nanoparticles [5]. Here, the parabolic increase in the band gap with decreasing particle size, as predicted by Equation (6.6), is clearly visible. However, these data also

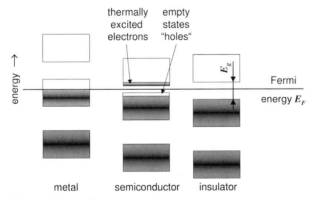

Figure 6.7 Energy bands in metals, semiconductors, and insulators. Metals are electric conductors because the highest occupied band is not completely filled allowing movement of electrons, which is equivalent to the possibility for electric conduction. In semiconductors and insulators, the last occupied band is filled completely; therefore, electric conduction is impossible. However, in semiconductors the energy gap is so narrow that the thermal energy of some electrons is sufficient to jump into the next band, creating "holes" in the originally filled band and free electrons in the next, initially empty, band. Therefore, conduction of electricity is possible. Additionally, the level of the Fermi energy, E_F (the energy level of the least tightly bond electron within a solid), is indicated.

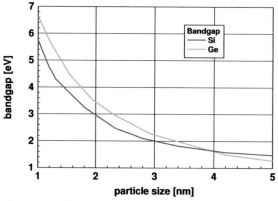

Figure 6.8 Band gap energy E_g for silicon and germanium nanoparticles according to Khurgin et al. [5]. Note the parabolic increase of the band gap with decreasing particle size. These data also show that the increase in band gap width is important only for particles well below 5 nm diameter.

show that these phenomena are only of any importance when in the particle size range below ~5 nm.

The influence of particle size is found not only in the band gap but also in the wavelength of the emitted photons. The emission spectra of CdSe nanoparticles with different sizes are shown in Figure 6.9 [6], where the blue shift of the emission for decreasing particle size is clearly apparent. The color of the spectra should indicate the color of the emitted light. Generally, it is possible to functionalize particles which emit at different wavelengths in different ways in order to attach them to various biological phases. The clear separation of the emission spectra of the particles shown

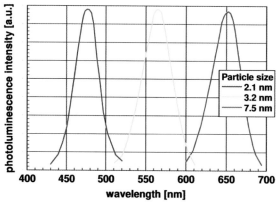

Figure 6.9 Emission spectra of CdSe nanoparticles as a function of the particle size according to Smith et al. [6]. The figure shows that particle size influences not only the band gap but, therefore, also the wavelength of the emitted photons. The blue shift of the emission for decreasing particle size is also evident.

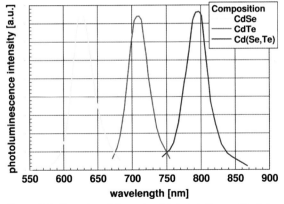

Figure 6.10 The color of the emitted light is influenced both by particle size and composition. Here, the emission of CdSe and CdTe of equal particle size ~5.5 nm is plotted; clearly, CdTe emits less energy-rich photons compared to CdSe. The exchange of selenium by tellurium leads to an additional red shift of the emission beyond that of pure CdTe [6].

in Figure 6.9 demonstrates the possibility of distinguishing different biological phases by the color of the emitted light. However, it is a necessary requirement that the particle size distribution is extremely narrow, as otherwise the emission spectra will overlap too much.

In addition to changing the particle size, the energy of the emitted light may also be influenced by the composition of the nanoparticles. The emission of CdSe and CdTe of equal particle size of ~5.5 nm is shown in Figure 6.10, and CdTe can be seen clearly to emit photons of less energy compared to CdSe. As both materials show some degree of solubility in the crystalline phase, it might be possible to produce solid solutions in this system, and thereby to alter the energy of the emitted light. Interestingly, the partial exchange of selenium by tellurium leads to a further red shift of the emission beyond that of pure CdTe.

As mentioned above, the change in band gap and quantum confinement phenomena led to a blue shift of any interaction with photons, and an excellent example is provided by Pratsinis and coworkers [7,8] (see Figure 6.11). Here, the absorbance of ZnO is shown as a function of wavelength and particle size, and the blue shift with decreasing particle size from 12 to 3 nm is clearly visible. This again demonstrates the possibility of tailoring the optical properties of nanoparticles simply by the particle size (as displayed in Figure 6.9).

Based on the *Tauc* relationship [9], Mills *et al.* [10] developed an empirical formula that allows the energy of the band gap to be estimated, by relating optical absorption α and band gap energy E_g.

$$(\alpha h\nu)^n = \mathrm{const}(h\nu - E_g) \tag{6.9}$$

where h is *Planck*'s constant and ν is the frequency of the light. According to Equation (6.9), a plot of $(\alpha h\nu)^n$ versus α can be used to determine E_g. According to

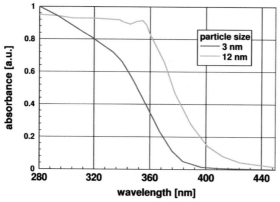

Figure 6.11 Absorption spectra of ZnO nanoparticles of different sizes according to Pratsinis *et al.* [7,8]. Reducing the particle size from 12 to 3 nm leads to a clearly visible blue shift of the absorption spectrum.

Tauc, the exponent n represents the type of electronic transition causing photon absorption; values for n, together with their interpretation, are listed in Table 6.2.

For practical applications, the steep increase in absorption can be used for this estimation. In relation to the example given in Figure 6.11, the energy-rich, linear part of the graph in Figure 6.11 is extrapolated, and the intersection with the abscissa $(\alpha h v)^n \rightarrow 0$ is then determined to give the optical band gap. In many cases, it is not unequivocal which exponent should be selected.

Whilst the application of Equation (6.9) supposes one distinct particle size, in reality there is always a more or less broad particle size distribution, and therefore the band gap obtained when using this method is only a rough estimation. Such a *Tauc* plot is shown in Figure 6.12, where the data from Figure 6.11 were applied.

Extrapolation of the (more or less) linear part of the graph leads to a gap width of 3.08 eV for the 3-nm particles, and 2.87 eV for the 12-nm particles. The strong deviation from linearity and Equation (6.9) is, most likely, caused by the particle size distribution.

Semiconducting nanoparticles are the classic example of quantum confinement systems. A wonderful example of the behavior of semiconducting nanoparticles is that of lead sulfide, PbS. Depending on the particle size, the band gap increases from 0.41 eV in bulk crystals up to a few electron volts in nanoparticles. Therefore, bulk

Table 6.2 Values for the *Tauc* exponent and its relation to electronic transitions causing absorption.

Tauc exponent n	Electronic transition
2	Direct allowed
2/3	Direct forbidden
1/2	Indirect allowed
1/3	Indirect forbidden

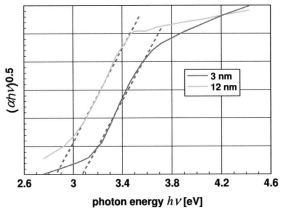

Figure 6.12 *Tauc* plot according to Equation (6.9) [9,10], using the absorption data for ZnO nanoparticles with different sizes shown in Figure 6.11. The intersection of the extrapolation with the abscissa $(\alpha h \nu)^{0.5} \rightarrow 0$ gives the width of the optical band gap. In this example, a gap width of 3.08 eV for the 3-nm particles and 2.87 eV for the 12-nm particles was determined.

PbS absorbs throughout the visible range and hence appears black. However, with decreasing crystal size, the color changes to dark brown, while suspensions of PbS nanoparticles are clear and reddish. In this context, interesting experimental results have been provided by Reisfeld [11], who prepared nanosized semiconducting PbS particles in glasses by using the sol–gel method. The optical properties of these materials are shown graphically in Figure 6.13. Reisfeld demonstrated a blue shift of nanoparticles with decreasing particle size; the absorbance of PbS nanoparticles as a function of the wavelength for particle sizes 4.8, 5.4, and 6.0 nm are shown in Figure 6.13a. The blue shift of the absorbance found with decreasing particle size is correlated to a widening of the band gap.

Figure 6.13b displays a plot according to Equation (6.9) to estimate the gap width. The intersection of the extrapolation of the linear part of the graph with the abscissa at $(\alpha h \nu)^2 = 0$ yields the band gap, which is seen clearly to widen, from 1.42 eV for the 6-nm particles to 1.92 eV for the 4.8-nm particles. A plot of the gap width versus inverse particle size squared is shown in Figure 6.14, and this demonstrates the validity of Equations (6.7) and (6.8). Clearly, the exponent –2 for the particle size is perfectly valid, although a minor deviation may be caused by the particle size distribution and experimental uncertainties. In any case, the validity of the simple considerations leading to the exponent –2 is very well justified.

Again, this example clearly demonstrates the possibility of adjusting the optical properties – in this case "color" – by the particle size which, in this example, is adjusted by annealing at different temperatures.

These elementary considerations are valid for isolated particles and composites with low concentrations of particles. Otherwise, the properties are defined rather by the distance between the particles than by their number. Often, the excited nano-particles form a dipole and, as in the case of organic lumophores, these dipoles may

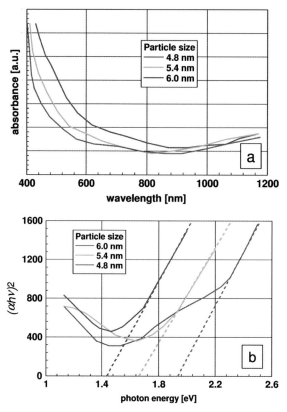

Figure 6.13 Absorbance and *Tauc* plot of PbS nanoparticles with sizes of 4.8, 5.4, and 6.0 nm according to Reisfeld [11]. (a) Absorbance of PbS nanoparticles as a function of the particle size. The blue shift of the absorbance found with decreasing particle size is correlated to a widening of the band gap. (b) *Tauc* plot for PbS according to Equation (6.9) allows an estimation of the gap width. The intersection of the extrapolation of the linear part of the graph with the abscissa at $(\alpha h\nu)^2 \rightarrow 0$ gives the band gap. There is a clear widening of the band gap, from 1.42 eV for the 6-nm particles to 1.92 for the 4.8-nm particles.

interact. Although the distance of interaction is, at maximum, 10 or 15 nm, the interacting dipoles give rise to *excimer* formation and this may lead to significant changes in the emission spectra. Zinc oxide, ZnO, is a good example of the formation of dipolar nanoparticles [12] which, in the excited state, exhibit a dipole moment leading to a permanent dipole-induced dipole interaction. This significantly influences the emission wavelength of the composites, and in this situation the size-dependency of the emitted wavelength is no longer than D^2; rather, a dependency on D^3 was theoretically predicted and verified experimentally [12,13]. As an example, the emission wavelength of ZnO coated with PMMA as a function of particle size is shown in Figure 6.15a [13]. However, perhaps more interesting is the rectified plot shown in Figure 6.15b, where the emission wavelength is plotted according to considerations of Monticone *et al.* [12] versus D^{-3}. Again, the minor deviations between the experimental and fitted lined are due to experimental variation, notably in relation to the particle size. (It should be noted here that, in contrast to Figure 6.14,

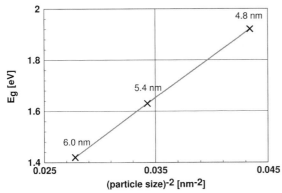

Figure 6.14 A plot of gap width versus inverse particle size squared for PbS data derived from Figure 6.13a and b. This perfect linearization clearly shows the validity of Equations (6.7) and (6.8). Here, the exponent–2, as predicted by theory for the particle size dependency, is perfectly valid.

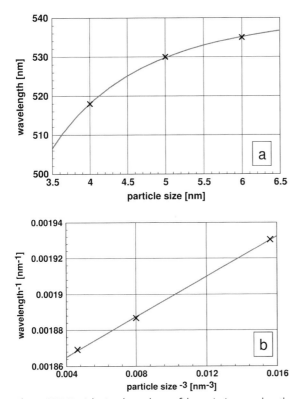

Figure 6.15 Particle size dependency of the emission wavelength of ZnO in the visible range [13]. (a) Emission wavelength of ZnO coated with PMMA as a function of particle size. (b) Rectification of the experimental data from Figure 6.15a. According to Monticone *et al.* [12], the proportionality $1/\lambda \propto 1/D^3$ was applied, indicating a dipole interaction between the particles in the excited state. The experimental data follow this relationship perfectly.

the ordinate is inverse wavelength rather than energy, these being proportional quantities.)

Further materials used for exploiting the quantum confinement phenomena include CdSe, CdS, GaAs, and GaN. In most cases, these nanoparticles are applied as isolated particles, especially in biotechnology, or as thin films.

In many cases, the properties of luminescent material depend heavily on the environment and the method of synthesis. As a typical example, the photoluminescence spectra of PbS in polystyrene in the infrared (IR) (Figure 6.16a) and visible

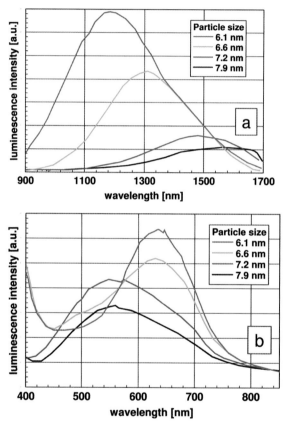

Figure 6.16 Photoluminescence of PbS/polystyrene nanocomposites with particle sizes from 6.1 to 7.9 nm [14]. The relationship between particle size and composition is shown in Table 6.3. (a) Emission of PbS/polystyrene nanocomposites in the IR range excited by 532 nm photons. As expected from the theory of quantum confinement, the wavelength of emission in the IR decreases with decreasing particle size. The explanation for the different luminescence intensities is found in the different particle size. As the luminescence intensity is related to the number of particles, at constant concentration, the number of small particles exceeds that of the larger particles; hence, the luminescence intensity increases with decreasing particle size. (b) Emission of PbS/polystyrene nanocomposites in the visible range. The excitation wavelength for this emission was 325 nm. The emission in the visible range does not follow the general rule of blue shift with decreasing particle size, and there is no simple explanation for this phenomenon. For the intensity, the same consideration as in the case of the IR emission may hold.

Table 6.3 Particle sizes, lead content, and PbSO$_3$/PbS ratio of specimen for which luminescence properties are shown in Figure 6.16a and b.

Particle size (nm)	Lead content (wt.%)	PbSO$_3$/PbS ratio	Emission peak in the IR (nm)
6.1	13.0	3.2	1190
6.6	16.8	2.7	1320
7.2	24.5	1.7	1480
7.9	31.2	1.2	1570

range (Figure 6.16b) are provided here [14]. Emission in the IR range was excited by 532-nm photons, whereas the excitation wavelength for emission in the visible range was 325 nm. Besides the formation of PbS in polystyrene, the applied process of synthesis inherently led to the formation of some PbSO$_3$. The occurrence of this byproduct was identified using spectroscopic methods; a noncritical view of the results of luminescence measurements conveys the impression of a strong influence of this byproduct, though a more critical analysis may change this opinion.

The lead content, particle size and position of the maxima of the emission peaks of samples used for these measurements (as shown in Figure 6.16a and b) are listed in Table 6.3, in which the PbSO$_3$/PbS ratio is also specified. The data show clearly that the particle size increases with increasing lead content in the composite. Additionally, the luminescence intensity increases with decreasing particle size. The authors assumed that SO$_3^-$ ions covered the surface, and therefore the content of PbSO$_3$ would increase with decreasing particle size, which was in turn related to an increasing particle surface. However, this assumption proved inadequate as the content of PbSO$_3$ was so large that it would cover the PbS particle surface with a 2-nm layer for the smallest particles, and a 1.3-nm layer for the largest particles. Thus, the explanation for different luminescence intensities must lie in the different particle size. Since, at constant concentration, the number of small particles exceeds that of the larger particles, the luminescence intensity is related to the particle number. As would be expected from the theory of quantum confinement, the wavelength of the emission in the IR decreases with decreasing particle size. Whilst this is a further example of the blue shift with decreasing particle size, the emission in the visible range does not follow this general rule.

6.4
Quantum Dots and Other Lumophores

At this point it might be pertinent to ask, "What is so special about light-emitting quantum dots in comparison to organic lumophores". An answer might be provided by Figure 6.17a and b, in which the absorption and emission spectra of CdSe quantum dots and an organic compound (fluorescein isothiocyanate, FITC), emitting roughly at the same wavelength, are compared [6]. For the quantum dots,

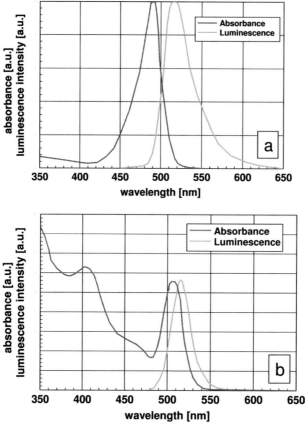

Figure 6.17 Comparison of absorption and emission spectra of organic lumophores (fluorescein isothiocyanate, FITC) (panel a) and CdSe quantum dots (panel b) emitting roughly at the same wavelength according to Smith *et al.* [6]. (a) Absorption and emission spectra of FITC. The characteristics of this class of compound are relatively broad emission spectra and a strong limitation of an effective absorption with respect to shorter wavelengths. Compared to quantum dots, the significantly smaller size of the organic lumophore molecules may be a significant advantage. (b) Absorption and emission spectra of CdSe nanoparticles. Compared to the spectra of the organic lumophore depicted in Figure 6.17a, the emission spectrum is more narrow and the absorption in the UV range is more effective. Therefore, the selection of an excitation source for luminescent nanoparticles is less critical as compared to organic lumophores.

provided that the particle size distribution is extremely narrow, the emission profile is more narrow and symmetric as compared to the organic lumophore. Therefore, nanoparticles which emit at different colors may be simultaneously excited with a single light source, making multiplexed detection of different biological targets possible. Additionally, the absorption spectra of organic compounds are relatively narrow in comparison to quantum dots exhibiting a broad absorption spectrum that range deep into the UV. Hence, the selection of an excitation source for luminescent

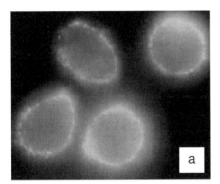
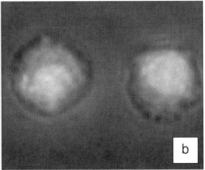

Figure 6.18 Application of (CdSe)(ZnS) quantum dots for luminescence staining of living cancer cells according to Smith *et al.* [6]. For this application, antibodies, peptides, or proteins are attached at the surface of the particles selected in such a way that they attach exactly at the intended cell receptor. (a) The nanoparticles are functionalized to attach at the cell surface. (b) Due to different functionalization, the quantum dots are localized in the cell nucleus. (Reprinted with permission from [6], Copyright: Elsevier 2003.)

nanoparticles is less critical as compared to organic lumophores. Furthermore, quantum dots can be tuned to emit over a broad range of wavelengths by changing size and composition. This tunability provides the possibility of adjusting the emission also in the near IR, thereby allowing fluorescence imaging to be conducted in living organisms. The only major disadvantage here is the significantly larger size as compared to organic lumophore molecules.

At present, the economically most attractive application of luminescent nanoparticles is found in biotechnology and diagnostics. For this application, it is necessary to attach antibodies, peptides, or proteins at the surface of the particles. These compounds must be selected in such a way that they attach exactly at the intended receptor. A typical example is shown in Figure 6.18 [6]. In this figure, CdSe·ZnS quantum dots were functionalized with two different compounds; one of these is attached at the surface (Figure 6.18a), and the other is localized in the cell nucleus (Figure 6.18b).

It is well known that the incorporation of nanoparticles in polymers may lead, in the case of chemical bonding at the particle surface, to additional optical phenomena. In this context, the most interesting area is the incorporation of luminescent particles into bulk polymers. One elementary approach is the protection of particles against degradation caused by interaction with the surrounding atmosphere [15], either by polymer coating the particles or by incorporating them into a polymer.

Perhaps of most interest is the creation of luminescent composites by coating oxide particles with a polymer. An example of luminescence based on the interaction of oxide particles with PMMA is shown in Figure 6.19a where, besides the excitation line at 325 nm, the photoluminescence spectra of alumina without and with PMMA coatings are shown. The suspension of nanocomposites in a liquid at either high or low concentration leads to identical spectra.

It is of interest to note that this emission is dependent only on the combination PMMA–oxide and not on the type of the oxide, as long it is an insulator. This is

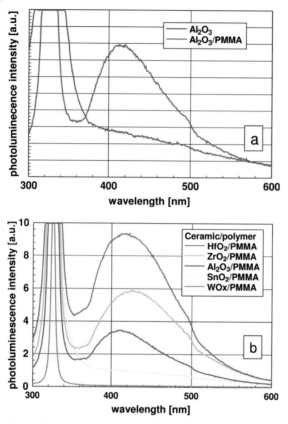

Figure 6.19 Photoluminescence of ceramic/ polymer nanocomposites according to Vollath *et al.* [13]. For excitation, an He−Cd laser with a wavelength of 325 nm was used. (a) The photoluminescence spectra of pure alumina nanoparticles and of Al_2O_3/PMMA composite particles of equal size. Uncoated alumina particles show no luminescence at all, whereas the PMMA-coated particles show a broad emission with a maximum around 425 nm.
(b) Photoluminescence spectra of oxide/PMMA nanocomposites for different oxides. While the oxide is an insulator, this emission depends only on the combination PMMA–oxide and not on the type of oxide. However, luminescence intensity depends heavily on the oxide core.

demonstrated in Figure 6.19b, where the different oxides clearly excel at different luminescence intensities of the powder. For example, HfO_2 nanoparticles exhibit the highest, and WO_x nanoparticles the lowest luminescence intensity. Except for tungsten oxide, the particle size of the ceramic cores was of the order of 5 nm, although in contrast to the intensity the position of the luminescence intensity maximum depended only weakly on the oxide core. With high probability, this is an effect of the particle size and does not depend on the oxide. Furthermore, it is obvious that PMMA-coated particles of the wide gap insulators HfO_2, ZrO_2, and Al_2O_3 show the highest photoluminescence intensity, while those of the semiconducting particles, such as SnO and WO_x, show the lowest. Additionally, intensity is increased with increasing UV absorption in the ceramic core; this relationship is shown

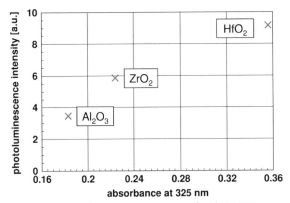

Figure 6.20 Photoluminescence intensity of oxide/PMMA nanocomposites as a function of absorbance at 325 nm, the excitation wavelength. Note the increase in photoluminescence intensity with increasing absorbance at the excitation wavelength.

quantitatively in Figure 6.20, where in aqueous suspension the absorbance at 325 nm (the excitation wavelength) is related to the luminescence intensity.

The strong correlation between UV absorption in the ceramic core and luminescence intensity is clearly apparent in Figure 6.20, and indicates that the primary process of photon absorption occurs – with high probability – in the ceramic core. Yet, the reason for this luminescence phenomenon remains unanswered, and to clarify this point attention should be turned towards the interface oxide polymer. It may be assumed that the polymer is bound chemically to the oxide surface (see Figure 6.21),

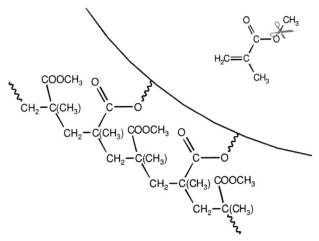

Figure 6.21 Model of the attachment and polymerization of MMA molecules at the surface of oxide particles, according to Meyer et al. [17] and Weng et al. [16]. The PMMA molecules touching the oxide surface are bound with an ester-like linkage to the surface; as shown in the insert, the methyl group (CH$_3$) adjacent to an oxygen ion is cut off. An oxide particle coated with PMMA may be described as one huge molecule, R–(C=O)–O–(oxide particle). The carbonyl groups (C=O) directly at the surface are essential for luminescence phenomena.

and it has been reported that PMMA which directly touches the oxide surface is bound with an ester-like linkage [16,17] to the surface. This means that the CH$_3$ group adjacent to an oxygen ion is cut off (see insert in Figure 6.21), and consequently the polymer at the surface becomes a modified PMMA (m-PMMA). Based on these results, nano-composite particles consisting of an oxide core bound to an m-PMMA may be described as one huge molecule, R−(C=O)−O−(oxide particle). The structural formula of MMA and a model of the connection of the polymer to the particle are shown in Figure 6.21, together with the position where the CH$_3$ group is cut off.

The most important feature in Figure 6.21 is the carbonyl group located directly at the surface of the nanoparticle. As with biacetyl, CH$_3$−(C=O)−(C=O)−CH$_3$, this carbonyl group is responsible for luminescence [18]. In the m-PMMA system, where a carbonyl group is close to the ceramic surface, the same mechanism is in operation, this having been proven by using the smallest molecule with a carboxylate group binding similarly to the ceramic surface. Here, formic acid methylester (FAME), H−(C=O)−O−CH$_3$, was selected for particle coating instead of MMA. The FAME molecules bind, in similar manner as m-PMMA, to the surface, thus forming H−(C=O)−O− (oxide particle). As oxide particles coated with polyhydroxypropyl-methacrylate (PHPMA) show only very weak luminescence, the particles were coated additionally with this compound. The difference from PMMA occurs because, in the PHPMA case, the (OH)$^-$ group binds with greater probability than the carbonyl group to the surface. The luminescence spectrum of these particles with zirconia core is shown in Figure 6.22, and is almost identical to that found with m-PMMA-coated materials (see Figures 6.19a and b). This is insofar remarkable, as aqueous solutions of FAME demonstrate luminescence with a few isolated lines in the UV range. These findings also indicate that the carbonyl group of these compounds, when bound to the particle surface, is responsible for the emission spectrum.

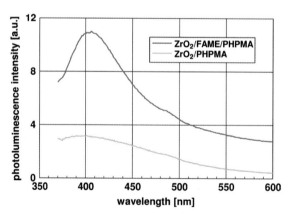

Figure 6.22 Luminescence of ZrO$_2$/FAME (formic acid methylester, H−(C=O)−O−CH$_3$)/ PHPMA nanocomposites in comparison to those without the FAME layer. As FAME molecules bind like PMMA to the surface forming H−(C=O)−O−(oxide particle), identical luminescence phenomena are observed. Oxide particles coated with PHPMA do not show luminescence because the (OH)$^-$ group binds to the surface with a greater probability than does the carbonyl group [13].

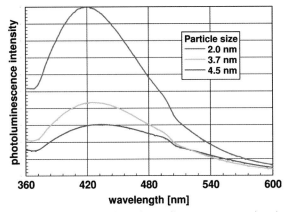

Figure 6.23 Particle size dependency of luminescence wavelength maximum and intensity of ZrO₂/PMMA nanocomposites with different sizes of the oxide core. These data indicate an increasing luminescence intensity with decreasing particle size, and a significant blue shift with decreasing particle diameter.

This is an interesting proof that the luminescence of oxide/m-PMMA nanocomposites is a surface-related phenomenon, stemming from the carbonyl group directly adjacent to the surface.

The influence of particle size on the emission spectra of ZrO₂/m-PMMA nanocomposites is shown in Figure 6.23, and demonstrates clearly that the luminescence intensity decreases with increasing particle size. Obviously, this is the reason why such a phenomenon was never observed with conventional particles. In addition, a remarkable blue shift is observed with decreasing particle diameter.

The following relationship fits the maximum of the luminescence intensities as it is shown as a function of the inverse particle size in Figure 6.24:

$$I = I_0 + \frac{b}{D} \qquad (6.10)$$

where I is the intensity, I_0 and b are the fitting parameters, and D the particle diameter. Equation (6.10) states that the luminescence intensity is directly proportional to the surface of the particles. Because for a given quantity of material, the number of particles is proportional to D^{-3}, the surface of one particle is proportional to D^2; thus, the surface of a given quantity of particulate matter is proportional to D^{-1}. The direct proportionality between luminescence intensity and particle surface offers additional proof of the idea that the interface ceramic/polymer is the source of the luminescence. Figure 6.24a shows diminishing small luminescence intensities for particle sizes larger than ca. 10 nm, while the data depicted in Figure 6.24b show a significant blue shift with decreasing particle size. The wavelength of the emission maximum as a function of the particle size follows the relationship:

$$\frac{1}{\lambda} = \frac{1}{\lambda_0} + bD^3 \quad \text{or} \quad \Delta\left(\frac{1}{\lambda}\right) = bD^3 \qquad (6.11)$$

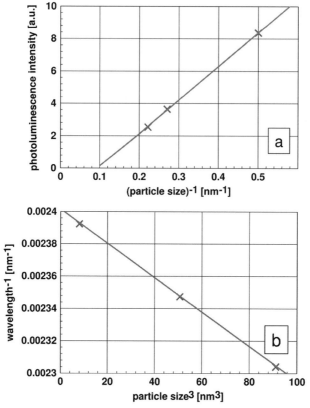

Figure 6.24 Dependency of luminescence intensity and wavelength maximum of ZrO$_2$/PMMA nanocomposites as a function of particle size. To show the dependencies, these graphs are rectified. (a) Luminescence intensity of ZrO$_2$/PMMA nanocomposites as a function of inverse particle size following Equation (6.10). The proportionality with $1/D$ indicates a proportionality of the intensity to the surface of the particles. (b) Wavelength maximum of ZrO$_2$/PMMA nanocomposites as a function of particle size. The mathematical description underlying this graph $1/\lambda = 1/\lambda_0 + bD^3$ is fundamentally different from that for quantum confinement. In contrast to all other mechanisms of blue shift, this shows a positive exponent of the particle size.

This blue shift is demonstrated graphically in Figure 6.24b, where the emission wavelength is plotted versus the particle size raised to the power of 3. Equation (6.11) describes a mechanism, expressed by a mathematical description, which is fundamentally different from the quantum confinement phenomena in semiconductor quantum dots. All mechanisms of blue shift – except for this one – show, in contrast to the one observed with oxide/polymer nanocomposites, a negative exponent of the particle size. The blue shift of the luminescence emission of oxide/m-PMMA nanocomposites exhibits the positive exponent 3.

Similar to the blue shift with decreasing particle size, a reduction in the line width at half-maximum intensity with decreasing particle size was observed [13,19].

Comparable phenomena have been described for many other combinations of oxides with polymers. Typical examples are the incorporation of oxide nanoparticles in poly(*p*-phenylene vinylene (PPV) and poly[2-(6-cyano-6β-methylheptyloxy)-1,4-phenylene] [20].

6.5
Metallic and Semiconducting Nanoparticles in Transparent Matrices

The free electrons in the incompletely filled conduction band of a metal move collectively in discrete waves. The metal particle is embedded in a cloud of electrons (see Figure 6.5). As with a quantized lattice, the vibrations are called *phonons* and the quantized waves of the free electrons are *plasmons*. This cloud oscillates relative to the lattice consisting of the positive charged atomic cores of the metal in different quantized modes, the *plasmons*. At the metallic surfaces, light with frequencies below the frequency of the surface plasmons is reflected. For most metals the plasmon frequency is in the UV range, and therefore these have a "metallic luster" in the visible range. Very few metals (e.g., copper, gold and a few alloys) have plasmon frequencies in the visible range, and this leads to their typical colors. Plasmon frequencies in the IR region are observed in highly doped semiconductors.

In Figure 6.5, it was shown that an electron cloud surrounds metallic nanoparticles; this in turn significantly influences the optical properties of metallic nanostructures, as these are controlled by the interaction of light with surface plasmons. The electron cloud which surrounds the positively charged lattice of the metal atom cores of a metallic nanoparticle oscillates relative to the positive charged lattice of the atom cores of the metal in different quantized modes. In Figure 6.25 this is shown for a spherical particle and the simplest oscillation mode.

Clearly, these oscillation modes depend heavily on the shape of the particle, and therefore the optical properties depend strongly on the particle shape. Usually, the oscillation frequency of the surface plasmons is in the visible region. In the case of resonance with incoming photons, one observes strong surface plasmon resonance absorption peaks; this is an entirely different mechanism of absorption as compared to semiconducting particles, where quantum confinement processes are predominant.

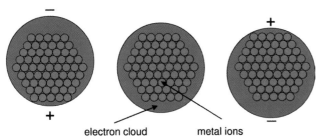

Figure 6.25 Oscillations of the cloud of free electrons surrounding a metallic nanoparticle. The quantized oscillations of the electron cloud are termed "plasmons".

The frequency of surface plasmon resonance absorption changes only slightly with particle size; therefore, the optical properties of metallic nanoparticles are, within a relatively broad range of particle sizes, almost uninfluenced by the particle size. The reason for the particle size independence is the mean free path length of the free electrons in metals of approximately 50 nm, which allows the electrons to travel undisturbed through a small nanoparticle. Therefore, all interactions will be with the surface, and scattering of light from the bulk is negligible. In the case of resonance interaction, the electrical field of the light moves the free electrons at the particle surface and causes oscillations with the frequency of the light; this process is called *surface plasmon resonance*. However, by adding shape anisotropy to the nanoparticle, such as the formation of ellipsoids or rods, the optical properties are changed dramatically. Nanorods or ellipsoids exhibit two different distinct plasmon resonances: (i) transverse oscillation of the electrons, independent of the aspect ratio, leading to the same absorption band as found for spherical nanoparticles; and (ii) longitudinal plasmon resonances at a longer wavelength, characteristic of the aspect ratio. Ellipsoid particles with an axis ratio close to one lead to a broadening of the absorption band.

Figure 6.26a and b display the resonance wavelength for surface plasmons of gold and silver spheroids with different axis ratios [21]. It can be seen that the transversal modes are almost independent of the particle size, whereas the longitudinal modes depend on particle diameter and axis ratio. In this figure, spherical particles are characterized with an axis ratio of one. As mentioned above, for particles up to dimensions of less than 50 nm, the wavelength of the absorption maximum is almost independent of the particle size.

The data in Figure 6.26a and b indicate that anisotropy adds an additional parameter to tune optical properties of metallic nanoparticles with respect to application. Additionally, the electron cloud around metal nanoparticles is the origin of strong electromagnetic fields, determined by the geometry of the nanoparticle, and affects the local environment. This electrical field influences the adsorption of other organic molecules, and is itself altered by such adsorption processes. The retroaction to the surface plasmon resonance frequency is caused by changing the dielectric constant of the surrounding material locally, thus influencing the oscillation modes to a significant degree. As this phenomenon is used for the detection and identification of other molecules, metallic nanoparticles – and especially highly anisotropic nanoparticles – may be used as extremely sensitive sensors. In addition, for dispersed metallic nanoparticles, entirely different absorption spectra as a function of the surrounding medium may be expected.

The absorption spectra of gold nanoparticles and nanorods are shown in Figure 6.27, with near-identical spectra being obtained for spherical particles of 15 or 30 nm diameter. In comparison, the absorption spectra of nanorods with aspect ratios of 2.25 and 6 are entirely different [22]. The relatively broad longitudinal peaks of the nanorods are caused by a broad distribution of aspect ratios. An electron micrograph of nanorods with mean aspect ratio of 6 is shown in Figure 6.28 [23], where the different lengths of the individual rods, resulting in a broad distribution of aspect ratios, are clearly visible.

In the spectra shown in Figure 6.27, the absorption spectra of the nanorods clearly show the well-separated absorption bands for the transversal resonances around

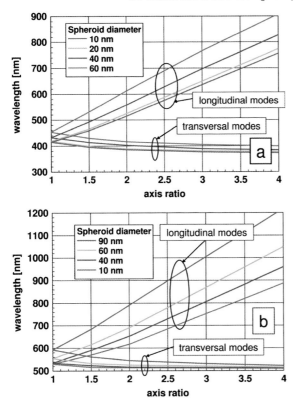

Figure 6.26 Wavelength of the absorption maximum of surface plasmons of silver and gold spheroids with different sizes and axis ratios [21]. (a) Wavelength of the surface plasmons of spheroidal silver nanoparticles with different sizes and axis ratios. The transversal modes are, up to a diameter of at least 20 nm, completely independent of the particle diameter. (b) Position of the wavelength of surface plasmons of spheroidal gold nanoparticles as a function of particle diameter and axis ratio. Up to a particle diameter of ∼40 nm, the transversal modes are independent of particle diameter.

520 nm, which is more or less identical with the absorption maximum for the spherical particles, and the broad absorption band attributed to the transversal plasmon resonances. Additionally, Figure 6.27 shows clearly that the relative contribution of transversal oscillation modes decreases with increasing aspect ratio.

As mentioned above, the energy of the plasmon resonance depends heavily on the surrounding medium and its composition and, especially when producing colored glasses and pigments, the composition of the matrix must be selected carefully. One of the oldest applications of nanoparticles is their use as pigment in glass. The first known application of these composites dates back to the Assyrians who, in approximately 700 BC, documented the composition of a red glass with gold nanoparticles as pigment. This was reinvented by Kunkel [24] during the seventeenth century in Leipzig. It was later found that, besides gold changing the color of glass to a characteristic red, the addition of silver leads to a yellow coloration. During the nineteenth century, *Faraday* attributed this color to very finely divided colloidal gold,

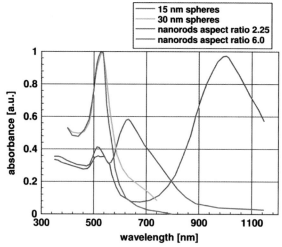

Figure 6.27 Absorption spectra of spherical and elongated gold nanoparticles. The absorption maximum of the transversal modes around 520 nm is almost independent of particle size for spherical particles and nanorods. The maxima for the longitudinal modes show a strong dependency on the aspect ratio.

the particle size of which was estimated by *Szigmondi* to be approximately 50 nm, well within the recent notation of nanoparticles.

During its long history, the composition of this "gold ruby glass" has not changed significantly to the present day. Currently, this colored glass is used not only for decoration but, after grinding, also as a pigment. Glasses containing metallic or semiconducting nanoparticles as colorants usually are composed of 50–60 wt.% SiO_2, 10–20 wt.% ZnO, and roughly the same amount of K_2O. Minor amounts of K_2O may be replaced by Na_2O, while some ZnO could be exchanged for CaO. In order to improve melting behavior, a few weight percent of B_2O_3 are sometimes added, while to adjust the index of refraction PbO or Sb_2O_3 may be added [25]. A typical example of such a glass containing gold nanoparticles is shown in Figure 6.29. The glass beaker

Figure 6.28 Electron micrograph of a specimen consisting of gold nanorods with an aspect ratio of about 6. The broad distribution of aspect ratios leads to a broad absorption maximum caused by the distribution of longitudinal surface plasmon modes (reprinted with permission from [23]; Copyright: American Chemical Society, 2007).

Figure 6.29 A beaker covered with gold ruby glass, a composite consisting of gold nanoparticles. Note the faint blue hue in the color, which is typical of pigments based on gold nanoparticles.

is coated with a thin layer of a glass that contains gold nanoparticles as pigment, and into this layer of gold ruby glass an artist has engraved an image. Even when this glass shows a deep red, a slight blue hue is both visible and unavoidable.

The red color results from narrow band absorption in the range of 500–600 nm. The absorbance of gold as nanoparticles dispersed in a polymer is shown in Figure 6.30 [26], where the nanoparticles ranged in size from 5 to 15 nm. Clearly,

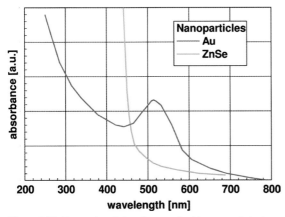

Figure 6.30 Narrow band absorbance of gold nanoparticles in polymer [26] with sizes in the range from 5 to 15 nm, in comparison to the edge-like absorbance of ZnSe nanoparticles [27].

the position – and also therefore the width – of the absorption peak depend heavily on the particle shape and shape distribution. To some extent, this allows the hue to be adjusted, for example by temperature treatment of the composite. However, as the spectral absorption shows a maximum around 520 nm and not an absorption edge, these glasses always have a more or less pronounced blue hue. This is avoidable by the application of a material showing an absorption edge, as found typically in many semiconducting nanoparticles. For comparison, the absorbance of ZnSe [27] is plotted additionally in Figure 6.30.

High-quality gold ruby glasses excel in a clear color which is obtained by well selected particle size distribution, of near-ideal spherical particles. In glasses, the particle size of the gold nanoparticles is stabilized by adding tin dioxide into the glass matrix. Besides gold, nanoparticles of silver, copper, and platinum are often used as colorants for glasses. The color can also be adjusted by replacing pure metal particles with alloys of gold with other metals; the absorbance spectrum of a typical example of a gold–silver alloy with a Au:Ag ratio of 1:2 is shown in Figure 6.31 [28], with the position of the absorption maxima of pure silver and gold nanoparticles clearly indicated. A complete theoretical description of the color of metal nanoparticles in glass has been provided by Quinten [29].

With respect to recent technological developments, nanoparticles made from semiconducting compounds are attracting an increasing amount of attention. Originally, nanoparticles made of solid solutions from the system CdSe−CdS −ZnS−ZnSe dispersed in glass were used as pigments, but the equivalent compounds of tellurium (either pure or as a solid solution) are also used. Whilst nearly all colors, from deep red to orange and yellow, are obtained with these systems, more recently the luminescence properties of these materials have also attracted attention,

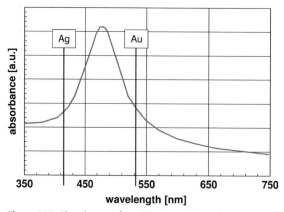

Figure 6.31 Absorbance of a gold–silver alloy with Au : Ag ratio of 1 : 2 [28]. The positions of the absorption maxima of pure silver and gold nanoparticles are also indicated. These data show the additional possibility of adjusting the position of the absorption maximum, which is equivalent to the color, by alloying.

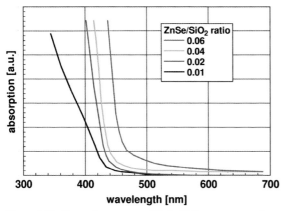

Figure 6.32 Absorption spectra of ZnSe/SiO$_2$ nanocomposites with different ZnSe contents [27]. The absorption is reduced with decreasing ZnSe content. The blue shift of the onset of absorption with decreasing concentration indicates a decreasing particle size with decreasing concentration.

and therefore these groups of semiconducting compounds move into the center of interest.

Semiconducting nanoparticles in a transparent matrix are extremely interesting with respect to their absorption and luminescence properties. In this context, Figures 6.32 and 6.33 show the absorption and photoluminescence spectra of ZnSe in SiO$_2$ [27]. In Figure 6.32, the absorption spectra of ZnSe in silica glass are shown for different concentrations, with the absorption seen to be reduced with decreasing ZnSe content, as might be expected. There is also a blue shift of the onset of the absorption with decreasing ZnSe content, reflecting a (readily understood) decrease in the particle size.

The absorption and the emission spectra of a specimen with a ZnSe/SiO$_2$ ratio of 0.04 is shown in Figure 6.33, the interesting point here being that the absorption and luminescence spectra are overlapping. Hence, the emission of one particle can excite a further particle of the same type, which allows the transport of information from one particle to the next.

Another interesting example of a quantum confinement system which is dispersible in silica glass matrix and perfectly tunable by particle size, is that of CdTe [30]. Interestingly, the optical properties of CdTe nanoparticles are essentially identical, independently of whether they are dispersed in a liquid or a glass, thus proving that the particles are embedded individually in the glass matrix. The absorption spectra of particles with two different sizes, namely 3.4 and 6.2 nm, suspended in a liquid and dispersed in glass, are shown in Figure 6.34a. When comparing the two different carriers the difference is not important, but in terms of absorbance the differences between the two particle sizes are clearly visible. As expected, a reduction in grain size leads to a significant blue shift. The same blue shift is visible in the emission lines, but there is no difference between the emissions of particles suspended in a liquid or

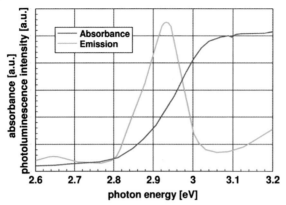

Figure 6.33 Absorption and the emission spectra for specimen with a ZnSe : SiO$_2$ ratio of 0.04 [27]. The absorption and luminescence spectra are overlapping; hence, the emission of one particle can excite a further particle of the same type. (Note that the energy units of the abscissa are an inverse of the wavelength.)

those dispersed in a glass. The relatively narrow emission lines indicate a narrow particle size distribution.

Even when the behavior of semiconducting nanoparticles in pure silica glass is – more or less – straightforward, the complexity increases dramatically in more complex glasses, as the dopants and nanoparticles will interact with the glass matrix. Thus, easily interpreted results cannot be expected. As an example, the optical properties of nanoparticles in the system CdSe−CdTe dispersed in an alkali-containing silica glass are shown [31]. The glass matrix was composed of SiO$_2$−CaO−A$_2$O (A = alkali metal), and as an alkaline a mixture of equal amounts of Na$_2$O, K$_2$O and Li$_2$O were used. Up to 1 wt.%, any composition of CdSe$_x$Te$_{1-x}$ dissolves in the selected melted glass, without causing it to crystallize. During cooling, CdSe$_x$Te$_{1-x}$ precipitates in uniform distribution as a nanoparticle. In order to understand the behavior of these materials, it is necessary to examine the phase diagram of the CdSe−CdTe quasibinary system (see Figure 6.35), where two phases may be observed: (i) on the selenium-rich side the particles crystallize in the wurtzite; while (ii) on the tellurium side they crystallize in the sphalerite structure [32]. However, it must be pointed out that this phase diagram is valid for materials of conventional grain size and that, due to the large surface of the nanoparticles, the phase diagram may be significantly different for nanoparticulate systems. Bodnar *et al.* [31] assumed that the two-phase region, which is quite narrow for bulk materials, broadens significantly with decreasing particle size; this suggestion may be explained by the difference in lattice and surface energy between the sphalerite and wurtzite phases.

The absorbance of a series of solid solutions in the system CdSe−CdTe dispersed in silicate glass is shown in Figure 6.36a, where the concentration of nanoparticles in the glass was 0.75 wt.%. Although this figure shows clearly the red shift with increasing tellurium content in the nanoparticles, this process is not

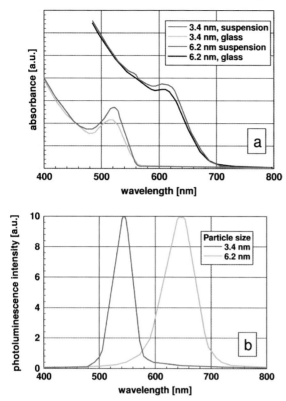

Figure 6.34 Absorption and emission spectra of CdTe quantum dots suspended in silica glass or in a liquid, according to Li and Murase [30]. The particles are dispersed individually in the glass matrix or the liquid. (a) Absorbance of CdTe nanoparticles of two different sizes, 3.4 and 6.2 nm. The different media used to suspend the particles are unimportant; however, the difference between the two particle sizes is clearly visible. Again, reducing the grain size leads to a blue shift. (b) Emission spectra of CdTe quantum dots embedded individually in silica glass or suspended in a liquid. The two dispersing media do not influence the emission behavior. Due to a relatively narrow particle size distribution, the emission spectra of the two sizes (3.4 and 6.4 nm) are well separated.

straightforward. Pure CdTe shows a maximum of absorption at about 570 nm, but this maximum does not appear with the particle composition $CdSe_{0.2}Te_{0.8}$, even when the composition is clearly in the sphalerite, one-phase field. CdSe does not show such a maximum of the absorption, and therefore it is not surprising that neither in the wurtzite one-phase field, nor in the two-phase field, is this maximum observed.

In the two-phase area, at a Se/Te ratio of about one, a minimum of the absorption is observed. As in the two-phase field, the same concentration is distributed into two different species of particles; one may assume the presence of extremely small particles in this field.

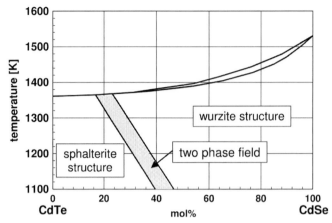

Figure 6.35 Phase diagram of the quasibinary system CdTe–CdSe [32]. As the two substances crystallize in different structures, there exists a miscibility gap. It is important to note that this phase diagram is for materials of a conventional grain size, and is not necessarily valid for nanoparticulate materials. Because of the difference in lattice and surface energy between the sphalerite and wurtzite phase, a broadening of the two-phase region may be assumed [31].

Furthermore, the spectra in Figure 6.36b show that the absorption decreases at higher concentrations. The nanomaterial used for this figure consisted of $CdSe_{0.2}Te_{0.8}$, and therefore lay clearly in the sphalerite one-phase field. It is of interest to note that the same maximum close to 570 nm is observed only at the lowest concentration; consequently, only at low concentrations of $CdSe_{0.2}Te_{0.8}$ nanoparticles will the absorption spectrum be similar to that of CdTe nanoparticles. In contrast, for other $CdSe_xTe_{1-x}$ solid solutions, this maximum is missing, independently of a concentration above 0.5 wt.%. It is most likely that, only under these conditions, $CdSe_{0.2}Te_{0.8}$ and CdTe nanoparticles are similar in structure.

When considering photoluminescence in glass, great care must be taken when associating a mechanism, as luminescence is not necessarily connected to nanoparticles and to quantum confinement. In this respect, a very instructive example for the possibility of such an erroneous attribution is provided in Figure 6.37 [33].

Here, the experiments were performed on thin silica glass films doped with 1 mol.% silver, prepared by sol–gel dipping of substrates, the sheets being either of pure silica glass or a soda lime glass. After depositing, the films were dried and then annealed at different temperatures. While the Ag-doped coatings deposited on silica substrates did not exhibit any luminescence, the coatings on soda-lime substrates excelled with intense emission in the UV, with a maximum at 320 nm. Up to 648 K, the intensity increased with annealing temperature, but at higher annealing temperatures the photoluminescence intensity decreased (see Figure 6.37). This emission is well known and ascribed to the presence of Ag^+ ions in glass matrix, and not to the precipitation of silver nanoparticles. The difference between the two substrates is explained by the

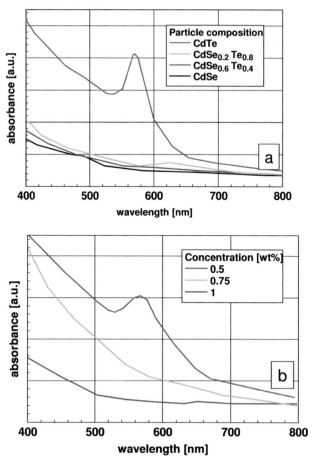

Figure 6.36 Absorption spectra of nanoparticles embedded in an SiO_2–CaO–A_2O (A = alkali metal) glass matrix, consisting of solid solutions of CdTe and CdSe as a function of composition and concentration [31]. (a) Absorbance of 0.75 wt.% nanoparticles of a series of solid solutions in the system CdSe–CdTe dispersed in silicate glass. Note the clear red shift with increasing tellurium content in the nanoparticles. Pure CdTe shows a maximum of the absorption around 570 nm, which is not evident in particles of any other composition. (b) Influence of concentration of $CdSe_{0.2}Te_{0.8}$ nanoparticles on optical absorbance. These particles are clearly in the one-phase field with a sphalerite structure. At the lowest concentration of 0.5 wt.%, the same maximum close to 570 nm as observed with pure CdTe is visible.

diffusion of silver into the soda-lime glass substrate, this process being based on an exchange between Ag^+ ions in the coating and Na^+ ions of the substrate. In the coatings on silica carriers, silver is always present as Ag^0; annealing of the specimen with soda-lime glass substrates above 523 K leads to a reduction of the Ag^+ ions by the precipitation of Ag nanoparticles, which show no photoluminescence.

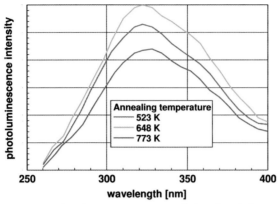

Figure 6.37 Photoluminescence spectra of silver-doped silica films on soda lime glass substrate as a function of annealing temperature, according to García *et al.* [33].

6.6
Special Luminescent Nanocomposites

Various specialized applications exist where the use of semiconducting quantum dots based on for example, cadmium, selenium, or tellurium, is not desirable due to potential problems of toxicity and carcinogenicity. Furthermore, most luminescent compounds made from these elements have only a limited stability against oxidation and hydrolysis. On occasion, such problems make the handling – and therefore also the application – of these materials extremely difficult, and this has led to a search for luminescent oxide nanoparticles comprised of nontoxic constituents that are stable in water. One possible way in which these problems might be overcome would be to dope the luminescent oxide nanoparticles with small amounts of rare earth ions.

A completely new approach [13,19] of overcoming the above-described problems is to use insulating oxide nanoparticles coated with a polymer. Although the lumophores produced are highly stable and made from nontoxic constituents, because their luminescence originates in the interface oxide/polymer the variability of emission colors is limited. Therefore, an additional design of luminescent nanocomposites, consisting of a ceramic core, a layer of an organic lumophore, and a polymer coating at the outside was developed. The design of these composites is shown schematically in Figure 6.38.

This concept makes use of the vast amount of organic lumophore molecules available. The ceramic core is selected according to the properties demanded and, as shown later, it may even be used to add an additional physical property besides luminescence to the particle. The polymer coating at the outer surface is selected according to the demands with respect to the surroundings during application. In general, PMMA is used for coating, but in order to obtain a highly hydrophobic layer fluorinated aliphatic compounds are used, while for hydrophilic layers PHPMA is

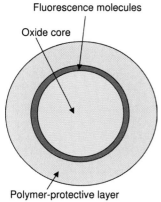

Fluorescence molecules

Oxide core

Polymer-protective layer

Figure 6.38 Luminescent nanocomposites, consisting of a ceramic core, a lumophore layer, and a polymer coating at the outside [13,19]. The ceramic core may be selected according to the desired properties, to add physical properties (besides luminescence) to the particle. The polymer coating at the outer surface is selected according to demands, and based on interaction with the outside world.

selected. As the amount of lumophore at the surface of the ceramic cores increases with the increasing surface to be coated, the size of the ceramic core is kept as small as possible. The best results are obtained using a monolayer of lumophores; moreover, to minimize parasitic absorption in the coating the outer polymer layer is made as thin as possible, usually less than 1 nm.

A typical spectrum of such a composite is shown in Figure 6.39a. For these particles, iron oxide was used as the ceramic core, PMMA as the outer coating, and pyrene was applied as the lumophore. The emission spectrum of pure solid pyrene is also plotted in the figure. Clearly, the spectrum of the nanocomposite is quite similar to that of the pure lumophore, with just a slight blue shift of ca. 15 nm; more prominently, the spectrum of the composite has lost the fine structure characteristic of pyrene. In contrast to pyrene, the situation with composites containing anthracene as lumophore is more complicated. Figure 6.39b shows the spectra of pure anthracene and a composite using hafnia as ceramic core. As shown in Figure 6.39b, anthracene shows a more structured spectrum as compared to pyrene. Each of the maxima in the spectrum can be associated with a vibration mode. On comparison, the spectra of pure anthracene and the anthracene–hafnia composite are essentially identical, except for the strongest line of the composite (positioned between vibration modes 5 and 6), which was not seen in the pure anthracene spectrum.

As described for the luminescent oxide–PMMA nanocomposites, there is a strong interaction between the oxide core and the lumophore coating. The influence of the ceramic core (which is selected according to its interaction with the exciting UV radiation) on the luminescence intensity of different oxide/pyrene/PMMA nano-composite powders, is shown in Figure 6.40. Absorption in the range of the 325-nm radiation used for excitation is increased (see Figure 6.40), from silica, SiO_2 over

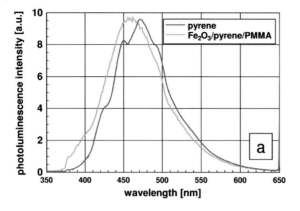

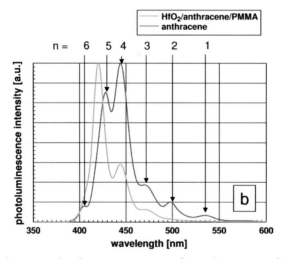

Figure 6.39 Photoluminescence spectra of nanocomposite powders according to Figure 6.38 [13,19]. (a) Photoluminescence spectra of pure pyrene and a composite consisting of Fe$_2$O$_3$/pyrene/PMMA, and of pure solid pyrene. When comparing the two spectra there is a slight blue shift of ca. 15 nm of the emission spectrum of the nanocomposite in comparison to the pure lumophore. However, more prominent is the loss of the fine structure characteristic of the pyrene spectrum.
(b) Luminescence spectra of pure anthracene and the composite HfO$_2$/anthracene/PMMA. Each of the maxima in the spectrum can be associated with a vibration mode. The main difference between these two spectra is the strongest line of the composite, positioned between vibration modes 5 and 6, that is not found in pure anthracene.

alumina, Al$_2$O$_3$ and zirconia, ZrO$_2$ to hafnia, HfO$_2$, with the highest absorption. In all of these examples the particle size and coverage of the surface of the oxide kernels with lumophore were kept constant.

The data in Figure 6.41 display maximum photoluminescence intensity for the different ceramic cores in Figure 6.40, versus absorbance at the excitation wavelength of 325 nm. As shown in Figure 6.20, for the PMMA/ceramic composites there is a

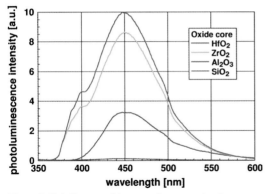

Figure 6.40 Influence of the ceramic core on luminescence intensity of different oxide/pyrene/PMMA nanocomposite powders, according to Figure 6.38. The strong influence of the ceramic core on luminescence intensity is striking. In all examples, particle size and coverage of the surface of the oxide kernels with pyrene were held constant.

strong correlation between luminescence intensity and UV absorption in the ceramic core. The intensities of the pyrene luminescence as a function of the ceramic core (which are visible in Figures 6.40 and 6.41) reflect the interaction of the lumophore and the core. This suggests a mechanism of excitation of the lumophore where the photon energy is absorbed in the ceramic core, after which the excitation is transferred to the lumophore.

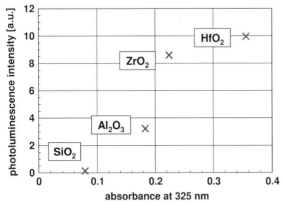

Figure 6.41 Maximum photoluminescence intensity for different ceramic cores (as shown in Figure 6.40) versus absorption at the excitation wavelength of 325 nm. The increase in photoluminescence intensity with increasing absorption of the excitation line suggests a mechanism of excitation where the photons are absorbed in the ceramic core; the excitation is later transferred to the lumophore (pyrene).

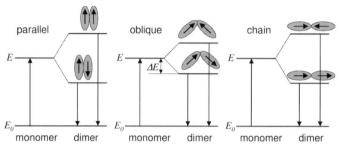

Figure 6.42 Excited organic lumophores, especially in solutions, may form intermediary excimers by dipole–dipole interaction, when they come close together. This changes the energy levels, depending on their configuration. The energy levels in turn depend on the orientation of the two dipoles. Configurations leading to higher energy levels are quite improbable. Additionally, some emission transitions are forbidden by quantum selection rules. Most probable are oblique configurations, where a change in energy depends on the angle between the two dipoles. As many angles are allowed, the dimer spectrum is, in most cases, quite broad and not structured.

Excited lumophore molecules often form dipoles which interact, provided that the distance between the molecules is sufficiently small. As distance of 10–15 nm is assumed as the maximum distance for interaction, it is possible in organic lumophores to distinguish between the spectra of the molecule and of the excimer. An *excimer* is a dimer which is formed when such an excited molecule comes into close contact to a second lumophore molecule. In this case, an intermediary excimer is formed by dipole–dipole interaction; the process is shown schematically, and greatly simplified, in Figure 6.42.

Coupling of two dipoles to form a dimer changes the energy levels; this depends on the new intermediate configuration. In extreme cases, the dipoles may be arranged either in parallel or as chain, when the energy level depends on the orientation of the two dipoles. As a general rule, it can be said that the configurations leading to higher energy levels are quite improbable. Additionally, some of the emission transfers are forbidden by quantum selection rules, but the most probable are oblique configurations. In this case, the change in energy depends on the angle between the two dipoles. As many angles are allowed, the spectrum of such a dimer is quite broad and, in most cases, not structured. The pyrene spectrum shown in Figure 6.39a is that of the excimer. As the formation of excimers depends on the distance between the molecules, the emission spectrum of a lumophore in a solution with low concentration is one of the molecule, with increasing concentration; the probability of excimer formation increases, leading at high concentrations, to the pure excimer spectrum.

In a nanocomposite according to Figure 6.38, the lumophore molecules are held tightly together, and therefore it is not surprising to see the excimer spectrum shown in Figure 6.39. Unfortunately, things are not that simple, as Figure 6.43 shows the spectra of a composite consisting of an alumina core (~5 nm diameter) coated with pyrene and PMMA. One of these spectra, measured with powder, is the excimer spectrum, while the other spectrum, measured in a suspension of low particle

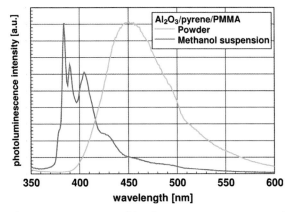

Figure 6.43 Comparison of the photo-luminescence spectra of an Al₂O₃/pyrene/PMMA nanocomposite as a powder and suspended in methanol. For the powder, the particles are close together, and therefore the excimer spectrum is emitted. In the methanol suspension with a low particle concentration, the distance between particles is large and therefore the probability for interaction is small. In such a diluted suspension, the molecule spectrum is emitted.

concentration in methanol, shows the molecule spectrum. Clearly, lumophore molecules located on one ceramic core are not interacting. Similar phenomena are found in nanocomposites with incompletely covered ceramic cores.

Although magnetism and luminescence are never found together in nature, for special applications in biotechnology and medical diagnosis a combination of these properties is quite often demanded. When such applications are connected to, for example, magnetic cell separation, identification, and quantification, this combination of properties is extremely valuable. The luminescence spectra and magnetization curves of some materials that are both luminescent and superparamagnetic (e.g., Fe₂O₃/anthracene/PMMA and Fe₂O₃/pyrene/PMMA nanocomposites) are shown in Figure 6.44a and b.

The difference in saturation magnetization for composites with anthracene and pyrene as lumophore is due to different sizes of the magnetic core, and not to the lumophore. The combination of inorganic nanoparticles with organic lumophores may be exploited for further modifications of the emission spectrum. Liu *et al.* [34] report on the combination of CdSe(ZnS) nanoparticles with an organic lumophore, the idea being to combine an organic lumophore capable of absorbing some of the light emitted by the inorganic nanoparticles. The emission of the organic lumophore modifies the total emission spectrum to more closely resemble a highly efficient materials combination that emits white light. For this application, a triplet iridium(III) complex [bis(4-trifluoro-methyl)-2-phenyl-benzothiazolatoacetylacetonate-iridium(III)] was selected as lumophore. The emission spectrum of the CdSe(ZnS) nanoparticles, together with the absorption spectrum of the organic iridium(III) complex, are shown in Figure 6.45. The emission spectrum of the complex depicted in Figure 6.46 clearly shows the emission peak of the inorganic nanoparticles and, in addition, a broad shoulder on the side of the longer

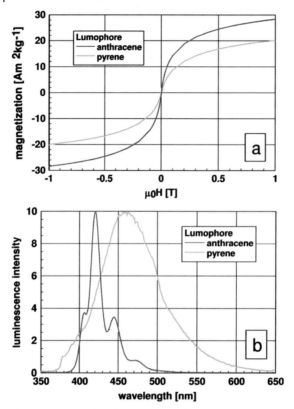

Figure 6.44 Examples of composites with combined properties. Here, the properties of luminescence and magnetism are combined – a situation not found in nature [13,19]. (a) Magnetization curves of Fe$_2$O$_3$/anthracene/PMMA and Fe$_2$O$_3$/pyrene/PMMA nanocomposites. The difference in magnetization is caused by different sizes of the ceramic core, and not by the different lumophores. (b) Luminescence spectra of the nanocomposites with superparamagnetic core as described in panel a. Note the typical spectra of anthracene and pyrene.

wavelength that is clearly associated with the emission of the iridium(III) complex. Additionally, it must be pointed out that the intensity of that part of the emission spectrum attributed to the iridium(III) complex has a higher intensity in combination to the nanoparticles as compared to the pure material. The CdSe(ZnS) particles with sizes around 1 nm, as used in this study, were covering gold nanoparticles of 5 nm diameter, and these agglomerates were embedded in the organic lumophore matrix. The light emission of Au–CdSe(ZnS) was slightly blue-shifted as compared to pure CdSe(ZnS) nanoparticles. Additionally, the photoluminescence intensity of the gold-containing agglomerates exceeded that of the CdSe(ZnS) nanoparticles. However, even when the increased emission intensity is of major technological importance, in this context only the modified emission spectrum is essential. Again, this is an example where the higher absorbance of nanoparticles is used to increase

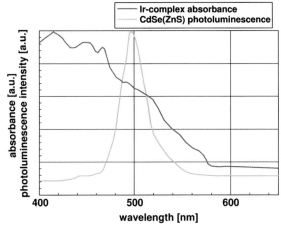

Figure 6.45 Absorbance of the iridium(III) complex (as described in the text) and photoluminescence spectrum of CdSe(ZnS) nanoparticles. The emission maximum of the nanoparticles is close to the maximal absorption of the iridium(III) complex. Therefore, the emission of CdSe(ZnS) nanoparticles excite the iridium(III) complex in a nanocomposite.

the efficiency of a nanoparticle/lumophor composite. This was different from the examples based on the particle design according to Figure 6.38 in the case described above, when energy was transferred by the emission and absorption of photons. The authors also showed that the described composite might be used for electroluminescent devices.

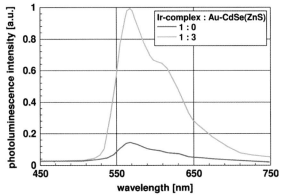

Figure 6.46 Photoluminescence of the pure iridium(III) complex and a nanocomposite with a ratio of 1:3 of the iridium(III) complex to the CdSe(ZnS) nanoparticles. Note the significant increase in photoluminescence intensity by energy transfer from the nanoparticles to the organic lumophore.

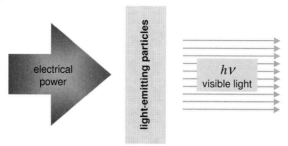

Figure 6.47 The basic principle of electroluminescence. Electrical energy is transformed into light; in an electroluminescence device, this is caused by the excitation of nanoparticles with electrical energy.

6.7
Electroluminescence

From an economics viewpoint, electroluminescence applications have an extremely high potential for technical applications among consumer products. In electroluminescent devices, light emission is stimulated by electric fields instead of energy-rich photons; the basic concept of electroluminescence is illustrated schematically in Figure 6.47.

The electrical stimulation may be provided either by the electrical field or by the injection of charge carriers; the latter process is deemed to have significantly better chances for broad industrial applications.

The main advantage of devices based on electroluminescent materials lies in the fact that, in contrast to liquid crystal devices (LCDs), an external light source is no longer necessary. Although today, most electroluminescent devices function with organic materials, these compounds (which are used as the starting materials for organic light-emitting diodes. OLEDs) currently face the problem of limited operation time. In addition, the three basic colors of red, magenta and yellow each age with a different time constant, and therefore a discoloration of the displayed image is observed well before the end of the device's lifetime. These problems have forced research teams to seek alternatives based on inorganic materials. However, as in most cases electroluminescent nanoparticles are embedded in organic matrices, this situation is – until now – only marginally improved. Primarily, nanoparticles made from semiconducting materials or doped insulating particles are applied as luminescent materials.

The general design of an electroluminescence device is shown in Figure 6.48. Here, the carrier glass plate is coated with indium tin oxide (ITO), an electric conductive transparent oxide. ITO has the additional advantage of injecting positively charged holes as it has a high work function for electrons, the energy necessary to emit electrons. (The work function is the minimum energy necessary to remove an electron from a solid. In most cases, the work function is about a half of the ionization energy of an isolated atom of the metal. In a first approximation, the work function is equal to the *Fermi* energy.) The next layer carries the nanoparticles. A layer of

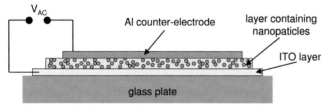

Figure 6.48 The set-up of an electroluminescence device. This normally consists of a glass carrier plate coated with an optically transparent electric conductor (ITO). The next layer contains the electroluminescent particles, which is coated with an aluminum counterelectrode.

sputtered aluminum (a material with a very low work function for injecting electrons into the system) is then applied as the counterelectrode.

The process of electroluminescence is shown in greater detail in Figure 6.49. When this device is connected to a direct electric current source, the cathode (aluminum) emits electrons (e^-), which jump from the conduction band of the metal into the lowest unoccupied band of the luminescence material. The anode, ITO, releases holes h^+; in reality, an electron jumps from the highest occupied band of the luminescence material into the conduction band of the anode, releasing a positively charged hole, in the luminescence layer. The energy necessary for this process is U_e at the side of the cathode, and U_h at the anode's side. The anode material must be selected in such a way that the energy to emit electrons – the work function – is significantly larger than that of the cathode material. Radiative recombination of the hole in the highest occupied band of the electroluminescent material and the

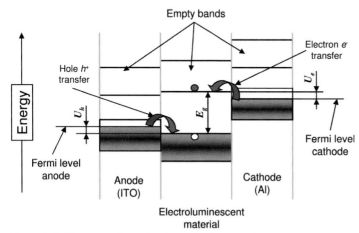

Figure 6.49 Charge transfer and excitation in the case of electroluminescence. The anode emits holes and the cathode electrons into the bands of the luminescent material. Light is emitted as result of the electron–hole recombination.

electron in the conduction band leads to the emission of light. In the simplest case, the wavelength of the emitted light λ can be calculated using the energy of the band gap E_g: $\lambda = ch/E_g$ (where c is the velocity of light; h is *Planck*'s constant). In order to emit light, one of the electrodes must be made from transparent or translucent material, usually ITO. Doping the luminescence material may modify the process described above, without changing the basic principles. Although the basic concept (as shown in Figure 6.49 and explained above) clearly utilizes a direct current system, experimental experience has shown that the application of an alternating current source increases the efficiency of the system.

In many cases, an additional organic layer (known as the charge carrier emitter layer) is applied between the nanoparticles and the counterelectrode. This is necessary if the active layer system is to be repeated a few times to increase the efficiency of the system. Based on this concept, such electroluminescence systems with semiconducting nanoparticles have several important advantages, the most important of which are:

- The emission wavelength of these quantum dots can be adjusted by the particle size.
- As a result, the emission color can be tuned according to the application, without changing the process chemistry and technology.
- Various intermediate polymer layers may be applied to optimize charge carrier transport.
- The polymer itself may be luminescent, thus improving the yield of light.
- Finally, an unmatched quantum efficiency of such inorganic/organic layered systems may be expected.

Certainly, one of the crucial problems of organic electricity-conducting compounds, namely the high sensitivity against oxidation, is reintroduced, and this may limit the lifetime of these devices.

Pioneering studies on such double layer systems were conducted by Colvin *et al.* [35], who applied CdSe semiconducting quantum dots and a poly(*p*-phenylene vinylene (PPV) layer, which showed luminescence and electric conductivity together. A typical example of a system using CdSe nanoparticles and a PPV layer was reported by Gao *et al.* [36]. The electroluminescence spectra of a CdSe/PPV multilayer system, consisting of 20 double layers, for different voltages, is shown in Figure 6.50. Here, a broad emission spectrum starting at 500 nm and reaching beyond 800 nm into the near-IR region, is achieved. In addition, it is obvious that the emitted intensity increases with the increasing voltage applied to the system. The intensity of the emission maximum at 657 nm as a function of the applied voltage is shown in Figure 6.51, where the need for a minimum voltage on the order of 3.5 V to obtain a first emission is clearly recognized.

Perhaps of most interest here is the near-perfect linear proportionality of the emitted intensity and the current density, indicating relatively small parasitic losses (see Figure 6.52). A detailed summary of the electric properties of an electroluminescence cell may be found in the report of Nelson and Fothergill [37].

Changing from nanoparticles to nanorods may lead to further improvements in optical luminescence and electroluminescence properties; this has been demonstrated for ZnS by Manzoor *et al.* [38].

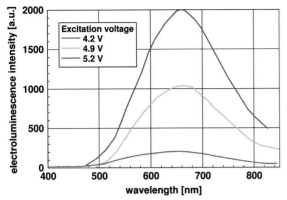

Figure 6.50 Electroluminescence spectra of a CdSe/PPV multilayer system, consisting of 20 double layers, for different voltages. The emitted intensity increases with increasing voltage; the spectral distribution, however, remains unchanged [36].

The spectral intensity distribution of ZnS nanorods in comparison to ZnS nanoparticles synthesized by an equivalent process is illustrated in Figure 6.53. Transmission electron microscopy studies revealed a mean diameter of 4 nm for the nanoparticles, and a length of 400 nm and diameter of approximately 35 nm for the rods. The nanorods were grown by aging in aqueous suspension by agglomeration of more or less spherical particles; this indicated that, under these conditions, the rods were thermodynamically more stable shapes. In addition, in order to obtain a broad emission band, the particles and rods were doped with 0.13% Cu^+ and 0.1% Al^{3+}. As zinc is Zn^{2+} in the lattice, the doping created positively charged holes and additional electrons in the energy bands. Sulfur vacancies, predominantly at the surface, were also the reason for the side bands extending the spectral range of the emitted light to a

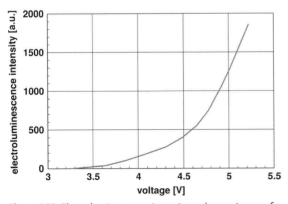

Figure 6.51 Electroluminescence intensity at the maximum of a CdSe/PPV multilayer system, consisting of 20 double layers, as a function of the applied voltage. A threshold voltage in the range of 3.5 V is clearly visible [36].

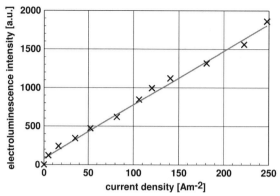

Figure 6.52 Electroluminescence intensity at the maximum of a CdSe/PPV multilayer system, consisting of 20 double layers, plotted versus current density. The experimental values show only small deviations from a strictly linear relationship [36].

shorter wavelength. Because of the smaller surface:volume ratio of the rods, such emission is less important for the larger nanorods. In total, the luminescence intensity of doped ZnS nanorods is significantly larger than that of the equivalent nanoparticles. Besides the emission related to sulfur vacancies, the red shift observed in the transition from nanoparticle to rod, caused by the larger size, was remarkable, and a near-identical behavior was found in the spectra of electroluminescence (see also Figure 6.54). When comparing the photo- and electroluminescence spectra, an interesting red shift in the case of electroluminescence was also visible.

The difference between ZnS nanoparticles and nanorods is similarly striking when luminescence intensity is plotted against voltage, the graph showing clearly

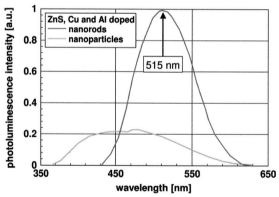

Figure 6.53 Comparison of the photoluminescence intensity of ZnS nanorods and nanoparticles. Both types of material were doped with Cu and Al. Note the significantly higher intensity obtained with nanorods. The wavelength shift of the maximum is caused by different sizes of particles and rods [38].

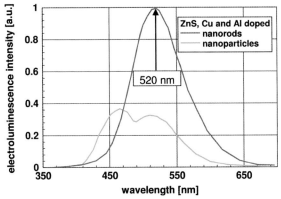

Figure 6.54 Electroluminescence spectra of doped ZnS nanorods and nanoparticles. Both types of material were doped with Cu and Al. As shown in Figure 6.53, a significantly higher intensity is obtained with nanorods. The shift of the maximum is caused by different sizes of particles and rods. Compared to photo-luminescence, the emission maximum of the rods shows a 5-nm red shift [38].

that the advantage for the rods occurs at higher excitation voltages. At lower voltages, the particles emit a higher intensity as compared to the rods.

The dependency of the emitted intensity of the applied voltage, as depicted in Figures 6.51 and 6.55, is important for technical applications in displays, as it allows the control of display brightness and analog or digital modulation of the brightness of each pixel to present pictures and other information.

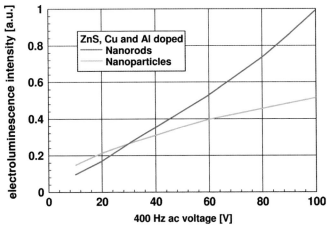

Figure 6.55 Electroluminescence intensity of doped ZnS nanorods and nanoparticles. Note that the improved performance of the rods starts at about 30 V [38].

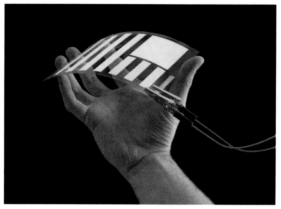

Figure 6.56 A commercially available electroluminescence device [39]. It is essential that these devices can be produced by printing on flexible carriers (Reprinted with permission from Merck KGaA [39]).

In the future, electroluminescence devices may gain enormous importance in computer display screens and television sets, as well as for other types of display. The crucial new point is that electroluminescent nanoparticles can be dispersed in liquids, which allows the structures to be printed. Clearly, there is no difference between printing on a rigid substrate such as glass, or on a flexible substrate, perhaps a foil made from polymers. Therefore, perhaps the most interesting future application will be seen in flexible display technologies, an example of which (produced by the company Merck KGaA) is shown in Figure 6.56. Although the object shown is only small and the procedure somewhat experimental, the figure demonstrates the broad field of possible applications.

6.8
Photochromic and Electrochromic Materials

6.8.1
General Considerations

Photochromic materials change color reversibly as a function of light intensity. In the dark, they are usually white or colorless, whereas in sunlight or UV radiation they change color. In most cases, the intensity of coloration is a function of the light intensity and, after removal of the light source, the material loses its coloration.

Electrochromic materials change color when connected to a source of electricity, but to change color only small electrical charges are needed in most cases.

Both, photochromic and electrochromic materials, have a huge economic potential for applications. However, from a technical viewpoint suffer from one crucial problem, namely that when they do change color the new color is distinct. Yet, in many large-volume technical applications a change from white (or colorless) to gray or even black would be adequate, and even preferred. A typical example of such an application would be that of sun-protecting windows.

A second problem is the time constant of the color change. Ideally, the consumer requires an immediate reaction to a change of external conditions, whether to sunshine intensity for photochromic windows or simply to change illumination levels in rooms by the use of electrochromic materials. Both types of material also have great potential to save energy as they can regulate not only illumination levels but also glare, and heat gain or loss. Windows with incorporated photochromic or electrochromic coated glasses save energy by keeping the heat out as they gradually darken when the sun rises. However, in the morning and evening, the windows should remain transparent while the sun is low in the sky. Hence, buildings fitted with these windows use less energy for air conditioning, and consequently save money and reduce air pollution associated with energy consumption. Further applications include large-scale electrochromic display panels, front and rear windows, and mirrors for cars and trucks, although the latter applications require the material to respond rapidly to changing conditions. An additional problem is that, for broader applications in electronic display systems, materials (or combinations of materials) which produce red and yellow colors are, at present, unavailable.

6.8.2
Photochromic Materials

The best-known photochromic materials are WO_3, MoO_3, and Nb_2O_5. WO_3 and Nb_2O_5 change from white to blue, whereas MoO_3 changes from white to green. An explanation for the described phenomena is found in the ability of these oxides to change stoichiometry. Considering the examples of WO_3 and MoO_3, it is known that in the colorless, bleached state, the metal ions are in the valency states 6^+ and 4^+, while in the colored state ions of valency 5^+ are present. (Strictly speaking, photochromic and electrochromic devices always use these oxides in the hypostoichiometric state; therefore, the exact formula is MeO_{3-x}. As the deviation from ideal stoichiometry is small, for reasons of brevity, in the following text, generally the shorter version MeO_3 is used.) The photochromic or electrochromic behavior performance of these oxides is related to electron/hole pairs intimately connected to deviations from the perfect stoichiometry. When these oxides are excited, positively charged holes and free electrons are formed. The holes react with adsorbed water to produce protons:

$$H_2O + 2h^+ \Rightarrow 2H^+ + O \tag{6.12a}$$

By reaction of these protons with the oxide,

$$MeO_3 + xH^+ + xe^- \Rightarrow H_x Me_{1-x}{}^{6+} Me_x{}^{5+} O_3 \tag{6.12b}$$

the material changes color to blue or green, while the oxygen radicals either occupy vacant sites inside the sample or escape from the particle. An immediate recombination of the charge carriers, $h^+ + e^- \Rightarrow$ heat, reduces the performance of the material. The reaction described above always require exchange with the surrounding atmosphere, and therefore the performance is difficult to control. When considering the rate-controlling steps, it can be assumed that the reaction according to Equation (6.12a) occurs at the surface, whereas diffusion processes control the step according to Equation (6.12b). The effects observed with protons are also obtained with alkaline metal doping. For alkaline metal-doped materials, Equation (6.12b) becomes

$$MeO_3 + xA^+ + xe^- \Rightarrow A_xMe_{1-x}^{6+}Me_x^{5+}O_3 \qquad (6.13)$$

where A^+ is a monovalent ion (in most cases an alkaline metal). Again, the number of free electrons that can react with the metal Me is most important for the efficiency of the system.

A typical example of the absorption spectra of MoO_3 nanoparticles indicating photochromic behavior is shown in Figure 6.57 [40]. This illustrates the absorption of MoO_3 in the bleached state (0 s illumination time) and after two different times, 360 and 2160 s, of illumination with a pulsed laser emitting 72 mW per pulse at 308 nm. Although it is clear that the intensity of coloration increases with time, the most interesting point is the comparison of the time-dependent coloration of nanoparticulate and coarse-grained MoO_3 at a wavelength of 750 nm, as shown in Figure 6.58. Here, the advantage of applying nanoparticles instead of coarse-grained material is clearly visible. Certainly, the reaction mechanisms indicated in Equations (6.12a) and (6.12b) require the transport of ions through the particle. However, as the diffusion time is indirectly proportional to the square of the particle size, this acceleration is to be expected.

Although a decrease in particle size by a factor of only 10 should increase the rate of coloration by a factor of 100, this is not the case as, experimentally, an increase in the coloration rate only by a factor of 13 was verified. Clearly, diffusion is no longer the

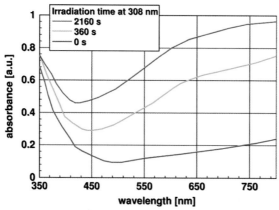

Figure 6.57 Absorption spectra of photochromic MoO_3; the parameter is the illumination time with a 308-nm pulsed laser. At 0 s, the material is in the bleached state, but at 2160 s the coloration has reached saturation level [40].

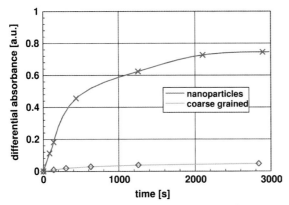

Figure 6.58 Increase of absorption of MoO₃ during illumination with a 308-nm pulsed laser at a wavelength of 750 nm. Note the significantly faster reaction of the nanoparticles as compared to conventional coarse-grained material [40].

rate-controlling step; rather, the rate-controlling process is more likely the surface reaction according to Equation (6.12a).

An entirely different mechanism has been reported for silver-containing photochromic layers by Okumu *et al.* [41], who described the photochromic properties of silver nanoparticles between two 30-nm TiO_2 layers. The new and crucial point of these nanocomposites is the multicolor photochromic property which states that, depending on the wavelength of the illuminating light, the absorption spectrum changes characteristically. These nanocomposites were produced by sputtering on a silicon carrier plate oxidized at the surface. In between the TiO_2 layers, the size of the silver nanoparticles ranged from 8 to 20 nm. Electron micrographs showing the silver nanoparticles in cross-sectional and top views are shown in Figure 6.59. The silver nanoparticles in the center of the TiO_2 layer were seen to exhibit a significant deviation from spherical geometry. This specimen was tempered at 573 K, and annealing at higher temperatures led to larger particles.

The absorbance of these composites depends heavily on the annealing temperature utilized; for example, the absorption spectra shown in Figure 6.60 were recorded with specimens annealed at 573 and 773 K. With an increasing annealing temperature, an increasing particle size and aspect ratio was observed. A red shift of the maximum of absorption was apparent with an increasing annealing temperature, which is equivalent to increasing the aspect ratio. When considering Figure 6.26a, in the particle size range from 10 to 20 nm, the aspect ratio of the spheroids needs to be in the range of 2 to 2.5 in order to obtain positions of the absorption maxima as are visible in Figure 6.60. In order to obtain absorption on the longer wavelength tail of the spectra shown in Figure 6.60, an aspect ratio of about 3 would be necessary.

In Figure 6.60 the absorption peaks are also very broad, this being equivalent to the broad distribution of the aspect ratios of silver nanoparticles. The illumination of such a photochromic layer with visible light leads to a reduction in absorbance; the difference in absorbance after illuminating with a 488-nm laser is shown in

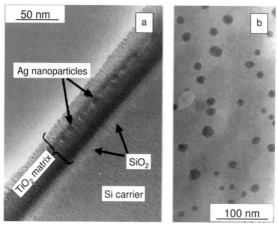

Figure 6.59 Photochromic material consisting of Ag nano-particles embedded in a TiO$_2$ matrix [41]. (a) Cross-sectional view of the photochromic layer system. (b) Top view of the layer system. Only the silver particles are visible. Note the broad distribution of particle sizes and aspect ratios, which is essential for the special properties of this material combination. (Reproduced with permission from [41]; Copyright: American Institute of Physics 2005.)

Figure 6.61. At this point it is important to note that, after illumination, the absorbance is reduced in the vicinity of the wavelength of the illuminating laser. This phenomenon, which is well known as "hole burning," leads to multicolor photochromic properties.

The difference in absorbance at the wavelength of the irradiating laser increases with increasing local power density. The differential absorbances of silver nanoparticles

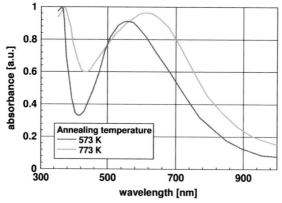

Figure 6.60 Absorption spectra of Ag/TiO$_2$ nanocomposites as a function of the annealing temperature. The red shift with increasing annealing temperature indicates increased grown grains and, more importantly, the aspect ratio of the ellipsoidal silver particles [41].

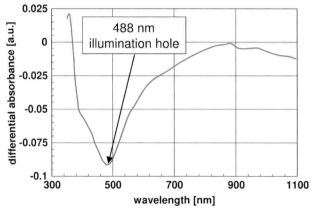

Figure 6.61 Change in the absorption spectrum relative to the unirradiated state as a consequence of irradiation with a 488-nm laser. The absorbance is changed predominantly around the laser wavelength. Due to the photon-activated interaction between the Ag nanoparticles and the TiO_2 matrix, a "hole" was burned into the absorption spectrum [41].

after irradiation at 532 nm for 1 min for different laser power densities at the surface of the specimen are shown in Figure 6.62.

As is observed for many photochromic materials, the differential absorbance moves towards a saturation level. Yet, the original absorbance is obtained again within several hours (occasionally a few days) after stopping the illumination. The recovery time can be reduced by tempering the samples at elevated temperatures in a range from 343 to 373 K. Additionally, removal of the spectral hole is accelerated by illumination with UV light of 270 nm (3.4 eV photon energy). The photochromic process in Ag/TiO_2 nanocomposites may be explained by electron emission from

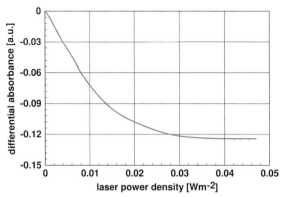

Figure 6.62 Change in absorbance of an Ag/TiO_2 nanocomposite relative to the unirradiated state in the spectral hole at the wavelength of the irradiation laser 532 nm as a function of laser power density. The irradiation time was 60 s [41].

those Ag nanoparticles, for which the particle plasmon resonance wavelength matches the laser wavelength, into the conduction band of the insulating matrix. After electron emission, the silver nanoparticles are charged positively. Subsequently, the electrons emitted into the TiO_2 conduction band are trapped by adsorbed oxygen near the silver nanoparticles. The electron depletion in the nanoparticles reduces the light absorption at the wavelength of irradiation, and this leads to the multicolor photochromic behavior. The whole process requires inherently a broad aspect ratio, or otherwise the photochromic wavelength range is extremely narrow. The recovery of the absorbance (which is equivalent to recovery of the particle plasmon band) observed after irradiation, may be explained by the thermal release of electrons from the trapping centers and subsequent recombination with positively charged silver nanoparticles. The acceleration of recovery by UV irradiation is explained by the capturing of photon-excited electrons of the TiO_2 by the metallic nanoparticles, thus restoring the absorption band.

6.8.3
Electrochromic Materials

A series of studies has reported the electrochromic properties of many materials such as MoO_3, WO_3, V_2O_5, Nb_2O_5, and TiO_2. The most efficient electrochromic properties of these materials are related to doping with small ions such as Li^+, H^+, or P^{5+} in their structures, and further interesting improvements are obtained with additions of silver. The general design of an electrochromic device is shown in Figure 6.63. This system consists of two transparent carrier plates each with one conductive surface. Between these two plates there is a layer of the electrochromic material and an additional electrolyte layer containing monovalent ions. The electrolyte layer may, for example, consist of propylene carbonate containing a few tenths of a mol.% of

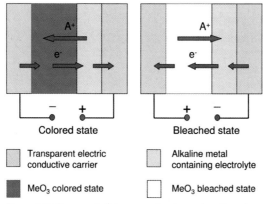

Colored state   Bleached state

Transparent electric conductive carrier   Alkaline metal containing electrolyte

MeO_3 colored state MeO_3 bleached state

Figure 6.63 Transport of charge carriers to produce the colored or bleached state. A^+ is an alkaline metal or a proton, MeO_3 may be WO_3 or MoO_3; however, the same mechanism is valid for all other electrochromic oxides consisting of a metal that can change valency.

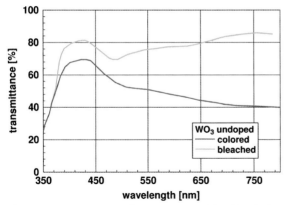

Figure 6.64 Transmittance of pure WO_3 at $+0.8\,V$ (bleached state) and at $-0.7\,V$ (colored state) after $10\,s$ [42].

$LiClO_4$ as electrolyte. In order to switch the system into the colored state, a voltage is applied across the system to move the monovalent ions into the photochromic layer. In the case of WO_3, the alkaline ions A^+ intercalate in the lattice to form $A_xMe_{1-x}^{6+}Me_x^{5+}O_3$, and the systems become colored. Changing the direction of the electrical current moves the alkaline ions back into the electrolyte layer, and the system bleaches.

The transmittance of an electrochromic cell according to Figure 6.63 is shown in Figure 6.64 [42]. The transmittance spectrum is structured, such that the difference between the colored and bleached states increases with increasing wavelength. Therefore, in the colored state, the device is blue.

It is important to compare these curves with those obtained from the same cell, but with lithium-doped WO_3. Interestingly, doping with lithium ions does not lead to any coloration of the cell in the bleached state. Transmittance in the bleached state is similar to that of pure WO_3, as illustrated in Figure 6.64, although coloration in the colored state is much more pronounced. Transmittance in the bleached and colored states for lithium-doped WO_3 is shown in Figure 6.65.

This is easily visible when comparing the difference in transmittance between these two states for both types of WO_3-based material. This is shown in Figure 6.66, where the difference in transmittance between the colored and bleached states is plotted for pure and lithium-doped WO_3. Here, the advantage of the doped material versus the undoped one is clearly visible; in addition, the significantly larger difference in the long wavelength range (yellow and red regions), leading to a deeper blue coloration in the colored state, is clearly seen.

Besides maximizing the difference in transmittance between the bleached and colored states, the time necessary to change color is an essential parameter when evaluating the quality and applicability of electrochromic devices. This goal is reached by specially targeted doping. In an interesting report, Avellaneda et al. [43] demonstrated the effect of P^{5+} additions to WO_3, and noted a difference in the spectral transmission of WO_3 and $WO_3(P)$ in the bleached and colored states at a P^{5+}

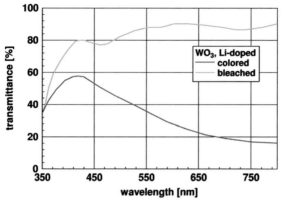

Figure 6.65 Transmittance of lithium-doped WO_3 in the colored and bleached states. Comparison with Figure 6.64 shows that there is almost no influence of doping in the bleached state, whereas transmittance of the colored state is significantly reduced [42].

content of 5 mol% (see Figure 6.67). Besides the different fine structures noted in these spectra, the most important difference was the higher transmission in the bleached state and, equally important, the significantly reduced transmission in the colored state. Again, this is shown in the differential transmission in Figure 6.67. In this context it is important to note that, in both cases, the color appears similar; it is simply the intensity of the coloration that is different.

As mentioned above, the electrochromic response time is extremely important. In this context, Figure 6.68 shows the response measured at a wavelength of 633 nm

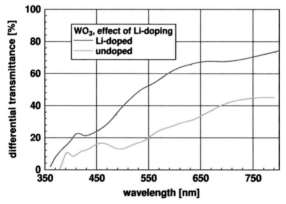

Figure 6.66 Difference in transmittance between the colored and bleached states for pure and lithium-doped WO_3. The advantage of the doped material over the undoped one is clearly visible. Note the larger difference in the range of the long wavelength (the yellow and red region), leading to a deeper blue coloration of the doped material in the colored state [42].

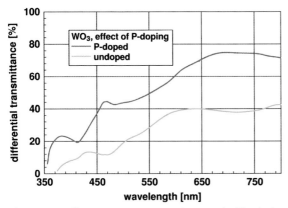

Figure 6.67 Difference in transmittance between the bleached and colored states of P^{5+}-doped and pure WO_3. These data demonstrate the superior performance of the doped material [43].

with alternating voltages between -0.8 V in the colored state and $+0.8$ V in the bleached state, when each was applied for 15 s. In both the cases, the bleaching process for both doped and undoped WO_3 films was completed, although bleaching and coloration of the doped film were significantly faster than for the undoped film. The difference in transmittance between both states was seen to be larger for the P-doped layer than for the undoped material (see Figure 6.67).

In the pure and doped states, essentially the same mechanisms are active in materials such as V_2O_5 and Nb_2O_5 [44]. Because of its high coloration efficiency, nickel oxide (NiO) is an interesting electrochromic material, as it shows good

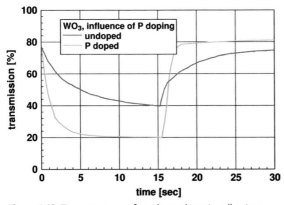

Figure 6.68 Time response of an electrochromic cell using pure and P^{5+}-doped WO_3. In both cases, transmittance was measured at a wavelength of 633 nm. The voltage was -0.8 V to obtain the colored state, and $+0.8$ V to obtain the bleached state. Note that the coloration and bleaching occurred significantly faster in the P^{5+}-doped material [43].

reversibility and low cost. Nanostructured materials containing the hydroxide phase, $Ni(OH)_2$, are cycled with the following coloring/bleaching process:

$$Ni^{2+}(OH)_2 \text{ (bleached, reduced)} \Leftrightarrow Ni^{3+}OOH + H^+ + e^- \text{ (colored, oxidized)}$$

Although the efficiency of this process may be improved by distributing the NiO nanoparticles in an amorphous Ta_2O_5 matrix [45], the most important point is the finding that the $NiO-Ta_2O_5$ nanocomposite withstands at least an order of magnitude more coloration–bleaching cycles than does pure NiO.

6.9
Magneto-optic Applications

One further interesting optical application of ceramic/polymer nanocomposites is their use as magneto-optical materials. Within this context, the phenomena under question include the *Faraday* effect (rotation of the plane of polarization of light in transmission), and the *Kerr* effect (rotation of the polarization plane after reflection at the surface of magnetic materials).

When discussing a phenomenon working in transmission, like the *Faraday* effect, it is necessary to look first at the wavelength range, where the absorption is sufficiently small for technical applications. Figure 6.69 displays the absorbance of Fe_2O_3 nanoparticles of different size.

Figure 6.69 shows the well-known blue shift with decreasing particle size. This phenomenon is very distinct for α-Fe_2O_3; in this case, the decrease of the particle size from 40 to ca. 25 nm brings a significant shift of the onset of the optical absorption to shorter wavelength. Therefore, the 40-nm material is red and the 25-nm material is brown. When looking at the γ-phase material, one realizes that a reduction in grain size from 12 to 3 nm has almost no influence. However, the transition from the 25-nm α-phase particles to the 12-nm γ-phase particles shows the expected

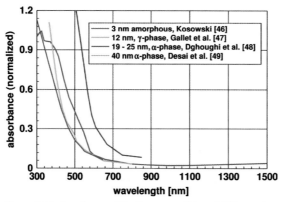

Figure 6.69 Absorbance of Fe_2O_3 nanoparticles of different size. The blue shift of the absorption edge with decreasing particle size is, at least in the α-phase, clearly visible [46–49].

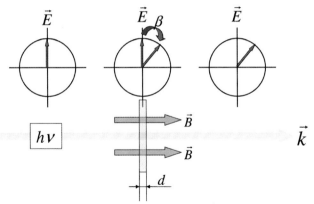

Figure 6.70 The magneto-optical *Faraday* effect. $\vec{E}$ is the polarization vector of the incoming light and $\vec{k}$ the wave vector of the light. The polarization of the incoming light is rotated for and angle β during passing a magneto-optical active layer with the thickness d in a magnetic field $|B|$.

significant blue shift. Additionally, one realizes from the absorbance data displayed in Figure 6.69 that, for a device working in transmission, only the γ-phase material is applicable. In any case, wavelengths shorter than ca. 700 nm should be avoided.

Figure 6.70 shows the general outline of an arrangement using the *Faraday* effect. The rotation angle β of the polarization plane of the incoming light in an transparent medium exhibiting *Faraday* rotation is given by

$$\beta = vd|B| \tag{6.14}$$

where v is the *Verdet* constant, d the thickness of the active medium, and $|B|$ the strength of the magnetic field.

As shown in Figure 6.70, the polarization plane of the incoming light is rotated for the angle β after passing the layer with the thickness d. Figure 6.71 shows experimentally determined values for the *Faraday* rotation of γ-Fe_2O_3 nanoparticles [50]. These data are given within a relative broad scattering band, as the absorption in the magneto-optic active layer of γ-Fe_2O_3 nanoparticles is quite high.

Using the data given in Figure 6.71, one can calculate the *Verdet* constant for γ-Fe_2O_3 to be in the range of $2 - 4 \times 10^6$ degree $T^{-1} m^{-1}$. These values of the *Faraday* rotation are significantly larger than those of the best magneto-optic active materials, which are garnets with the compositions $Tb_3Ga_5O_{12}$ or $(Tb_x,Y_{1-x})_3Fe_5O_1$ which are in the range of a few degree $T^{-1} m^{-1}$. However, in contrast to the highly transparent garnets, because of the relatively large absorbance, only thin films of iron oxide can be used, in the range of a few micrometers, or equivalently thin suspensions stabilized by surfactants. Lastly, as maghemite is significantly cheaper than the above-mentioned garnets, this material has a great potential for these applications. However, as also mentioned above, the efficient transmission of light through γ-Fe_2O_3 is possible only in the red and infrared range of the optical spectrum.

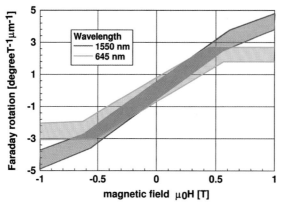

Figure 6.71 *Faraday* rotation of thin layers of γ-Fe$_2$O$_3$ nanoparticles prepared by sputtering for light with the wavelengths of 645 nm and 1550 nm as function of the applied magnetic field [53].

Although, for optomagnetic devices thin films of ferromagnetic materials are normally used, various attempts have been made to use nanoparticulate ferrites in a polymer matrix [51]. However, to date better results have been obtained with ferrite–silica nanocomposites. All of these materials are hampered by the high absorption of visible light by ferrites, and therefore useful applications are possible only in the range of red light. Zhou *et al.* [52] synthesized ZnFe$_2$O$_4$ nanoparticles in silica; this material was shown to be superparamagnetic, and the magnetization curves were – at least down to a temperature of 78 K– free of hysteresis. The optical absorption coefficient as a function of wavelength for particle concentrations of 30 and 5 wt.% are shown in Figure 6.72.

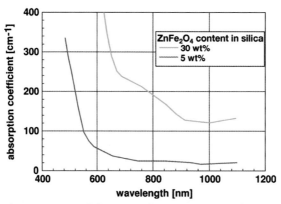

Figure 6.72 Spectral absorption of ZnFe$_2$O$_4$ nanoparticles in silica as a function of the ferrite content. The blue shift of the absorbance with decreasing ferrite content is a consequence of the smaller particle size [47].

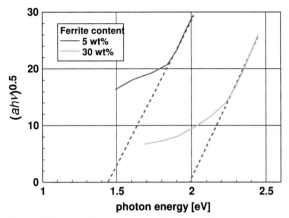

Figure 6.73 *Tauc* plot according to Equation (6.9) of the absorbance data shown in Figure 6.72. The intersection of the extrapolated data with the abscissa gives the value of the energy gap, clearly indicating a wider gap for the composite with the higher $ZnFe_2O_4$ concentration.

As might be expected, a higher concentration of ferrite particles clearly increases optical absorption. Furthermore, the strong increase in absorption for shorter wavelengths is shifted significantly to shorter wavelengths when the concentration is reduced. This blue shift may be attributed to a smaller particle size or surface phenomena on the band gap.

By using the experimental data shown in Figure 6.72, Figure 6.73 depicts the *Tauc* plot $(\alpha h\nu)^{0.5}$ versus $(h\nu)$ according to Equation (6.9), thus providing the relationship between optical absorption α and band gap energy E_g. In this case, when extrapolating the linear region towards $(\alpha h\nu)^{0.5} = 0$ for the specimen with 30 wt.% ferrite, the band gap was found to be significantly narrower as compared to that of the specimen with a 5 wt.% ferrite content.

With respect to the application of the *Kerr* effect, data relating to optical reflectance in the UV-visible range of γ-Fe_2O_3–SiO_2 nanocomposites are provided by Moreno *et al.* [53].

References

1 Nussbaumer, R.J., Caseri, W.R., Smith, P. and Tervoort, T. (2003) *Macromol. Mater. Eng.*, **288**, 44–49.

2 Böhm, J., Haußelt, J., Henzi, P., Litfin, K. and Hanemann, T. (2004) *Adv. Eng. Mater.*, **6**, 52–57.

3 Wu, M.K. (1995) KEMIRA Pigments, Inc., Portland. *Nanostructured Materials, and Coatings*, Gorham/Intertech Consulting.

4 Nussbaumer, R.J., Caseri, W.R., Smith, P. and Tervoort, T. (2003) *Macromol. Mater. Eng.*, **288**, 44–49.

5 Khurgin, J. *et al.* (1995) *Mater. Res. Soc. Symp.*, **358**, 193.

6 Smith, A.M., Gao, X. and Nie, S. (2004) *Photochem. Photobiol.*, **80**, 377–385.

7 Mädler, L., Stark, W.J. and Pratsinis, S.E. (2002) *J. Appl. Phys.*, **92**, 6537–6540.

8 Tani, T., Mädler, L. and Pratsinis, S.E. (2002) *J. Mater. Sci.*, **37**, 4627–4632. (www. ptl.ethz.ch/research/res_top_Qdots).

9 Tauc, J. and Menth, A. (1972) *J. Non.-Cryst. Solids.* **8–11**, 569.

10 Mills, G., Li, Z.G. and Meisel, D.J. (1988) *J. Phys. Chem.*, **92**, 822.

11 Reisfeld, R. (2002) *J. Alloys Compd.*, **341**, 56–61.

12 Monticone, S., Tufeu, R. and Kanaev, A.V. (1998) *J. Phys. Chem. B*, **102**, 2854–2862.

13 Vollath, D. and Szabo, D.V. (2004) *Adv. Eng. Mater.*, **6**, 117–127.

14 Lim, W.P., Low, H.Y. and Chin, W.S. (2004) *J. Phys. Chem.*, **108**, 13093–13099.

15 Mahamuni, S., Bendre, B.S., Leppert, V.J., Smith, C.A., Cooke, D., Risbud, S.H. and Lee, H.W.H. (1996) *Nanostruct. Mater.*, **7**, 659.

16 Weng, Y.X., Li, L., Liu, Y., Wang, L. and Yang, G.Z. (2003) *J. Phys. Chem.*, **107**, 4356–4363.

17 Meyer, T.J., Meyer, G.J., Pfennig, B.W., Schoonover, J.R., Timson, C.J., Wall, J.F., Kobusch, C., Chen, X., Peek, B.M., Wall, C.G., Ou, W.W., Erikson, B.W. and Bignozzi, C.A. (1994) *Inorg. Chem.*, **33**, 3952–3964.

18 Parker, C.A. (1968) *Photoluminescence of Solutions*, Elsevier, Amsterdam, p. 21.

19 Vollath, D. and Szabo, D.V. (2004) *Nanoparticle Res.*, **6**, 181–191.

20 Musikhin, S., Bakueva, L., Sargent, E.H. and Shik, A. (2002) *J. Appl. Phys.*, **91**, 6679–6683.

21 Abdolvand, A. (2006) Dissertation, Martin-Luther-Universität Halle-Wittenberg.

22 Eustis, S. and El-Sayed, M.A. (2006) *Chem. Soc. Rev.*, **35**, 209–217.

23 Eustis, S. and El-Sayed, M. (2005) *J. Phys. Chem. B*, **109**, 16350–16356.

24 Kunkel, J. Ars Vitraria, Leipzig, Frankfurt, 1679.

25 Vogel, W. (1992) *Glas Chemie*, 3rd edn, Springer, Berlin.

26 Park, J.-E., Atobe, M. and Fuchigami, T. (2005) *Electrochim. Acta*, **51**, 849–854.

27 Wanga, Y., Yaoa, X., Wanga, M., Konga, F. and Heb, J. (2004) *J. Cryst. Growth*, **268**, 580–584.

28 Cattaruzzaa, E., Battaglina, G., Calvellia, P., Gonellaa, F., Matteib, G., Mauriziob, C., Mazzoldib, P., Padovanib, S., Pollonia, R., Sadab, C., Scremina, B.F. and D'Acapitoc, F. (2003) *Sci. Technol.*, **63**, 1203–1208.

29 Quinten, M. (2001) *Appl. Phys.*, **73**, 317–326.

30 Li, C. and Murase, N. (2004) *Langmuir*, **20**, 1–4.

31 Bodnar, I.V., Gurin, V.S., Molochko, A.P. and Solovei, N.P. (2004) *Inorg. Mater.*, **40**, 115–121. [Translated from *Neorganicheskie Materialy*, **40** (2004) 158–165].

32 Mizetskaya, I.B. *et al.* (1986) *Physicochemical Principles of Crystal Growth of Solid Solutions Between II–VI Semiconductors*, Naukova Dumka, Kiev.

33 García, M.A., García-Heras, M., Cano, E., Bastidas, J.M., Villegas, M.A., Montero, E., Llopis, J., Sada, C., De Marchi, G., Battaglin, G. and Mazzoldi, P. (2004) *J. Appl. Phys.*, **96**, 3737–3741.

34 Liu, H.-W., Laskar, I.R., Huang, Ch.-P., Cheng, J.A., Cheng, S.-S., Luo, L.-Y., Wang, H.-R. and Chen, T.-M. (2005) *Thin Solid Films*, **489**, 296–302.

35 Colvin, V.L., Schlamp, M.C. and Alivisatos, A.P. (1994) *Nature*, **370**, 354–357.

36 Gao, M., Richter, B., Kirstein, S. and Möhwald, H. (1998) *J. Phys. Chem.*, **102**, 4096–4103.

37 Nelson, J.K. and Fothergill, J.C. (2004) *Nanotechnology*, **15**, 586–595.

38 Manzoor, K., Aditya, V., Vadera, S.R., Kumar, N. and Kutty, T.R.N. (2005) *Solid State Commun.*, **135**, 16–20.

39 Merck KGaA (2007). Darmstadt, Germany

40 Li, S. and El-Shall, M.S. (1999) *Nanostruct. Mater.*, **12**, 215–219.

41 Okumu, J., Dahmen, C., Sprafke, A.N., Luysberg, M., von Plessen, G. and Wuttig, M. (2005) *J. Appl. Phys.*, **97**, 094305-1–094305-6.

42 Bueno, P.R., Faria, R.C., Avellaneda, C.O., Leite, E.R. and Bulhoes, L.O.S. (2003) *Solid State Ionics*, **158**, 415–426.

43 Avellaneda, C.O. and Bulhoes, L.O.S.
(2003) *J. Solid State Electrochem.*, **7**,
183–186.

44 Melo, L., Avellaneda, C.O., Caram, R.,
Sichieri, E. and Pawlicka, A. (2002) *Mater.
Res.*, **5**, 43–46.

45 Ahn, H.-J., Shim, H.-S., Kim, Y.-S., Kim,
Ch.-Y. and Seong, T.-Y. (2005) *Electrochem.
Commun.*, **7**, 567–571.

46 Kosowsky, B. (2008) MACH I, King of
Prussia, PA private communication.

47 Gallet, S., Verbiest, T. and Persoons,
A. (2003) *Chem. Phys. Lett.*, **378**, 101–104.

48 Dghoughi, L., Elidrissi, B., Bernede, C.,
Addou, M., Lamrani, M.A. and Regragui,
M. (2006) *Appl. Surf. Sci.*, **253**, 1823–1829.

49 Desai, J.D., Pathan, H.M., Min, Sun.-Ki.,
Jung, Kwang.-Deog. and Joo, Oh.-Shim.
(2006) *Appl. Surf. Sci.*, **252**, 8039–8042.

50 Tepper, T., Ilievski, F., Ross, C.A., Zaman,
T.R., Ram, R.J., Sung, S.Y. and Stadler,
B.J.H. (2003) *J. Appl. Phys.*, **93**, 6948–6950.

51 Ziolo, R.F., Giannelis, E.P., Weinstein,
B.A. and O'Horo, M.P. (1992) *Science*, **257**,
219–223.

52 Zhou, Z.H., Xue, J.M., Chan, H.S.D. and
Wang, J. (2002) *Mater. Chem. Phys.*, **75**, 181.

53 Moreno, E., Zayat, M., Morales, M., Serna,
C., Roig, A. and Levy, D. (2002) *Langmuir*,
18, 4972–4978.

7
Electrical Properties of Nanoparticles

7.1
Fundamentals of Electrical Conductivity in Nanotubes and Nanorods

Conventional electrical conductors are ruled by *Ohm*'s law:

$$U = IR = \frac{I}{G} \quad \text{or} \quad G = \frac{I}{V} \tag{7.1}$$

where V is the applied voltage, I the electrical current, R the resistance, and G the electrical conductance. *Ohm*'s law implies that the electrical resistance depends only on the geometry and material of the conductor. The conductance G of a wire depends on the geometric parameters of the length L and the cross-section A; hence $G = \sigma L/A$. The electrical conductivity σ is a material-dependent property which, for metallic conductors, is independent of the applied voltage or the flowing electrical current. In contrast, for semiconductors or insulators the electrical conductivity usually increases with increasing applied voltage. When reducing the geometric dimensions of a wire to nanometer or molecular dimensions, *Ohm*'s law is no longer valid in any case. Rather, the strictly linear relationship between current and voltage is replaced by a nonlinear, nonohmic characteristic. In order to understand these phenomena, it is necessary first to consider the mechanism of electrical conductivity, the conventional, macroscopic case of which is shown is Figure 7.1.

Here, an electrical conductor – such as a metallic wire – is connected to an electrical circuit, and electrons start to move, driven by the electrical field. Within the wire there is a huge number of electrons, and these move slowly from one end of the wire to the other end. In this way, the electrons experience scattering processes that lead to a change in the momentum by interactions with electrons, phonons, impurities, or other imperfections of the lattice, which are responsible for the electrical losses. In metallic wires, electrical conductivity is characterized by the mean free path of the electrons. In an electrical field the electrons exhibit a type of "drift movement", and such a process of electrical conductivity is termed "diffusive conductance". Reducing the size of the conducting wire changes the mechanism of electrical conductivity; hence, when the geometric dimensions reach the mean free path length of the electrons, the mechanism of conduction changes from a "diffusive" to a "ballistic"

Nanomaterials: An Introduction to Synthesis, Properties and Application. Dieter Vollath
Copyright © 2008 WILEY-VCH Verlag GmbH & Co. KGaA, Weinheim
ISBN: 978-3-527-31531-4

Diffusive conductance

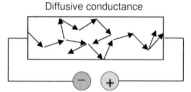

Figure 7.1 Model of diffusive electric conductance, as observed in conventional metallic conductors. Diffusive conductance is characterized by a scattering of free electrons in the conductor. The electrical current is transported by a slow drift movement of the electrons.

mechanism. The principle of ballistic conductivity is shown schematically in Figure 7.2.

As the scattering phenomena are no longer observed, classically zero resistivity would be expected, but this is not observed because now quantum mechanical phenomena are occurring. In order to understand this ballistic conductivity, Equation (7.1) must be rewritten in a form which takes into account the transport of electricity by electrons. The electrical current I transports within a time interval Δt the charge Q. As one electron carries the charge e, the charge Q is transported by $N = Q/e$ electrons. The time interval Δt is estimated from the length L of the wire and the velocity of the electrons v_e: $\Delta t = L/v_e$.

$$G = \frac{I}{V} = \frac{Q}{\Delta t V} = \frac{N e v_e}{V L} \tag{7.2}$$

Under the influence of the voltage V, the electrons are accelerated and obtain the energy $E = eV$. When considering electrons, it is necessary to apply *Planck*'s equation, $E = h v_e/\lambda$, to express the energy of the electrons. Inserting this into Equation (7.2), one finally obtains

$$G = \frac{N e v_e}{V L} = \frac{N e^2 v_e}{E L} = \frac{N e^2}{h} \frac{\lambda}{L} = \frac{N e^2}{h} \frac{1}{n} \tag{7.3}$$

Here, $L/\lambda = n$ is the electron wave mode number. Each electron wave mode can have two modes (spin up and spin down) leading to $N = 2n$; therefore, one finally obtains for the conductance of a short, thin wire with one mode $G = 2e^2/h$. Assuming m active modes in a wire, the conductance is $G = 2me^2/h$. In this formula, there are no

Ballistic conductance

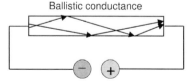

Figure 7.2 Ballistic conductivity of an electrical current in a small electrical conductor. Ballistic conductivity is not characterized by scattering of the free electrons in the lattice, as the geometric dimensions of the conductor are smaller than the mean free path length of the electrons.

longer any variables depending on the material or the geometry of the wire. It is clear that the electrical conductance of a small, thin wire increases with the increment $G_0 = 2e^2/h = 7.72 \times 10^{-5}$ S. It is important to note, again, that in the ballistic case the electrical conductance is independent of the material and geometry of the wire. The inverse value $h/e^2 = 26\,k\Omega$ is called the *resistance quantum*.

According to Equation (7.3), the conductance decreases in steps with increasing voltage. However, this is valid only at low temperatures or for extremely small wires; otherwise, the thermal energy is larger or in the range of the energy difference between two neighboring electron wave modes. In this case, the different modes may be activated thermally and not by the electrical field. This leads to a smearing of the distinct steps, such that the current–voltage (*I–V*) diagram resembles that which follows *Ohm*'s law. A more rigorous treatment of quantized electrical conductance may be found in the review of Datta [1].

One interesting method which can be used to demonstrate the steps in conductance is the mechanical thinning of a nanowire and measurement of the electrical current at a constant applied voltage. Usually, these measurements are performed at room temperature. A good example of such an experiment is shown in Figure 7.3 [2], where a short gold nanowire of 5 nm length was pulled with a piezo device in steps of 0.2 nm. After each step, the electrical current was measured at a constant voltage of 32 mV.

In Figure 7.3, the conductance of the wire is plotted as a function of the elongation, and only the steps in conductivity are displayed. Additionally, the conductance displays characteristic dips which stem from the deformation. These are local minima or maxima of the conductance caused by disorder–order transformation, resulting in a part change between ballistic and diffusive conduction mechanisms, on an atomic scale, during the elongation process. The important point here is that, with

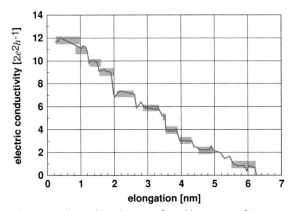

Figure 7.3 Electrical conductivity of a gold nanowire of 5 nm length, determined at a constant applied voltage of 32 mV. The gold wire was elongated during this experiment in 0.2-nm steps. The conductivity decreases stepwise with increasing elongation, which is equivalent to a reduction of the diameter. The shaded areas show the range of experimental scatter [2].

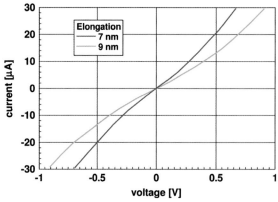

Figure 7.4 *I–V* characteristics of a gold nanowire of 9.5 nm length after two different elongations. The non*Ohmic* behavior, as indicated by the more than linear increase in current with increasing voltage, is clearly visible [2]. At increasing elongation, the cross-section becomes smaller and the wire longer, leading to an increased resistance.

decreasing diameter of the wire, the number of electron wave modes contributing to the electrical conductivity is becoming increasingly smaller by well-defined quantized steps, the height of which is $2e^2/h$. Even when taking the experimental scatter (characterized by the shaded areas in Figure 7.3) of the measured values into account, the quantized decrease in conductivity is clearly visible. Measurement of the *I–V* characteristics as a function of the applied voltage at constant length does not show the quantized steps. However a significant nonlinearity is observed, which increases with rising elongation of the wire (see Figure 7.4). In this figure, the two *I–V* curves, determined on a 9.5 nm-long gold wire, are measured at elongations of approximately 7 and 9 nm. It is important to realize that the more than linear increase of the electric current with increasing voltage will indicate an increasing conductance of the wire. It should be noted that these experimental data were obtained at room temperature.

The electrical conductance determined from the characteristic given in Figure 7.4 leads to the interesting result that the constant value, as predicted from the simple consideration about metallic or ballistic conductance, is valid only within a narrow range of low voltages. At higher voltages, as expected, the conductance increases. In Figure 7.5, the transition from a range with constant conductance at low voltages to a range where the conductance increases with voltage can be visualized, and this was interpreted by the authors as a transition from metallic behavior to a more semiconductor- or insulator-like characteristic. Because of the slight scattering of experimental values, the *G–V* characteristic is indicated only by shaded areas.

As mentioned above, deviations from *Ohm*'s behavior are observed only for very thin objects, and to demonstrate this the results of an *I–V* measurement for a 50-nm nanowire made from silver and with a length of approximately 10 μm are shown in Figure 7.6 [3]; the *I–V* characteristic is strictly linear.

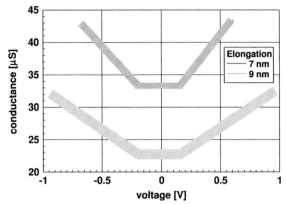

Figure 7.5 Electrical conductance of an originally 9.5 nm-long gold wire after elongation as a function of the applied voltage. These data were calculated using the values from Figure 7.4. Note the transition of a metallic behavior with voltage-independent conductance, to a semiconductor- or insulator-like behavior, where the conductance increases with the applied voltage.

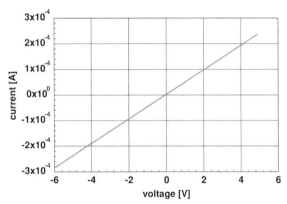

Figure 7.6 Current–voltage characteristics of a 50 nm-diameter, 5 μm-long silver nanowire measured at 4.2 K [3]. The current offset in the original data was removed.

A second fact may be derived from Figure 7.6, notably that even in the range where *Ohm*'s law is valid, the ability of nanowires to carry electrical currents is enormous, and any comparison with macroscopic electrical conductors is absolutely impossible. The current density applied during measurement of the *I–V* characteristic was up to $10^{12} \, \mathrm{A \, m^{-2}}$ ($= 10^6 \, \mathrm{A \, mm^{-2}}$) – a current density which would be unthinkable for a macroscopic wire. In addition, it should be mentioned that this current density did not lead to failures; Aherne *et al.* [4] measured the current at failure for gold nanowires as a function of the diameter in a range from 65 to almost 120 nm at room temperature, and the results are depicted in Figure 7.7.

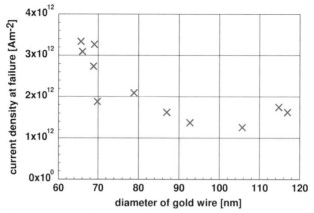

Figure 7.7 Current density to failure for gold nanowires of different diameters, determined at room temperature. Note the increasing maximum current density with decreasing wire diameter [4].

In Figure 7.7 it can be seen that, for gold nanowires, the current to failure is above $10^{12} \, A \, m^{-2}$. It is interesting to note that the current density leading to failure increases significantly for wire diameters below 70 nm; therefore, for thinner wires even higher current densities might be expected that would lead to failure. In carbon nanotubes, current densities in the range of $10^{13} \, A \, m^{-2}$ were observed.

7.2
Carbon Nanotubes

In electrically conducting carbon nanotubes, only one electron wave mode is observed which transport the electrical current. In an interesting experiment, Poncharal *et al.* [5] demonstrated the quantized nature of the electrical conductivity of carbon nanotubes. In order to reduce the problems with contacts, these authors measured the electrical resistance of bundles of multiwalled carbon nanotubes that had been pushed into a droplet of mercury. In this experiment (the set-up of which is shown in Figure 7.8), a bundle of carbon nanotubes of different length and orientation is fixed onto a sample holder which is moved slowly in the direction of a mercury droplet. As the lengths and orientations of the carbon nanotubes are different, they touch the surface of the mercury at different times, which provides two sets of information: (i) the influence of carbon nanotube length on the resistance; and (ii) the resistances of the different nanotubes.

As the nanotubes have different lengths, then with increasing protrusion of the fiber bundle an increasing number of carbon nanotubes will touch the surface of the mercury droplet and contribute to the electrical current transport (see Figure 7.9). With the four successive steps, stemming from four different nanotubes being inserted into the mercury droplet, the conductance increases for approximately

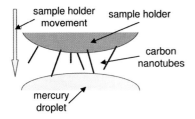

Figure 7.8 Experimental set-up to measure the electric conductivity of carbon nanotubes. A bundle of carbon nanotubes of different length and orientation is fixed onto a sample holder and moved in the direction of a mercury droplet. The lengths and orientations of the carbon nanotubes are different; hence they touch the surface of the mercury at different times. This provides information on the influence of carbon nanotube length on resistance, and the resistance of different nanotubes [5].

one G_0. Most likely, the minor deviations from the exact values of multiples of G_0 are caused by contact resistance. These measurements were performed at a voltage of 100 mV. Because of the small electrical current flowing during these measurements, the conductivity values were blurred by noise. Two important facts can be deduced from the data in Figure 7.9: (i) that each nanotube contributes equally to the conductance; and (ii) that the conductance is independent of the length of the nanotube since, at each step, with increasing submersion of the nanotubes, the conductance remains unchanged.

Beyond a voltage of approximately 100 mV, the I–V characteristic of multiwalled carbon nanotubes and individual graphene layers is more complex. Poncharal *et al.* [5] measured this relationship for multiwalled nanotubes, and Shklyarevskii *et al.* [6]

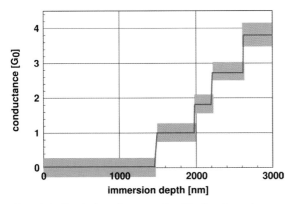

Figure 7.9 Electrical conductance of multiwall nanotubes determined at a voltage of 100 mV. As shown in Figure 7.8, for this experiment, four multiwall carbon nanotubes were immersed successively into a mercury drop. The graph shows that the conductivity is independent of the carbon nanotube length, in this case characterized by the immersion depth. The experimental scatter of the measured values is indicated by the shaded areas. Contact resistance causes the deviations from the multiples of G_0 [5].

for individual graphene layers. In both cases the experiments revealed that, above ca. 100 mV, the conductivity increased with increasing voltage. For voltages below or above ca. 100 mV, the conductance is G_0

$$G = G_0 \qquad\qquad V < 100\,\text{mV}$$

$$G = G_0(\alpha + \beta|V|) \quad V > 100\,\text{mV}$$

(7.4)

where α and β are specimen-dependent factors ranging from 0.25 to a maximum of 1.0. Due to the unavoidable contact resistance, the constant contribution to conductance is usually smaller than G_0. The function of the conductance is symmetrical against $V = 0$ V, according to Equation (7.4). Equation (7.4) leads to a I–V characteristic such as

$$I = VG_0(\alpha + \beta|V|)$$

(7.5)

The same relationship was observed for graphene layers, although the relationship to the conductance quantum G_0 is not well defined. The I–V plot for a multiwalled nanotube assuming $\alpha = 0.5$ and $\beta = 0.25\,\text{V}^{-1}$ is shown in Figure 7.10, and is similar to that described for gold nanowires.

Consequently, the electrical conductance of a multiwall nanotube as a function of the applied voltage has an appearance as shown in Figure 7.11, according to Poncharal *et al.* [3].

The I–V characteristic of multiwalled carbon nanotubes and individual graphene layers is entirely different from that of single-walled nanotubes. Likewise, in the case of single-walled nanotubes, above 100 mV the electrical conductivity depends heavily

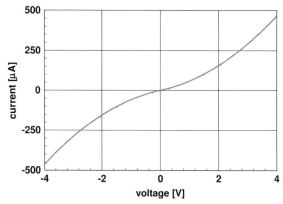

Figure 7.10 Typical I–V characteristic of multiwall nanotubes according to Poncharal *et al.* [5]. The I–V characteristic of graphene platelets is similar. The graph was calculated using Equation (7.5), setting $\alpha = 0.5$ and $\beta = 0.25\,\text{V}^{-1}$. The characteristic is similar to that shown in Figure 7.4 for gold nanowires.

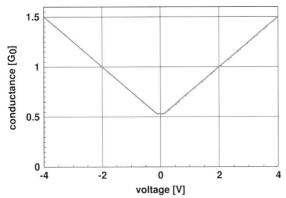

Figure 7.11 Electric conductance of multiwall nanotubes according to Poncharal *et al.* [5]. For voltages above ca. ± 100 mV, the electric conductance follows Equation (7.4); below that limit the conductivity has a constant value. Graphene layers show a similar behavior [6].

on the applied voltage. Yao *et al.* [7] described the *I–V* curves over a range of higher voltages up to ± 5 V, and the experimental results obtained on a single-wall nanotube with a length of approximately 1 µm are shown in Figure 7.12.

Again, as in all cases discussed previously, a nonlinear relationship was observed between the voltage and current. However, the *I–V* characteristic was entirely different to that of multiwalled nanotubes since, below approximately 100 mV, the electrical resistance was constant R_0. However, beyond a range of approximately ± 100 mV the resistance R of a single-walled nanotube is described by

$$R = R_0 + \frac{|V|}{I_0} \qquad (7.6)$$

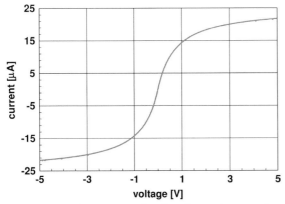

Figure 7.12 *I–V* characteristic of single wall nanotubes according to Yao *et al.* [7]. This graph is characterized by a saturation value of the electric current.

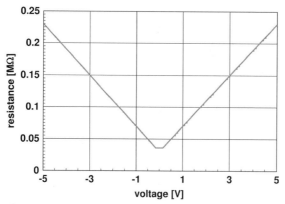

Figure 7.13 Dependency of the electrical resistance of a single-walled nanotube as a function of the applied electric voltage [7]. In contrast to multiwalled nanotubes, the resistance increases with increasing applied voltage.

where R_0 and I_0 are material-dependent constants (R_0 may also include some influence from the contact resistance). Experimental values for the resistance of single-walled carbon nanotubes, as determined by Yao *et al.* [7], show very clearly the constant value of resistance at low voltages and the increasing electrical resistance with increasing applied voltage (see Figure 7.13).

Derived from Equation (7.6), above 100 mV the *I–V* curve in Figure 7.12 is described by

$$I = \frac{V}{R_0 + |V|/I_0} \tag{7.7}$$

Equation (7.7) leads, for large values of the applied voltage, to a saturation current of I_0, and this has interesting consequences for the conductance:

$$G = \frac{I}{V} = \frac{1}{R_0 + |V|/I_0} \Rightarrow \lim_{|V| \to \infty} G = 0 \tag{7.8}$$

For infinite voltages, the conductance vanishes, and experimentally this saturation current was observed. Experiments led to a value of 25 μA for the saturation current I_0, which is independent of the individual nanotube, and expected from theory. Specifically, the value is independent of the nanotube length but, when considering this saturation current and the dimensions, a single-walled nanotube can carry a current density in the range of 10^{13} A m^{-2} ($=10^7$ A mm^{-2}).

When comparing Figures 7.10 and 7.11 with Figures 7.12 and 7.13, as expressed by Equations (7.5) and (7.7), an almost inverse *I–V* (respectively *R–V*) behavior between single-walled and multiwalled nanotubes is apparent. Although to date the reason for this strange difference is not clear, it is important to note again that isolated graphene layers behave like multiwalled nanotubes.

Potentially, one of the most important applications of carbon nanotubes lies in electronics, as it is possible to produce a field-effect transistor (FET) with just one

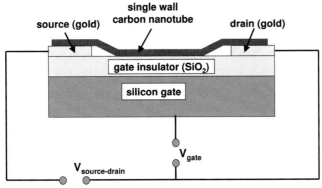

Figure 7.14 A metal oxide field effect transistor (FET) made from a carbon nanotube [8]. Such a simple device shows much promise as it is operated with voltages and currents that can be handled without major problems.

carbon nanotube which lies between two gold contacts (that act as the source and drain) and is in touch with the gate insulator. The set-up of such an FET, as realized at the IBM laboratories, is shown in Figure 7.14 [8]. Whilst it is astonishing that this simple device will act as a transistor, it is equally surprising that the voltages and currents controlled by such a nanotube FET (see Figure 7.15) are within a range that can be handled without major problems. This is clearly just the beginning of a potentially interesting development since, by examining Figure 7.12, it is clear that electrical currents in excess of 10 µA can flow through a single-wall nanotube, without destroying it. Others have also shown that even complete logical circuits can be realized using just one nanotube.

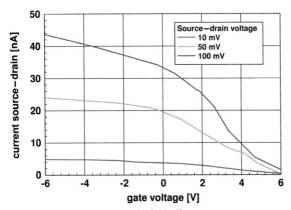

Figure 7.15 Characteristics of a field-effect transistor (FET) according to Figure 7.14 [8].

7.3
Photoconductivity of Nanorods

GaN is a semiconducting material which not only demonstrates photoluminescence but also, under UV irradiation, shows photoconductivity which is essential in terms of the compound's electrical conductivity. Calarco *et al.* [9] determined the photoconductivity of GaN as a function of nanorod diameter, and showed that it increased exponentially up to a critical diameter value of 100 nm. Beyond this limit, in a second range, the photoconductivity increased linearly with rod diameter. These two ranges are shown in Figure 7.16, where each of the experimental points was determined at a single nanorod. The photocurrent during UV illumination with a light intensity of 15 W cm^{-2} versus nanorod diameter is also shown in Figure 7.16.

When considering the time response to illumination, it is important to distinguish two different ranges. The first range, with rod diameter less than 100 nm, shows a fast response where the electrical conductivity goes immediately to the value of the dark current after switching off the UV illumination. However, in the second range, there is a significant persistent photoconductivity after switching off the illumination; after illumination, the time constant of the current decrease is almost two orders of magnitude larger. Therefore, for applications in optoelectronics, only the first regime is of interest. The conductivity of GaN nanorods, 70 nm in diameter, as a function of the applied voltage, is shown in Figure 7.17. At a constant voltage, under UV illumination, the current increases for more than an order of magnitude. The measured dark current indicated in Figure 7.17 is a parasitic background current, and is not related to transport through the GaN nanorod. As the dark current is extremely low, a significant noise level characterizes the voltage–current characteristic, and therefore a scattering band represents the dark current.

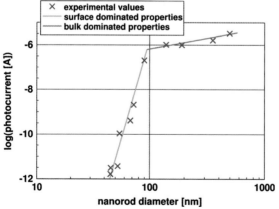

Figure 7.16 Photocurrent during UV illumination of 15 W cm^{-2} as a function of the diameter of the GaN nanorod. The edge at 85 nm indicates transition from the smaller diameters where the photocurrent increases exponentially to the range of larger diameter [9].

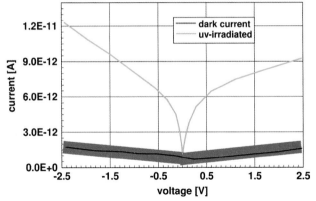

Figure 7.17 Conductivity of 70 nm GaN nanorods as a function of the applied voltage. During UV illumination, the current increases by more than an order of magnitude. The dark current (indicated by the shadowed area) is a parasitic background current and not related to transport through the GaN nanorod. According to Figure 7.16, this 70-nm nanorod operates in the range where the properties are surface-dominated [9].

The difference in behavior of GaN nanorods in these two size ranges, below and above 100 nm, is explained by a size-dependent surface recombination mechanism acting predominantly at specimens of small diameter.

The importance of surface phenomena on photoconductivity was also demonstrated for CdS nanorods. For example, Pan *et al.* [10] studied the properties of CdS nanorods with diameters ranging from 35 to 45 nm and lengths from 2 to 6 μm. As a semiconducting material, CdS shows photoluminescence, the spectrum of which is shown in Figure 7.18. The spectrum shows the band-edge emission around 505 nm and, in addition, a second luminescence peak at 548 nm, which is attributed to

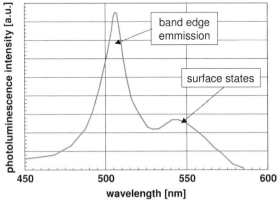

Figure 7.18 Photoluminescence spectrum of CdS nanorods [10]. Note the two clearly different features. The higher peak is related to band edge emission, whereas emission of the surface states is represented by the smaller peak at longer wavelength.

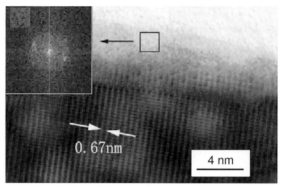

Figure 7.19 High-resolution transmission electron micrograph close to the surface of a CdS rod. Note the well-crystallized interior and an amorphous surface layer. The inset shows the Fourier-transform of the surface area indicated by the square. This Fourier transform also indicates a lack of crystallinity, and a weak preference for the distance to the next neighbors (with permission from [10]; IOP Publishing).

surface states. When examining the structure, transmission electron microscopy reveals an amorphous layer at the material's surface (see Figure 7.19).

This surface layer is clearly visible in the small areas indicated within the square, while the two-dimensional *Fourier* transformation of the indicated area is shown in the upper left corner inset. Here, a faint ring is just visible, this being characteristic for a material that has a only minor preference for the distance to the next neighbor. The thickness of this amorphous layer is in the range of 3 nm.

The voltage–current characteristic, both in the dark and under illumination, of an individual CdS fiber characterized by the amorphous surface layer is shown in Figure 7.20. At low voltages, this plot is characterized by a linear increase in the

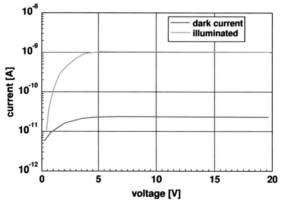

Figure 7.20 Dark current and photocurrent of CdS nanorods as a function of the applied voltage. Note that the current is in a logarithmic scale. Illumination increases the current by more than an order of magnitude [10].

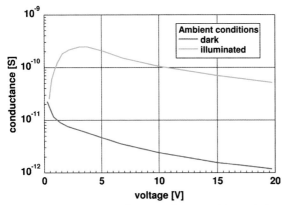

Figure 7.21 Conductance of CdS as a function of the applied voltage for illuminated and nonilluminated nanorods. Above a voltage of 1 V, the conductance in the illuminated state is more than one order of magnitude higher than in the dark [10].

current with increasing voltage, but beyond 5 V a saturation is observed. The difference between the current in the dark and illuminated states is almost two orders of magnitude. This remarkable response suggests that CdS would be a good material for the fabrication of photoelectric and sensing devices.

Calculating the conductance from the values in Figure 7.20 for CdS, whether illuminated or in the dark, provides interesting insights. In the dark, a decreasing conductance is observed with an increasing applied voltage, whilst at low voltages the illuminated material displays an increasing conductance with increasing applied voltage. In contrast, at higher voltages (above ca. 3 V) the phenomenon of decreasing conductance with increasing voltage, as described by Equation (7.8), is observed. Above an applied voltage of approximately 1 V, the ratio of conductance in the illuminated over the nonilluminated state is more than an order of magnitude (Figure 7.21). At 5 V, this ratio reaches its maximum of almost 50.

7.4
Electrical Conductivity of Nanocomposites

For technical applications outside of electronics, the electrical conductivity of nanomaterials can best be exploited as the electric-conducting phase in nanocomposites. In order to obtain electrical conductivity in a composite consisting of conducting and nonconducting phases, the conducting particles must touch each other to form a continuous series of conducting elements. Clearly, the probability of forming a continuous electrical conducting system increases with increasing concentration of the conducting phase. The concentration, at which such a continuous system is formed is termed the *percolation threshold*.

The electrical conductivity of nanocomposites depends on percolation, the theory of which treats the properties of two-phase mixtures consisting either of conducting and insulating phases, or of a solid and pores. In the latter case, percolation leads to the formation of a network of open pores. Detailed theories of percolation consider the shape and concentration of the constituents, but in all cases the crucial question relates to the critical concentration p_c, the percolation threshold, where the minority phase of the mixture forms a continuous network. Assuming an electric-conducting nanocomposite – a two-phase mixture consisting of an insulating and an electrical conductive phase – the percolation threshold describes, in simple terms, the concentration of conductors required for the onset of electrical conductivity. At concentrations below the percolation threshold, there is no electrical conductivity, whereas above the threshold conductivity is observed. Above the percolation threshold p_c, the electrical conductivity σ is described by [11]

$$\sigma = \sigma_0 (p - p_c)^t \tag{7.9}$$

where σ_0 is the conductivity of the conducting phase and p is the volume fraction of the conducting phase. The exponent t reflects the dimensionality of the network; usually, this is found not to be an integer, and experimental values of t are found to range between 1 and 3. Equation (7.9) is linearized in a double logarithmic graph, where $\log(\sigma)$ is plotted against $\log(p-p_c)$.

As mentioned above, percolation is extremely sensitive to shape and the aspect ratio of the second phase's particles. Therefore, any theory describing percolation must of necessity consider the aspect ratio of the second phase. For this, many stochastic theories have been devised, although most are so complex that their technical use is nearly excluded. As an example, a description of percolation for a fiber-shaped, second phase was reported by Balberg [12], and this theory led to the following equation for the percolation threshold:

$$p_c = 0.7 \frac{\langle L \rangle^3}{\langle L^3 \rangle} \frac{D}{\langle L \rangle} \tag{7.10}$$

where L is the length of the particles and D the particle diameter. This theory assumes a second phase with constant diameter and a distribution of lengths. The brackets $\langle \, \rangle$ denote mean values. It is an important characteristic that the percolation threshold increases with the mean value of the aspect ratio $a = \frac{\langle L \rangle}{D}$. Therefore, the concentration of particles necessary for the onset of percolation is for fibers up to many orders of magnitude less as compared to spherical particles. To obtain optically transparent electric conductive composites, one must apply extremely thin, long fibers, which are usually achieved with nanotubes or nanowires. In any real length distribution, $\frac{\langle L \rangle^3}{\langle L^3 \rangle} < 1$ is valid. Assuming fibers of equal length L, then Equation (7.10) boils down to

$$p_c = 0.7 \frac{D}{L} \tag{7.11}$$

In simple terms, Equation (7.11) states that using long fibers reduces the percolation threshold, as they have a huge aspect ratio. Furthermore, as the nanomaterial used as electric conductive filler is significantly more expensive than the polymer matrix, the application of nanotubes or nanowires with large aspect ratio is the most economic.

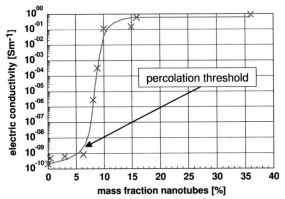

Figure 7.22 Experimental results on the electric conductivity of a carbon nanotube–PMPV (poly(*p*-phenylenevinylene-*co*-2,5-dioctoxy-*m*-phenlyenevinylene)) nanocomposite [13]. Note the sudden increase in conductivity at the percolation threshold.

Some typical experimental results for the electrical conductivity of a carbon nanotube–PMPV (poly(*p*-phenylenevinylene-*co*-2,5-dioctoxy-*m*-phenlyenevinylene)) nanocomposite are shown in Figure 7.22 [13]. Here, there is a clear and sudden increase in electrical conductivity at the percolation threshold between 7 and 8 vol.% carbon nanotubes. Additionally, a second characteristic feature of this type of nanocomposite is apparent, namely that the electrical conductivity shows saturation. This means that, for each combination of electrical conductive filler and insulator, a characteristic maximal conductivity is observed.

The percolation threshold and electrical conductivity at saturation level are heavily dependent on the fabrication process. In contrast to Figure 7.22, Figure 7.23 shows, as an example of a low percolation threshold, the electrical conductivity of a carbon nanotube–epoxy composite [14]. It is remarkable that the percolation threshold is as low as 2.5×10^{-3} wt.%, which is equivalent to a volume fraction of 1×10^{-5}. The electrical conductivity of these composites is in the range of $1 \, \text{S m}^{-1}$ in the case of 1 wt.% nanotubes in the composite. According to Equation (7.11), the low-percolation threshold indicates a large aspect ratio, which is expected for nanotube composites. However, such low-percolation thresholds are not always obtained. The percolation threshold of the carbon nanotube–PMPV nanocomposite, as shown in Figure 7.22, was in the range from 7 to 8 wt.% nanotubes. This huge difference, compared to the electrical properties of the composite shown in Figure 7.23, may be explained by insufficient singularization of the fibers in the composite. According to Equation (7.11), in this case the percolation threshold goes to higher concentrations because the diameter of fiber bundles is larger than that of one fiber.

The double logarithmic plot of electrical conductivity against the reduced weight fraction $p–p_c$, according to Equation (7.9), is shown in Figure 7.24. It is clear that, up to a volume content of almost 0.1 above the percolation threshold, the experimentally

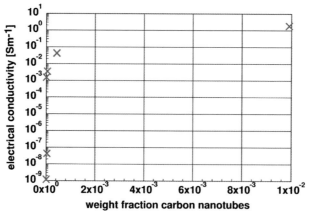

Figure 7.23 Electrical conductivity of a carbon nanotube–epoxy composite [14]. The percolation threshold of these well-distributed carbon nanotubes is in the range of 2.3×10^{-3} wt.%, which is equivalent to a volume fraction in the range of 10^{-5}.

determined data follow exactly Equation (7.9). The exponent t describing the dimensionality of this composite is 1.2 [14].

As a further example of the importance of the processing parameter, Figure 7.25 shows the electrical conductivity of carbon nanofiber-filled epoxy [15] with different viscosities of the polymer before curing. The vapor-grown carbon fibers were 150 nm in diameter and ranged in length from 10 to 20 μm. This led to an aspect ratio in the range of 100. The epoxy for the matrix was prepared using two different processes. In the first method, which led to the product denominated as "low viscosity" in

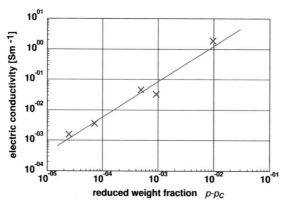

Figure 7.24 Double logarithmic plot of the electric conductivity of the material, as displayed in Figure 7.23, versus the reduced weight fraction $p - p_c$. The experimental data follow exactly Equation (7.9), up to a volume content of almost 0.1 above the percolation threshold [14].

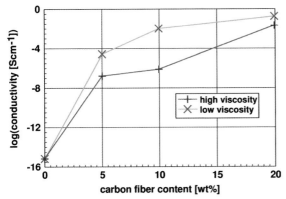

Figure 7.25 Electrical conductivity of carbon nanotube-filled epoxy as a function of fiber loading for low- and high-viscosity epoxy nanocomposite sheets. Data according to Choi et al. [15].

Figure 7.25, the epoxy resin was dissolved in acetone to reduce the viscosity, while in the second method a commercial epoxy resin was used. In both cases, the carbon fibers were dispersed in the liquid by stirring and sonication at room temperature. From Figure 7.25, it is clear that the electrical conductivity increases in the low-viscosity case by more than ten orders of magnitude over a concentration range from 0 to 5 wt.% carbon fibers, whereas the increase in conductivity is less significant in the case of the high-viscosity process.

A significantly more advanced system was introduced by Murphy et al. [16] in which, instead of carbon fibers, $Mo_6S_{4.5}J_{4.5}$ fibers were used. This material crystallizes linearly in wires with a diameter of ca. 1 nm, although it must be pointed out that these fibers form bundles. It is possible to reduce the diameter of the bundles by sonication, but during this treatment the length of the wires is also reduced. After sonication, the length of the wires was reduced to approximately 1000 nm, and this resulted in an aspect ratio greater than 1000. The electrical conductivity of a composite of PMMA and these fibers is shown in Figure 7.26, where the electrical conductivity is plotted against the volume

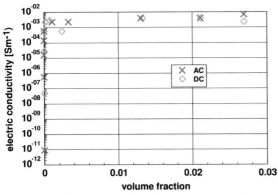

Figure 7.26 Electrical conductivity of a composite of PMMA and $Mo_6S_{4.5}J_{4.5}$ fibers; after Murphy et al. [16].

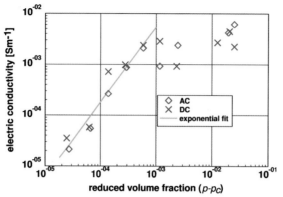

Figure 7.27 Double logarithmic plot of the experimental data displayed in Figure 7.26. The solid line is a fit according to Equation (7.9).

fraction of nanowires. Although the conductivity was measured by both DC and AC methods, interestingly no difference was found. The composites showed a saturation value of electrical conductivity of $5 \times 10^{-3}\,S\,m^{-1}$. Figure 7.27, which is based on an analysis according to Equation (7.9), shows the electrical conductivity plotted against the reduced volume fraction of the conducting phase $(p-p_c)$. In this double logarithmic plot, a linear relationship up to a reduced volume fraction of 10^{-3} is apparent. Further analysis of these experimental data led to a percolation threshold of $p_c = 1.3 \times 10^{-5}$, which was an extremely low value and similar to those found for well-prepared composites of carbon nanotubes. Such a small percolation threshold is possible with an aspect ratio of more than 10 000. The discrepancy between the aspect ratio determined from micrographs and from electrical conductivity is not yet clear. The conductivity of the isolated $Mo_6S_{4.5}J_{4.5}$ fibers is estimated to be about $80\,S\,m^{-1}$, and the dimensionality of the system was determined as 1.4. The data in Figures 7.26 and 7.27, taken together, show that in this case, over an extremely narrow concentration range from 10^{-5} (percolation threshold) to 10^{-3}, the electrical conductivity increased from 10^{-5} to almost $10^{-2}\,S\,m^{-1}$. Additionally, the data in Figure 7.27 indicate that, in this case, Equation (7.9) is valid only up to a volume fraction of nanowires of approximately 10^{-3}. Although this is a significantly smaller value than is often found for carbon nanotubes, it is most likely caused by an insufficient debundling of the nanotubes at higher concentrations.

References

1 Datta, S. (2004) *Nanotechnology*, **15**, 433–451.
2 Pascual, J.J., Gómez-Herrero, J., Méndez, J., Barón, D.A.M., García, N., Landman, U., Luedke, W.D., Bogachek, E.N. and Cheng, H.P. (1994) Proceedings of the NATO Advanced Research Workshop: "Scanning Probe Microscopies and Molecular Materials", Schloss Ringberg, Tegernsee.
3 Graff, A., Wagner, D., Ditlbache, H. and Kreibig, U. (2005) *Eur. Phys. J. D*, **34**, 263–269.

4 Aherne, D., Satti, A. and Fitzmaurice, D. (2007) *Nanotechnology*, **18**, 125–205.

5 Poncharal, Ph., Frank, St., Wang, Z.L. and de Heer, W.A. (1999) *Eur. Phys. J.*, **9**, 77–79.

6 Shklyarevskii, O.I., Speller, S. and van Kempen, H. (2005) *Appl. Phys. A*, **81**, 1533–1538.

7 Yao, Z., Kane, C.L. and Dekker, C. (2000) *Phys. Rev. Lett.*, **84**, 2941–2944.

8 Martel, R., Schmidt, T., Shea, H.R., Hertel, T. and Avouris, Ph. (1998) *Appl. Phy. Lett.*, **73**, 2447–2449 www.research.ibm.com/nanoscience/fet.html.

9 Calarco, R., Marso, M., Richter, T., Aykanat, A.I., Meijers, R., Hart, A.v.d., Stoica, T. and Lulth, H. (2005) *Nano Lett.*, **5**, 981–984.

10 Pan, A., Lin, X., Liu, R., Li, C., He, X., Gao, H. and Zou, B. (2005) *Nanotechnology*, **16**, 2402–2406.

11 Stauffer, D. and Aharoni, A. (1994) *Introduction to Percolation Theory* 2nd edn. Taylor and Francis London.

12 Balberg, I. (1985) *Phys. Rev. B*, **31**, 4053–4055.

13 Coleman, J.N., Curran, S., Dalton, A.B., Davey, A.P., McCarthy, B., Blau, W. and Barklie, R.C. (1998) *Phys. Rev. B*, **58**, R7492–R7495.

14 Sandler, J.K.W., Kirk, J.E., Kinloch, I.A., Shaffer, M.S.P. and Windle, A.H. (2003) *Polymer*, **44**, 5893–5899.

15 Choi, Y.-K., Sugimoto, K.-I., Song, S.-M., Gotoh, Y., Ohkoshi, Y. and Endo, M. (2005) *Carbon*, **43**, 2199–2208.

16 Murphy, R., Nicolosi, V., Hernandez, Y., McCarthy, D., Rickard, D., Vrbanic, D., Mrzel, A., Mihailovics, D., Blau, W.J. and Coleman, J. (2006) *Scripta Mater.*, **54**, 417–420.

8
Mechanical Properties of Nanoparticles

8.1
General Considerations

The huge interest in nanomaterials began with a report on the plastic deformation of CaF_2 at room temperature [1]. The discussion of mechanical properties of nanomaterials is, in to some extend, only of quite basic interest, the reason being that it is problematic to produce macroscopic bodies with a high density and a grain size in the range of less than 100 nm. However, two materials, neither of which is produced by pressing and sintering, have attracted much greater interest as they will undoubtedly achieve industrial importance. These materials are polymers which contain nanoparticles or nanotubes to improve their mechanical behaviors, and severely plastic-deformed metals, which exhibit astonishing properties. However, because of their larger grain size, the latter are generally not accepted as nanomaterials. Experimental studies on the mechanical properties of bulk nanomaterials are generally impaired by major experimental problems in producing specimens with exactly defined grain sizes and porosities. Therefore, model calculations and molecular dynamic studies are of major importance for an understanding of the mechanical properties of these materials.

It is common practice to characterize the mechanical properties of a material by its stress–strain diagram, which can be determined either in tension or compression experiments. A typical example of such a stress–strain diagram is shown in Figure 8.1, where elongation of the specimen during increasing deformation is plotted on the abscissa, while the stress is plotted on the ordinate. To be independent of the geometry of the specimen, elongation of the specimen is plotted as strain $\varepsilon = \Delta l / l$, where l is the length of the specimen and Δl the elongation under load. Similarly, the stress σ is given by $\sigma = P/A$, where P is the load and A the cross-section of the specimen. At low stresses, the deformation of a specimen starts with a linear, elastic range, where the deformation is fully reversible. After the elastic regime, plastic deformation begins; the stress of the onset of plastic deformation is called "yield stress." As experimentally it is almost impossible to determine the yield stress exactly, in general the stress where 0.2% plastic deformation is observed is defined as the yield stress. After reaching the yield point, characterized by the yield stress, the

Nanomaterials: An Introduction to Synthesis, Properties and Application. Dieter Vollath
Copyright © 2008 WILEY-VCH Verlag GmbH & Co. KGaA, Weinheim
ISBN: 978-3-527-31531-4

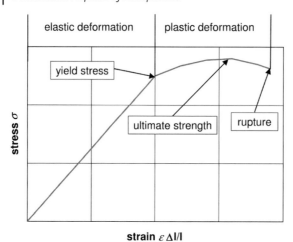

Figure 8.1 A typical stress–strain diagram, as obtained in a tension experiment. The important ranges are those where elastic and plastic deformation occurs. Yield stress, ultimate strength, and rupture are also indicated.

range of plastic deformation begins. Plastic deformation is not reversible, and during any further increase in deformation the increasing stress passes the point of the ultimate strength, after which the specimen will break. In many cases, the ultimate strength and the rupture stress are identical.

In the elastic region, *Young*'s modulus E or the modulus of elasticity (elasticity modulus) is determined by

$$E = \frac{\sigma}{\varepsilon} \tag{8.1}$$

Equation (8.1) is referred to as *Hooke*'s law; however, it should be noted that many materials do not demonstrate this linear elastic range.

For the sake of completeness, it should be noted that a differentiation can be made between an engineering stress–strain diagram – where the original cross-sectional area of the specimen is used to determine the stress for every value of applied force – and the true stress–strain diagram, where the applied force is divided by the actual value of the cross-section of the specimen. However, within this chapter, such differentiation is not made.

The stress–strain diagram depicted in Figure 8.1 is the most common, although not the only possible, form. The three most important types of stress–strain diagram are shown in Figure 8.2.

In Figure 8.2, the stresses are not plotted according to their actually possible values; rather, they are equalized to a constant level to demonstrate their characteristic behaviors. The most common form, denoted by "metal" is found primarily in metals and ceramic materials. A second type, denoted as "rubber", is characteristic of highly elastic materials, and this is an important example where *Hooke*'s law is not

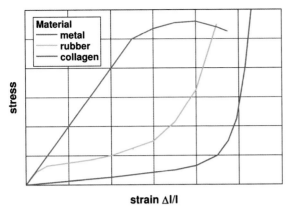

strain Δl/l

Figure 8.2 Different types of stress–strain diagram. As the three materials used as examples have significantly different strengths, the curves are normalized for comparison.

applicable. The last type, labeled as "collagen", is typical of many biological materials. In reality, it is quite rare to find an experimentally determined stress–strain diagram which follows exactly the curves plotted in Figure 8.2, and in most cases a "mixed type" diagram is observed.

At a very early stage, Siegel and Fougere [2] postulated that a change in the deformation mechanism of metals from dislocation processes to grain boundary processes when the grain size is reduced should be expected. This transition in deformation mechanism, which is equivalent to a transition of the behavior from metal-like to ceramic-like, is illustrated graphically in Figure 8.3.

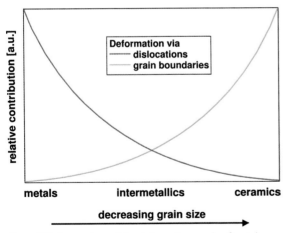

Figure 8.3 Comparison of the deformation mode of metals, intermetallic compounds, and ceramics with the mechanical behavior of metals with decreasing grain size [2]. Note the clearly decreasing importance of dislocation processes for plastic deformation with decreasing grain size; in contrast, the importance of the grain boundary processes increases.

This first model of the overall behavior of materials with respect to grains becoming increasingly smaller was well confirmed. Additionally, it should be noted that a significant part of the interpretation of the mechanical properties of nanocrystalline materials is based on model calculations, not least because experiments are often impaired by secondary influences such as porosity or grain growth at elevated temperatures.

8.2
Bulk Metallic and Ceramic Materials

8.2.1
Influence of Porosity

When considering mechanical properties, it is essential to study deformation mechanisms as a function of materials' structure. Important structural features in this context are grain size and porosity. As bulk nanomaterials are not in thermodynamic equilibrium, grain growth at elevated temperature or during deformation must be taken into account.

The influence of porosity on mechanical properties has been the subject of many studies, and in this context the porosity p is defined as

$$p = 1 - \rho_{\text{specimen}}/\rho_{\text{theor}} \tag{8.2}$$

where ρ_{specimen} is the actual, experimentally determined density of the specimen and ρ_{theor} is the theoretical density of the full, dense material. Most important is the influence on elastic properties, and one of many formulae describing this was provided by MacKenzie [3], who applied a *Taylor* series development for *Young's* modulus E to estimate the influence of the porosity p:

$$E = E_0(1 + \alpha_1 p + \alpha_2 p^2 + \cdots) \tag{8.3}$$

In Equation (8.3), E_0 is *Young's* modulus of the full dense material, $p = 0$, and α_1 and α_2 are fitting parameters. For small values of porosity the linear term is sufficient, but for higher porosities an increasing number of series elements are necessary. Besides this simple series development, a huge number of theoretically developed equations are described. However, theoretically well-based formulae also require fitting parameters, and Figure 8.4 depicts *Young's* modulus for nanocrystalline copper and palladium as a function of the porosity [4]. This is a good example of small porosity, where the linear term is sufficient. Additionally, it demonstrates the unavoidable scattering of experimental values which, most likely, is due to the different grain sizes for specimens with different porosities. When examining these heavily scattered experimental data, it is clear that a fit using more than the linear approximation is not justified.

Evaluation of the experimental data depicted in Figure 8.4 led to the following fitting parameter for *Young's* moduli (in GPa):

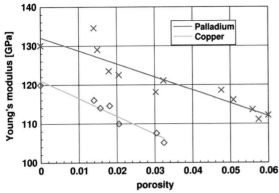

Figure 8.4 *Young's* modulus of palladium and copper according to Sanders *et al.* [4]. The wide scattering of experimental values highlights the problem in preparing a consolidated nanocrystalline specimen and measuring the porosity. The severe scattering of the experimental values is probably caused by different grain sizes.

$$E_{Cu} = 121(1 - 3.8p)$$
$$E_{Pd} = 132(1 - 2.5p)$$
(8.4)

The values for E_0 in Equation (8.4) are close to the bulk values found in coarse-grained materials. The yield stress is influenced by the porosity in a similar way; Figure 8.5 shows the yield stress of nanocrystalline copper and palladium again as function of porosity [5].

However, in most cases, a varying density is connected to a varying grain size. As the yield strength is also grain size-dependent, correlations between porosity and yield strength are usually quite poor. For the samples used for Figure 8.5, Figure 8.6 provides a correlation between grain size and porosity. The remarkable increase in

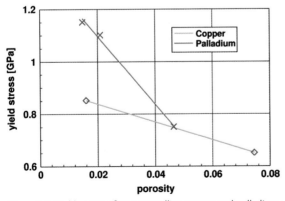

Figure 8.5 Yield stress of nanocrystalline copper and palladium as a function of the porosity [5].

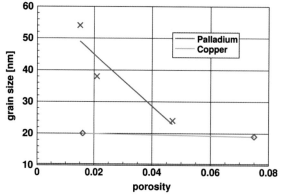

Figure 8.6 Relationship between grain size and porosity for the specimen used for the mechanical measurements plotted in Figure 8.5. It is clear that, especially for palladium, the grain size was far from constant [5].

grain size with decreasing porosity indicates that the data in Figure 8.5 demonstrate, for the example of palladium, only a trend and not the actual correlation.

8.2.2
Influence of Grain Size

A comparison of the strength of nanocrystalline and conventional materials with grain sizes in the micrometer range shows that the yield stress of nanocrystalline materials is found at significantly higher values. A comparison of the stress–strain diagrams for palladium with 50 µm and 14 nm grain sizes is shown in Figure 8.7 [6]. At first glance, there is clearly a higher yield stress of 259 MPa for the nanocrystalline

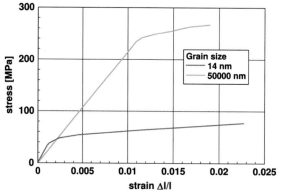

Figure 8.7 Comparison of the stress–strain diagram of palladium with different grain sizes. Note the significant increase in yield stress and reduction in *Young*'s modulus for the nanocrystalline material in comparison with the coarse-grained material [6].

material compared to the significantly lower value of 52 MPa for the coarse-grained material.

On examination of Figure 8.7, two features are striking:

- The elastic modulus (*Young's* modulus) of the nanocrystalline specimen is smaller than that of the coarse-grained material. This is highly probable to have been caused by residual porosity in the specimen as, according to Equation (8.3), *Young's* modulus is reduced with increasing porosity. As the nanocrystalline specimen was produced by the compaction and sintering of nanocrystalline palladium powder, a residual porosity is unavoidable.

- The increase of the strength cannot be overseen. This is also an effect of the grain size, although the increase is less distinct than might be expected. Again, this is an influence of the residual porosity.

The correlation between grain size D and yield stress σ_y, the *Hall–Petch* relationship, is described by:

$$\sigma_y = \sigma_0 + \kappa D^{-0.5} \tag{8.5}$$

where σ_0 and κ are constants. The grain size D is not necessarily defined by wide-angle grain boundaries; rather, it is related to the smallest units that may also be limited by small-angle grain boundaries. The exponent -0.5 is in theory well established; however, when analyzing experimental data in general the exponents are found in a wide range, perhaps from -0.3 to -0.7. It is important to note that the *Hall–Petch* relationship is also valid for the dependency of hardness as function of grain size.

The yield stress of nickel as a function of grain size is shown in Figure 8.8 [7], the so-called "*Hall–Petch* Plot", where yield stress or hardness is plotted against the square root of the inverse grain size. Such a plot visualizes immediately the validity of

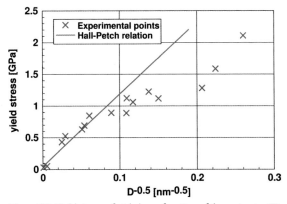

Figure 8.8 Yield stress of nickel as a function of the grain size [7]. Note the deviation from the straight line representing the *Hall–Petch* relationship below grain sizes of 200 nm ($D^{-0.5} = 0.07$). The deformation mechanism, leading to the *Hall–Petch* relationship, is no longer valid.

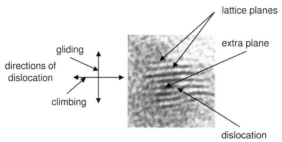

Figure 8.9 A dislocation in a nanoparticle made from WS$_2$. The inserted extra lattice plane is clearly visible. The possible directions of dislocation movement are also indicated. Movement of dislocation is possible only for mobile dislocations.

the *Hall–Petch* relationship for the material in question. The minimum grain size of these experiments was 15 nm. Clearly, there is a deviation of the experimental points from the yield stress versus (grain size)$^{-0.5}$ straight line for grain sizes below approximately 200 nm.

Obviously, at small grain sizes the deformation mechanism is changing, and in order to understand these changes it is necessary to analyze the deformation mechanisms.

Generally, plastic deformation is related to the generation and movement of dislocations, and within a grain there may be both immobile and mobile dislocations. However, for plastic deformation only the latter dislocations are of importance.

An electron micrograph of an edge dislocation in a WS$_2$ nanoparticle, demonstrating all of the important features of such a one-dimensional lattice defect, is shown in Figure 8.9. Between two lattice planes, an extra lattice plane is inserted, around the edge of which there is a stress field. Depending on the lattice plane, a dislocation is either mobile or immobile. A mobile dislocation has two possibilities to move: for example, it may either slip perpendicularly to the extra plane, a process known as "dislocation gliding" (Figure 8.9), or it may move in the direction of the extra plane, a process known as "dislocation climbing". Dislocation gliding always occurs after an increase of the stress beyond the yield stress, whereas dislocation climbing processes are observed during creep deformation, and are connected to diffusion. The change of shape of a single crystal specimen deformed by dislocation gliding and dislocation climbing is shown in simplified form in Figure 8.10.

Besides edge dislocations, screw dislocations are of major importance. However, as they are less important with respect to nanomaterials, the interested reader should seek specific information from the many textbooks on this subject.

A dislocation near a surface produces a stress in the surface plane, which in turn pulls the dislocation to the surface. Provided that the dislocation is mobile, it will begin to move and, on reaching the surface, it will be annihilated. As in nanomaterials with sufficiently small grain size any point is close to a surface, mobile dislocations are impossible. Therefore, dislocation-generating systems, which are pinned at nodes, must be discussed in view of plastic deformation. The most

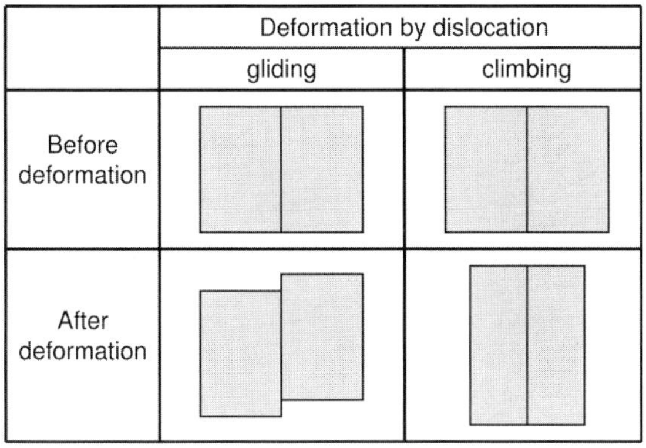

	Deformation by dislocation	
	gliding	climbing
Before deformation		
After deformation		

Figure 8.10 The influence of dislocation gliding and climbing on the shape of a single crystalline specimen.

common mechanism to generate dislocations is the *Frank–Reed* source which, to operate within a slip plane of one grain, requires a dislocation to be anchored at two nodes. A *Frank–Reed* source in its temporal sequence is shown in Figure 8.11.

A *Frank–Reed* source of dislocations starts at a dislocation which is pinned at two fixed nodes within one grain (an interior source), or at one node and a point at the grain boundary (a surface source). Because of their high elastic energy, dislocations are connected to a line tension, and therefore any dislocation has the tendency to shorten. The applied stress bows the dislocation out, and this leads to a decrease in the radius of curvature until the line tension is in equilibrium with the applied stress. Increasing the line stress beyond a point, where the dislocation is semicircular, creates a situation where the dislocation no longer has an equilibrium position. Consequently, the dislocation expands rapidly and rotates around the nodes until the loops meet each other to form a complete dislocation loop and a new line source

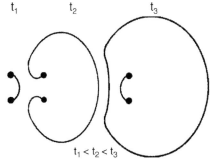

Figure 8.11 A *Frank–Reed* source of dislocations in its temporal sequence. Note the extension of a dislocation fixed at two pinning points under the influence of a stress and, finally, the separation from the pinning points. This leads to the formation of the next dislocation, and the cycle then re-starts. For the indicated times $t_1 < t_2 < t_3$ is valid.

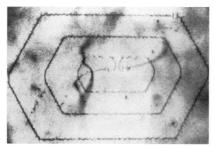

Figure 8.12 Electron micrograph of a *Frank–Reed* source [8] in single crystalline silicon. Note the dislocation loops in their different stages of development. (Reprinted with permission; Copyright: Springer 2005.)

between the nodes. The process may then start again. This generation and movement of dislocations is connected to plastic deformation. Acting *Frank–Reed* sources can be visualized using electron microscopy, and a typical electron micrograph is shown in Figure 8.12 [8]. Here, two closed dislocation loops and one which is starting to close can be visualized. The *Frank–Reed* source shown is in the micrometer size range. Even when the size of this example source does not fit into nanoparticles, the geometry and mechanisms are independent of the actual size.

Frank–Reed sources are not usually found in nanoparticles. This situation is more easily understood if the stresses necessary to activate a *Frank–Reed* source are considered, these being

$$\tau = \frac{Gb}{l} \tag{8.6}$$

where τ is the shear stress in the plane of the dislocation, G is the shear modulus, and l the distance between the two nodes. The *Burgers* vector b, which is characteristic of the type of dislocation and the crystal lattice, has a length of a few tenths of a nanometer. From Equation (7.6) it is clear that the shear stress to activate a *Frank–Reed* source increases with decreasing distance between the pinning points. The maximum possible distance between the pinning points is the grain size. Lastly, this is one of the reasons for increasing strength or hardness with decreasing grain size as represented by the *Hall–Petch* relationship. Even when it is assumed that l may reach the grain size, for nanocrystalline materials the necessary shear stress exceeds the maximal achievable strength of a technical body. This may be explained with the following simple estimations. Assuming, the maximal shear strength $\tau_{max} = \alpha G$ of a polycrystalline specimen is in the range from $10^{-3}G$ to $10^{-2}G$, and the *Burgers* vector is in the range of 10^{-10} m, then the minimal size l_{min} of a grain with an active *Frank–Reed* source may be estimated by:

$$l_{min} = \frac{Gb}{\tau_{max}} = \frac{b}{\alpha}$$

This estimation leads to a minimum grain size for dislocation deformation in the range from 10 to 100 nm. This result fits well with experimental findings, where

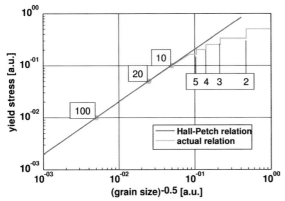

Figure 8.13 *Hall–Petch* plot of results of model calculations on the correlation between yield stress and grain size [9]. The numbers on the Hall–Petch line indicate dislocations in one grain; below ca. 10 dislocations per grain, the model calculations predict a stepwise increase of the yield stress with decreasing grain size.

grain sizes in the range between 10 and 50 nm serve as limits for deformation processes via dislocations. Therefore, as in most nanocrystalline materials, *Frank–Reed* sources for dislocation generation are impossible; moreover, deformation of these materials via dislocations is not possible.

Figure 8.13 shows, in a significantly simplified manner and interpretation, results from model calculations on the interaction of grain size, number of dislocations in a grain, and yield stress [9]. The boxed numbers indicated in the figure give the numbers of dislocations in one grain. It can be seen easily that, for up to 10 dislocations, the deviations from the *Hall–Petch* relationship are insignificant. However, when examining the scatter found in the experimental data (as given in Figure 8.8), a clear and significant deviation from the straight line may be detected at a later point. Interestingly, below about five dislocations in a grain each reduction in the number of dislocations is accompanied by a step in the yield stress. These phenomena, which are found in model calculations, cannot be proven experimentally as in any specimen used for experiments a distribution of grain sizes will be necessary. Hence, it is highly improbable that these steps can be verified.

Experimental results on yield stress and hardness show a decrease in strength at very small grain sizes. One of the very few experiments where mechanical properties were measured over a large range of grain sizes is depicted in Figure 8.14, where the hardness of TiAl with grain sizes between 3 and 30 nm is shown at 30 and 300 K [10]. Interestingly, for grain sizes above ca. 20 nm an increase was found in hardness with decreasing grain size, but below such grain size there was a decrease in hardness. In the transition range, insufficient data were available for any valid discussion. (It should be noted that to have more than eight different grain sizes in the range of 10 and 100 nm was a remarkable achievement by these authors.) Within the range of increasing hardness, the *Hall–Petch* relationship was fulfilled quite well, but in the

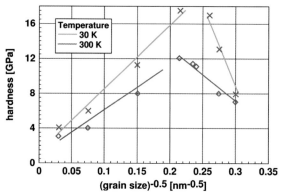

Figure 8.14 *Hall–Petch* plot of the hardness of TiAl, an intermetallic compound, at 30 and 300 K [10]. At both temperatures, there is an increase in hardness with decreasing grain size, as expected according to the *Hall–Petch* relationship. Below ca. 20 nm, the hardness decreases with decreasing grain size; this is the so-called inverse *Hall–Petch* range.

range of the smallest grains an inverse *Hall–Petch* relationship following $D^{0.5}$ was observed.

The existence of a range with extremely small grains, where the strength or hardness is decreasing with decreasing grain size, is very general, being found also in pure metals. Typical examples for copper and palladium are shown in Figure 8.15, where values of *Vickers* hardness for grain sizes ranging from 6 to 25 nm are displayed [11]. The decrease in hardness can be seen to follow more or less distinctly the inverse *Hall–Petch* relationship which, for copper, is quite well fulfilled. However,

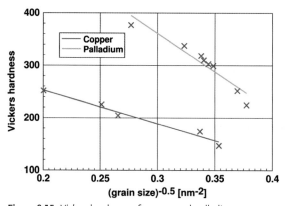

Figure 8.15 *Vickers* hardness of copper and palladium as a function of grain size. This *Hall–Petch* plot shows for both metals the inverse *Hall–Petch* relationship. The very wide scattering of the experimental data shows that the exponent 0.5 is only an approximation to the actual correlation [11].

in the case of palladium the fit was quite poor. Overall, this indicates that the exponent 0.5 is a quite rough approximation.

This inverse *Hall–Petch* relationship demands other deformation mechanisms as discussed above. In conventional materials, at high temperatures – and especially for ceramics – plastic deformation processes via grain boundary mechanisms are active, although a few of these processes have been described in the literature. For technical materials, the most important are the grain boundary deformation processes described by *Nabarro–Herring* or *Coble*. Deformation processes according to *Nabarro – Herring* act via volume diffusion when, under constant stress σ, the deformation rate $\dot{\varepsilon}$ is given by:

$$\dot{\varepsilon} \propto \frac{D_{vol}\sigma}{D^2} \tag{8.7}$$

where D_{vol} is the volume diffusion coefficient and D is the grain size. The *Coble* mechanism uses grain boundary diffusion with the diffusion coefficient D_{GB} as the rate-controlling mechanism; hence, the deformation rate is in that case:

$$\dot{\varepsilon} \propto \frac{D_{GB}\sigma}{D^3} \tag{8.8}$$

As a rule of thumb, the *Nabarro–Herring* mechanism acts at higher temperatures, perhaps close to the melting point, while the *Coble* mechanism is active at lower temperatures. Therefore, for nanocrystalline materials, the *Nabarro–Herring* mechanism can be excluded as in the temperature range where this mechanism operates the nanocrystalline materials are no longer stable and grain growth occurs. When considering the exponent at the grain size, it is clear that the grain boundary diffusion mechanism according to *Coble* has a higher exponent describing the grain size dependency as compared to the *Nabarro–Herring* mechanism. In both deformation mechanisms, the shape of the grains will be stretched during deformation; such behavior is shown in Figure 8.16, in an idealized manner.

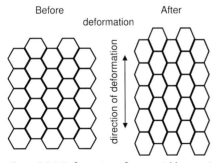

Figure 8.16 Deformation of a material by a grain boundary mechanisms. In an idealized manner, the grains are depicted as hexagons of equal size. It is important to note (as depicted in the figure) that the numbers and arrangement of the grains remain unchanged.

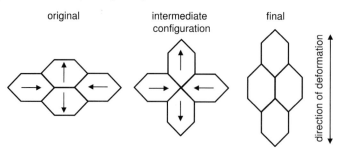

Figure 8.17 Deformation of a material by the *Ashby–Verall* mechanism. Shown here are the original, intermediary, and final states. The grain size dependency of this mechanism is D^{-3}.

In Figure 8.16 and all subsequent figures, the grains are depicted as hexagons of equal size. Although extremely idealized, however, this is the configuration best accessible for the theoreticians. From Equations (8.7) and (8.8), it is clear that grain boundary mechanisms become more prominent for decreasing grain sizes when, as found in nanomaterials, the contribution derived from the grain boundary processes is increased. Ashby and Verall [12] added a grain deformation mechanism to accommodate the grains with their new shapes in their new configuration. Also in this case, the deformation rate is proportional to D^{-3}. The principle of this mechanism is depicted in Figure 8.17, which also shows clearly why this deformation mode is called the "grain switching" mechanism.

Chang *et al.* [13] developed a concise experimental proof of the validity of the *Ashby–Verall* process for the deformation process of nanocrystalline intermetallics by comparing the experimental results of hardness measurements on TiAl with model calculations. The results of this study are shown in Figure 8.18, where the size of the

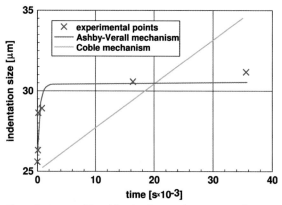

Figure 8.18 Size of the *Vickers* hardness indentation as a function of the indentation time, together with the results of model calculations assuming the *Ashby–Verall* and *Coble* mechanisms. In this case, TiAl deformation follows the *Ashby–Verall* mechanism [13].

Vickers hardness indentation is plotted versus time. The length of the impression diagonal was selected as the size of the indentation and, for the theoretical calculations, the models of *Coble* and *Ashby–Verall* were used. Subsequently, the experimental points were described quite well by the *Ashby–Verall* mechanism, whereas the assumption of the *Coble* process led to a time-dependency far from experimental reality.

Like the *Coble* mechanism, the mechanism according to *Ashby–Verall* is proportional to D^{-3}. Although this mechanism is the most probable for nanocrystalline materials, molecular dynamics calculations on the deformation of nanocrystalline aluminum in the size range from 3.15 to 9.46 nm by Kadau *et al.* [14] did not lead to any conclusive results on the exponent of the grain size in the range of the inverse *Hall–Petch* relationship. The results of these calculations, plotted assuming a grain size dependency D^{-n}, with $n = 1, 2$, and 3, are shown in Figure 8.19a–c. These figures

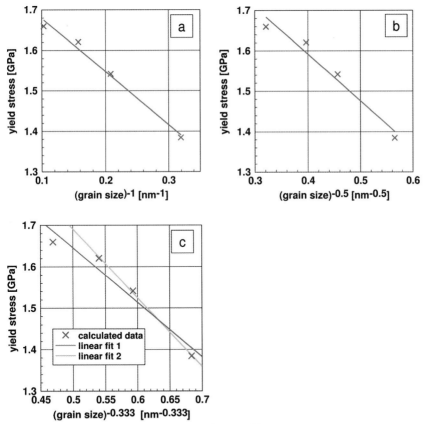

Figure 8.19 Results of molecular dynamic calculations of the deformation of nanocrystalline aluminum as a function of grain size, according to Kadau *et al.* [14]. The different plots assume a dependency of D^{-1} (a), D^{-2} (b), and D^{-3} (c). The grain sizes were selected in the range from 3.15 to 9.46 nm.

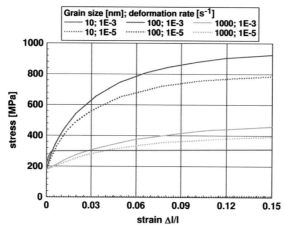

Figure 8.20 Results of model calculations on the influence of grain size and deformation rate on the stress–strain diagram of copper. The general observation is that, as expected intuitively, an increasing deformation rate requires higher stresses. Furthermore, the highest strength is observed at an intermediary grain size of 100 nm. The stress to deform a specimen with 10-nm grain size at a deformation rate of $\dot{\varepsilon} = 10^{-5}\,\mathrm{s}^{-1}$ is so small that it is no longer visible in this figure. Note the increase in *Young's* modulus with decreasing grain size [15].

show clearly that all of the plots are equally good or poor, with Figure 8.19c being the most interesting in this context. Here, fit 1 uses all of the data points, but is worse compared to the fits in Figures 8.19a and b. However, the fit is perfect when the experimental point relating to the largest grain size is omitted; clearly, this point is outside of the size range where the *Ashby–Verall* mechanism is acting.

With decreasing grain size, the influence of the deformation rate becomes dominant, and this has been experimentally very well proven and described by many theoretical models. As examples of the results of extensive model calculations, stress–strain graphs of copper with different grain sizes and deformation rates are shown in Figure 8.20 [15].

In Figure 8.20, three features are striking. First, the stress necessary for deformation decreases with decreasing deformation rate; such an observation is found to be independent of the grain size. Second – and most importantly – there is no simple correlation between grain size and stress necessary to obtain the same strain. As might be expected from previous discussions, the maximum stress necessary for deformation is found at an intermediary grain size of 100 nm (shown as blue curve in Figure 8.20). Most important is the stress–strain behavior at a grain size of 10 nm and a deformation rate of $10^{-5}\,\mathrm{s}^{-1}$. According to the data in Figure 8.20, only a very low stress is necessary for deformation, from which the important conclusion may be drawn that materials consisting of sufficiently small grains have almost no resistance against slow deformation. In general, this may be considered as a major disadvantage, but it may in fact be used for shaping by plastic deformation via the application of relatively small forces. The third point which is clearly visible in Figure 8.20 is the increase in *Young's* modulus with decreasing grain size.

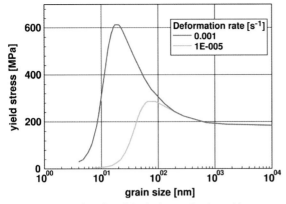

Figure 8.21 Results of model calculations for the yield stress
of copper as a function of grain size and deformation rate.
The grain size, where the maximum strength is observed, depends
on the deformation rate [15].

Based on the content of Figures 8.14 and 8.15, the result shown in Figure 8.20 was
to be expected. Finally, many experimental indications have suggested that there
exists a grain size where strength is maximal, and that beyond this grain size the yield
stress is reduced with decreasing grain size. This phenomenon is one of the clear-cut
results of these calculations. The yield stress of nanocrystalline copper as function of
the grain size and deformation rate is displayed in Figure 8.21, where these results
are from the model calculations performed by Kim *et al.* [15]. In this graph, because of
the broad range of grain sizes covered in the calculations, a logarithmic scale was
selected as the abscissa, and the results confirmed, perfectly, the tendency of the
experimental results as depicted in Figure 8.14.

The calculations leading to Figure 8.21 clarify that a maximum strength is found
with larger grain sizes, when the deformation rate is reduced. This figure also shows
that the inverse *Hall–Petch* effect is limited to a very narrow range of grain sizes; when
the grain size reaches a lower limit, any further reduction of the yield stress is
negligible. This point is of major importance for the near-net-shape forming of
ceramic parts by plastic deformation.Calculations which led to the graphs depicted
in Figures 8.20 and 8.21 also provided information on the contributions of
different mechanisms for deformation; these contributions as a function of grain
size at a strain rate of $10^{-5}\,s^{-1}$ are shown in Figure 8.22, which also indicates the
contributions of dislocation and grain boundary sliding processes. The contribution of lattice diffusion processes is so small that it was not plotted. It is
important to realize that, at a strain rate of $10^{-5}\,s^{-1}$, below a grain size of
approximately 35 nm practically the whole deformation occurs via grain boundaries. According to these authors, such calculations reveal that the deformation
rates at small grain sizes are related to the grain boundary surface-to-volume
ratio. A comparison of the results depicted in Figure 8.22 with those in
Figure 8.21 reveals that the maximum yield stress is correlated to the transition
between the two deformation mechanisms.

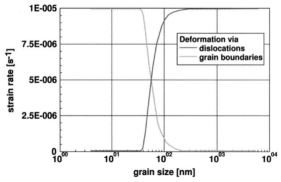

Figure 8.22 Contributions of the deformation mode via dislocation and grain boundary processes for copper as a function of grain size and deformation rate. In this example, at a deformation rate of $10^{-5}\,\mathrm{s}^{-1}$, only grain boundary processes contribute to deformation at grain sizes below approximately 35 nm [15].

Nanocrystalline bodies are, in thermodynamic terms, far from the minimum of free enthalpy, and consequently the tendency to come closer to equilibrium promotes grain growth. During a deformation process (such as that depicted in Figure 8.23), the orientation of the grains is changing. Assuming that the orientation of the slip system of neighboring grains is quite similar, then during such a rotation it may so happen that their slip systems are almost equally oriented. As plastic deformation continues, the neighboring grains might rotate in a way so as to bring their orientation closer together and, as a result, the grain boundaries in-between will be eliminated. In this way a new extended path for dislocation movement is opened,

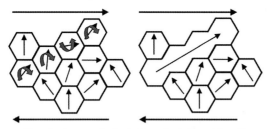

The short arrows indicate the orientation of the slip system in the grains.

Figure 8.23 Neighboring grains with similar orientation of the slip systems may rotate during plastic deformation, leading to an equal orientation of the grains. This allows the reduction of energy by eliminating the grain boundaries, opening an additional path for dislocation movement; this results in a softening of the material [16].

leading to a softening of the material [16]. The mechanism involved is depicted in Figure 8.23.

8.2.3
Superplasticity

The phenomenon of *superplasticity* is found especially in connection with small grain sizes. Superplastic materials allow deformation in tensile tests of a few hundred percent; indeed, in special cases, deformations of more than 1000% were found in metals. Although restricted to a certain, material-dependent range of grain sizes and deformation rates, superplasticity is typically found at around $0.5\,T_m$, where T_m is the melting temperature in Kelvin. Superplasticity is observed in metals, alloys, and ceramic materials. Superplastic specimens neither narrow locally nor form internal cavities, as both are local sites where fractures start. To date, no generally accepted theory has been proposed of superplasticity which explains all of the available experimental results. As with other deformation processes, the deformation rate of a superplastic specimen can be described by

$$\dot{\varepsilon} \propto \frac{\sigma^n}{D^2} \tag{8.9}$$

In Equation (8.9) the stress exponent n is 1 for diffusion and grain boundary processes, or 2 for dislocation processes.

In order to demonstrate superplasticity in nanocrystalline metals, two different types of material were selected. Figure 8.24 displays the stress–strain curves of superplastic Ni3Al, an ordered intermetallic compound [17], and of Ti6Al4V, an alloy [18]. The intermetallic compound Ni3Al had a grain size in the range from 80 to 100 nm, whereas for the Ti6Al4V alloy the grain size was smaller, in the range of 30 to 50 nm. The deformation rate during the experiments displayed in Figure 8.24 was $10^{-3}\,\mathrm{s}^{-1}$.

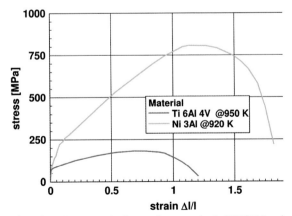

Figure 8.24 Stress–strain diagram for superplastic Ni3Al [17] and Ti6Al4V [18]. The large strain, exceeding 1 (>100%) is remarkable.

Over a quite narrow range of grain sizes and temperatures, both materials display mechanical properties that allow plastic deformations of more than 100%. In both cases, the specimens were produced by the severe plastic deformation of conventional materials. In spite of the essential differences between these two types of material, their mechanical properties with respect to superplasticity have several points in common. In both cases, superplasticity of the nanocrystalline material is observed at lower temperatures as compared to their coarse-grained analogs. In nanocrystalline metallic materials, superplasticity is often observed at the transition from dislocation to grain boundary processes for plastic deformation. This is a very general and very important point. When considering Figures 8.21 and 8.22, it is apparent that the transition from dislocation to grain boundary processes proceeds, with decreasing deformation rate, to larger grain sizes. Insofar, these findings are equivalent, since in both examples shown in Figure 8.24 the temperatures of superplastic deformation are relatively low, and the nanocrystalline structure is quite stable.

Superplasticity is found not only in metallic materials but also in ceramic materials, especially in oxides. However, as the ductility of ceramic materials is significantly lower than that of ceramics, such huge plastic deformations as are found in metals would not be expected. In addition, ceramic parts are produced from powders by a sequence of pressing and sintering such that, unavoidably, fully densified bodies are not obtained, especially as the pressing behavior of nanoparticulate powders is extremely poor. These poor pressing properties are not compensated by the excellent sintering behavior of nanoparticulate powders. Rather, the individual pores of residual porosity act as failure points, where cracking starts. This occurs because, around a flaw, stress concentrations are observed which depend on the aspect ratio of that flaw. As a consequence, the strength of a material containing pores is less than might be expected from a reduction of the bearing cross-section. In fact, the maximal strength σ_{flaw} of a specimen containing flaws of the size c follows the proportionality

$$\sigma_{flaw} \propto \frac{1}{c^{0.5}} \tag{8.10}$$

Increasing the size of the flaws necessarily leads to premature cracking, with the consequences of reduced strain and strength. The successful production of ceramic specimens exhibiting superplasticity is a reference for excellent abilities in ceramic technologies. Densities close to theoretical values may be obtained with specimens produced from submicron powders, and these materials have the potential for superplastic deformation of a few hundred percent.

As typical example of superplastic ceramics, Figure 8.25 shows a stress–strain diagram of yttria-doped zirconia [19]. In this example, 5% yttria was added to stabilize the tetragonal phase of zirconia; otherwise, the material would transform into the monoclinic phase. The investigations were performed as tensile tests and not (as is usually done for ceramics) in compression. The density of the material ranged from 84 to 94% of the theoretical density, the grain size ranged from 45 to 75 nm, and sintering was performed at 1420 K. As the testing temperature was not significantly

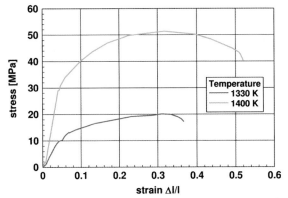

Figure 8.25 Superplastic deformation of ZrO$_2$ (5 wt.% Y$_2$O$_3$) with grain size in the range from 45 to 75 nm and densities of ca. 90% theoretical density. The tests were performed under tension. As an elevated temperature was used, unavoidable grain growth was observed during the experiments [19].

lower than the sintering temperature, substantial grain growth was observed during deformation. A process of grain boundary sliding was identified as the deformation mechanism; hence, the experiments indicated that the deformation was directly connected to grain growth and a reduction in density.

8.3
Filled Polymer Composites

8.3.1
Particle-Filled Polymers

Filling polymers with nanoparticles or nanorods and nanotubes, respectively, leads to significant improvements in their mechanical properties. Such improvements depend heavily on the type of the filler and the way in which the filling is conducted. The latter point is of special importance, as any specific advantages of a nanoparticulate filler may be lost if the filler forms aggregates, thereby mimicking the large particles. The stress–strain diagrams of filled polymers are shown in Figure 8.26, where the least strength is found at the unfilled polymers but, at least in the idealized case, the strain at rupture is largest. Particulate-filled polymer-based nanocomposites exhibit a broad range of failure strengths and strains. This depends on the shape of the filler, particles or platelets, and on the degree of agglomeration. In this class of material, polymers filled with silicate platelets exhibit the best mechanical properties and are of the greatest economic relevance. The larger the particles of the filler or agglomerates, the poorer are the properties obtained. Although, potentially, the best composites are those filled with nanofibers or nanotubes, experience teaches that sometimes such composites have the least ductility. On the other hand, by using

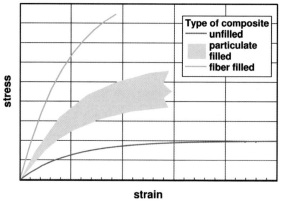

strain

Figure 8.26 Simplified stress–strain diagrams for different types of nanoparticulate-filled polymer. Generally, pure polymers exhibit the largest strain at rupture and the least strength, while fiber-filled polymers have the highest potential for high-strength composite materials. The stress–strain curves of particle- or platelet-filled nanocomposites lie within in a broad range between the unfilled and fiber-filled polymers.

carbon nanotubes it is possible to produce composite fibers with extremely high strength and strain at rupture.

It should be noted that many materials of biologic origin are high-strength nanocomposites, consisting of a mineral phase bound together by proteins. Typical examples are bones and nacre (mother-of-pearl), both of which apply well-ordered inorganic platelets to increase strength. Interestingly, these composites are, to a large extent, insensitive against flaws [20]. Technical polymer–ceramic nanocomposites may be produced in different ways. The simplest approach is simply to knead the ceramic powder together with the polymer. However, this does not lead to isolated particles in the polymer matrix; rather, agglomerates are obtained that are distributed in the polymer. More advanced processes start with a suspension of the nanoparticulate powder in a liquid; this may either be a solvent for the polymer or a liquid precursor compound, such as a monomer. Subsequently, the liquid phase is either evaporated or polymerized. Processes which start with liquid suspensions lead to products of the highest quality.

Experimental results indicate that the increase in strength obtained by utilizing a constant amount of second phase increases with decreasing particle size. This may occur for either of two reasons:

1. The size of the flaws, which are extended under load, is smaller when using nanoparticles as compared to the application of conventional ceramic powders. In extending a graph produced by Jordan *et al.* [21], this situation is shown schematically in Figure 8.27, where panels a and b relate to isolated larger or smaller ceramic particles in the polymer matrix, respectively. Clearly, by using fillers with a smaller grain size the size of the cracks that may occur during

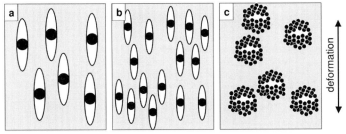

Figure 8.27 Influence of second phase particle size in ceramic–polymer nanocomposites on behavior under tension. Larger particles lead to larger failures under load (compare panels a and b), reducing the maximum stress before cracking. Under load, agglomerates of small particles may also lead to large failures and, therefore, to early cracking (c).

mechanical deformation are also smaller. Therefore, according to Equation (8.11), a higher strength might be expected. Such behavior might be expected primarily in cases where the binding between the polymer matrix and the filler is poor. However, in most cases, it will be difficult to fill the polymer matrix with isolated grains; rather, it is to be expected that nanopowders will be introduced as agglomerates, as shown in Figure 8.27c. Certainly, to some extent, these agglomerates are filled with the matrix polymer, although most probably this is not the case, and in this situation the agglomerate itself will break. It is highly probable that the flaw introduced by the broken agglomerate will be smaller than that in a particle of comparable size. One further essential point is the interaction of the filler particles with the polymer matrix, as when binding is insufficient the filler particles act as a flaw and not as a strengthening element.

2. A higher strength is found in the larger surface of the nanoparticulate filler with nanoparticle-filled polymers as compared to filling with conventional ceramic powder. However, this argument is valid only in composites, where the particles are bound firmly together with the polymer matrix.

Figure 8.28, which displays stress–strain diagrams of pure and filled polyamide-6, confirms this intuitive relationship by using experimental data [22]. It is of interest to note how the strength increases with filling; when silica with a different particle size was selected as the filler, the composite filled with 17-nm particles showed a higher strength compared to that filled with 80-nm particles, as might be expected.

As mentioned above, one essential property of ceramic–polymer nanocomposites is the interaction between the polymer matrix and the filler particles. In contrast to Figure 8.28, Figure 8.29 shows experimental results where the filler reduces the strength and fracture strain [23]. Here, the bonding between the polymethyl methacrylate (PMMA) matrix and the alumina filler was clearly insufficient, and therefore the filler particles acted as flaws. This is an interesting example of a composite with minimal interfacial interaction between the matrix polymer and the nanoparticles, as reported by Ash *et al.* [23]. The data in Figure 8.29 show a

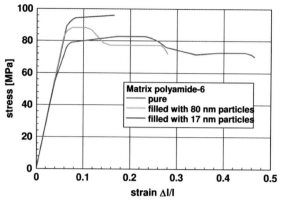

Figure 8.28 Stress–strain diagram of pure and nanoparticulate silica-filled polyamide-6 nanocomposite. Note that the largest strain and least strength is achieved with the pure polymer. Filling with nanoparticles improves strength; the influence increases with decreasing particle size of the filler [22].

significantly reduced yield stress and, interestingly, an eight-fold increased strain to failure in composites consisting of PMMA as matrix and 40-nm Al$_2$O$_3$ particles as filler. The results for the composite with 5 wt.% filler, which is equivalent to ca. 1.5 vol.% filler content, are shown graphically in Figure 8.29. In addition, in contrast to the example shown in Figure 8.28, there was a decrease in the *Young's* modulus of the filled material, in contrast to the pure PMMA.

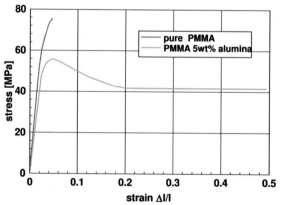

Figure 8.29 Stress–strain diagram of polymethyl methacrylate, PMMA, in the pure state, and filled with 5 wt.% alumina particles. The particle size was 40 nm. In contrast to expectation, the pure polymer exhibited the highest strength and least strain at rupture when compared to the filled material [23]. In this case, the bonding between the PMMA matrix and the filler was clearly insufficient, and therefore the filler particles acted as flaws.

8.3.2
Polymer-Based Nanocomposites Filled with Platelets

Among the most exciting nanocomposites are the polymer–ceramic nanocomposites, where the ceramic phase is platelet-shaped. This type of composite is preferred in nature, and is found in the structure of bones, where it consists of crystallized mineral platelets of a few nanometers thickness that are bound together with collagen as the matrix. One prominent example is that of nacre, which exhibits a "bricks-and-mortar" structure where the thickness of the aragonite bricks is a few hundred nanometers. Even when nacre does not exhibit a "real" nanostructure, there are crack-resistant elements in this structure that are essential in order to understand the properties of these composites and provide direction for further development. The structure of these composites, consisting of ceramic platelets (the bricks) and an organic matrix (the mortar) is shown schematically in Figure 8.30a.

Figure 8.30b shows the composite in a maximally deformed state. Here, the flow of the stresses is marked to demonstrate the stress distribution. The stress was assumed to be so high that the soft binder in between the ceramic platelets is broken; however, the bonding between the polymer and ceramic remains intact such that the part itself is not broken. This simplified model shows that the ceramic structure carries most of the load, which is transferred via the high-shear zones between the ceramic platelets. Consequently, the part will break when the stress reaches a level where either the ceramic platelets will break or, as is more probable, the shear stress between the organic binder and the ceramic filler leads to debonding [20]. An electron micrograph of this structure (nacre) is shown in Figure 8.30c. Clearly, there is not too much idealization in the model structure shown in Figure 8.30a but, according to the authors, there exists an optimum aspect ratio of the ceramic bricks which

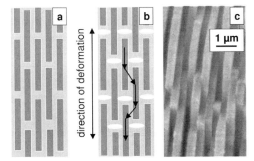

Figure 8.30 Structures of idealized and natural nanocomposites consisting of a binding polymer and a platelet-shaped filler. (a) Idealized arrangement of the ceramic building blocks and polymer filler. (b) Flow of stress in a composite according to (a). Here, it was assumed that the stress was so high that the soft binder in-between the ceramic platelets was already broken. The part itself was not broken, as the bonding between the ceramic platelet and binder remained perfect. (c) An electron micrograph of nacre [20] (mother-of-pearl), a naturally occurring, high-strength nanocomposite, the structure of which closely approaches that of the idealized structure (a) (reprinted with permission from [20], Copyright: National Academy of Sciences).

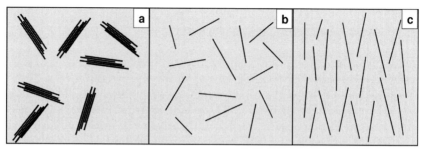

Figure 8.31 Different variations for the arrangement of filler particles in a nanocomposite, using particles crystallizing in layers as filler. (a) Conventional composite; the particles, which consist of stacks of layers, are distributed randomly in the matrix. (b) In this composite, the layers are individualized, defoliated, and the distribution of their orientation is random. (c) A nanocomposite according to (b), but with more or less equally oriented platelets; such a structure is very similar to that of nacre.

corresponds to the condition that protein and mineral fail at the same load. However, this design of a platelet-enforced nanocomposite is optimized for only one load direction. When assuming mechanical loads in directions perpendicular to the platelets, the structure shown in Figure 8.30a is not necessarily the best. Considerations such as these are of major importance when discussing the optimal structures of composites with platelet-shaped fillers.

In order to produce such a platelet–polymer nanocomposite, small particles of ceramic materials crystallizing in layered structures are applied. The best results are obtained when these particles are defoliated, which means that they are split into individual layers. The three fundamental types of man-made nanocomposite using platelets as filler, are shown in Figure 8.31.

Figure 8.31a shows the conventional type of nanocomposite, where just small platelet-shaped particles are used. However, more progressive and more successful types of nanocomposite utilize defoliated particles, which may either be arranged in random (as shown in Figure 8.31b) or they may be more or less oriented (see Figure 8.31c). As the latter structure comes close to that of nacre, the best mechanical properties may be expected as compared to other structures.

Defoliated platelets are obtained from compounds which are crystallizing in layered structures. The term "defoliating" means to separate the individual layers of a layered compound, and the process is explained schematically in Figure 8.32, using a layered silicate (phyllosilicate) as the ceramic starting material.

The layered silicates consist of negatively charged layers (these are gray-colored in Figure 8.32a) consisting of silicate tetrahedrons and aluminum or magnesium, and some lithium in varying quantities. These layers are approximately 1 nm thick, and the distance between two layers is 0.2 nm. The lateral size of the platelets is in the range of a few hundred nanometers. Depending on their composition, these silicates are referred to as montmorillonite, hectorite, or saponite. In between these layers there are positively charged alkaline ions (predominantly sodium); these are shown

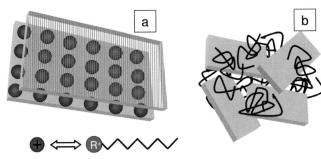

Figure 8.32 Defoliation of layered silicates to produce nanocomposites. (a) Model of a layered silicate. This consists of negatively charged silicate layers bonded together with positively charged alkaline ions. For simplicity, the upper layer is shown transparent. To start defoliating, the alkaline ions are exchanged with organic molecules carrying equal charges. (b) Following exchange of the alkaline ions with equally charged organic molecules, the crystal defoliates. The individualized layers then become embedded in the polymer matrix.

as red spheres in Figure 8.32a. In an initial step these intercalated cations are substituted by positively charged organic molecules, and this leads to a swelling of the particle, where the distance between the silicate layers increases to values between 2 and 3 nm. In a second step, defoliation of the crystal occurs. In addition to natural-layered silicates, various synthetic forms are currently in use; hence, synthetic layered silicates such as hydrotalcite are produced in a very pure form. In contrast to natural phyllosilicates, some synthetic counterparts may also carry positive charges on the platelets and negative charges on the intercalated ions.

Composites consisting of a polymer matrix and defoliated phyllosilicates exhibit excellent mechanical and thermal properties and, when developed by Toyota in 1989, were originally intended for use in the automotive industry. Here, nylon-6 was used as the polymer, and 5 wt.% montmorillonite as the layered silicate. In the meantime, polymer–phyllosilicate nanocomposites have become a multibillion dollar business and may represent one of the most successful nanomaterials at present. The electron micrographs of some example composites are shown in Figure 8.33.

A polypropylene (PP)–silicate nanocomposite with a 4 wt.% montmorillonite addition is shown in Figure 8.33a [24]. A variety of different features is immediately apparent as inclusions, and one nondefoliated silicate particle is also visible, where the defoliation process is already starting at the ends. The shape of this particle demonstrates perfectly the process of defoliation. In addition, defoliated single, double, and triple silicate sheets are visible, the lengths of which may be approximately 150 nm. A near-perfect defoliated specimen of a composite with 5.6 wt.% montmorillonite in polystyrene is shown in Figure 8.33b [25]. It is of interest to note that this is a more or less perfectly oriented composite, at least within the frame of the micrograph. Unfortunately, in this image it is quite difficult to estimate the size of the silicate sheets, although they may be in the range of a few hundred nanometers.

Figure 8.34 shows the stress–strain diagrams of the nylon-6–montmorillonite nanocomposite in comparison with the unfilled polymer [26]; these measurements

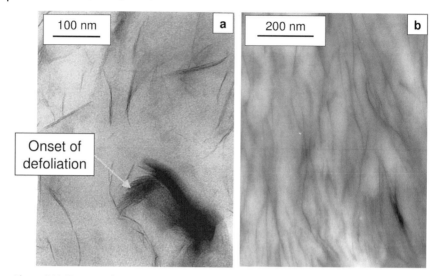

Figure 8.33 Two typical composites consisting of a layered silicate in a polymer matrix. (a) Nanocomposite with 4 wt.% montmorillonite as ceramic phase and polypropylene as matrix [24]. The distribution of the defoliated layers is random. The start of the defoliation process is visible on one particle. (Reproduced with permission from [24], Copyright: American Chemical Society, 2008.) (b) A near-perfect 5.6 wt.% montmorillonite–polystyrene composite. The defoliated layers are aligned almost in parallel (reproduced with permission from [25], Copyright: Elsevier, 2007).

were all performed at room temperature and 350 K. Here, two striking features are apparent. First, the silicate-containing nanocomposite exhibits a significantly higher strength as compared to the unfilled polymer, although at least at room temperature, the maximal strain before the specimen breaks is significantly lower. Second, the

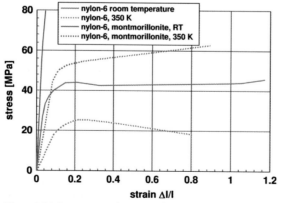

Figure 8.34 Stress–strain diagrams of pure nylon-6 and a nylon-6–montmorillonite nanocomposite, measured at room temperature and 350 K [26]. Note the dramatic increase in strength due to the silicate addition. The higher *Young*'s modulus of the composite, compared to the pure polymer, is clearly apparent.

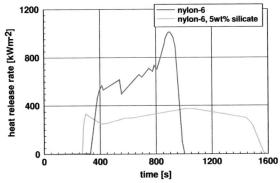

Figure 8.35 Heat release rate of pure nylon-6 and a nylon-6–5 wt.% silicate nanocomposite at a heat flux of 35 kW m^{-2}. The reduction in maximum heat release rate by ca. 60% is remarkable. For technical applications (e.g., in the automotive industry) this is an important safety feature [27].

Young's modulus of the filled material is, independently of the temperature, higher than that of the pristine material. The data in Figure 8.34 show that the polymer matrices filled with layered silicate are stiffer and exhibit a significantly higher strength as compared to unfilled material. This relatively high strength at elevated temperatures is of particular importance for uses in automobiles.

A further major property of these composites is equally important, namely that their flammability is significantly reduced as compared to the pure polymer. Advantages in this direction are realized in the released heat and maximum temperature of flames during burning. Heat release during burning is shown graphically in Figure 8.35 [27], where a specimen of the material was heated at 35 kW m^{-2}. It is clear from the data in Figure 8.35 that, for a nylon-6–5 wt.% silicate nanocomposite, the maximal heat release rate is almost one-third that of the unfilled material. In addition, the maximal flame temperature is reduced, as shown in Figure 8.36 [27].

The data in Figure 8.36 show clearly that the maximum flame temperature during a heat input of 35 kW m^{-2} is reduced, from 820 K to less than 750 K, and hence the burning time is extended. The reason for this observed improvement is the formation of a ceramic insulating layer at the surface that reduces the heat input to the residual material. The reduced flammability shown in Figures 8.35 and 8.36 represents a safety feature which is extremely important in the automotive industry.

Polypropylene, PP, has many advantages in its processing, but burns very well. Hence, the addition of layered silicates to PP will also reduce the heat release rate. Improvements in mechanical properties is less significant than when nylon-6 is used as the matrix material. The heat release rate for a PP filled with 4 wt.% layered silicate is shown in Figure 8.37 [24], where the reduction in heat release rate is much more dramatic than for the composite containing nylon-6 as the matrix material.

According to the data in Figure 8.37, PP–4 wt.% silicate nanocomposites have no higher a rate of heat release than do equivalent nanocomposites with nylon-6 as the

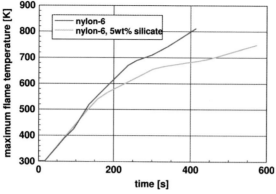

Figure 8.36 Maximum flame temperature of pure nylon-6 and a nylon-6–5 wt.% silicate nanocomposite at a heat flux of 35 kW m^{-2}. The temperature maximum is reduced from 820 to less than 750 K. Again, this is an important safety feature for technical applications [27].

matrix. However, whilst this leads to a much wider variety of materials for technical applications, the strength of the PP matrix composites is significantly lower, even when it is improved by the addition of phyllosilicates.

8.3.3
Carbon Nanotube-Based Composites

Nanocomposites with carbon nanotubes as filler have shown great promise, mainly because the carbon nanotubes are extremely stiff, with a *Young*'s modulus of about 1 TPa and an exhibited maximum tensile strength close to 30 GPa [28]. Yet, these values are currently under dispute, as other authors have reported *Young*'s moduli to

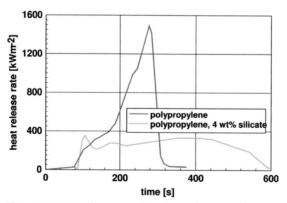

Figure 8.37 Heat release rate of a polypropylene–4 wt.% silicate nanocomposite under a heat flux of 35 kW m^{-2}. The reduction in maximal heat release rate, from almost 1500 to 330 kW m^{-2}, is remarkable [24].

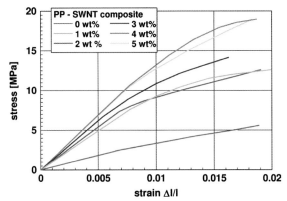

Figure 8.38 Stress–strain diagrams of polypropylene (PP), single-wall nanotube (SWNT) composites compared to a pure PP fiber. Remarkably, not only the strength but also *Young*'s modulus increase dramatically with the addition of SWNT. However, the addition of more than 4 wt.% SWNT has only a minor influence [29].

lie in the range from 0.64 to 1.8 TPa. In any case, whatever the correct values might be, carbon nanotubes are stiffer and exhibit a higher strength than any other material available in large quantities and for reasonable prices. In addition, it is not too difficult to distribute carbon nanotubes within a polymer matrix, a process which makes these composites a highly promising class of materials. At present, the technical realization is heading in two different directions: (i) composites with relatively small additions of nanotubes; and (ii) materials where the binding polymer is the minor phase.

The stress–strain diagrams for composites with polypropylene as the matrix and the addition of 0 to 5 wt.% single-wall carbon nanotubes (SWNT) are shown in Figure 8.38 [29]. These specimens were fibers with a diameter of 1.6 mm. The figure exhibits several interesting features, notably that the addition of 1 wt.% SWNT (which is equivalent to ca. 0.75 vol.%) increases the *Young*'s modulus and strength by more than a factor of two, while the maximum strain is almost unchanged. This indicates a dramatic improvement in the fiber's mechanical properties, with higher SWNT concentrations leading to increasing strength and *Young*'s moduli. Whilst in terms of *Young*'s modulus such an increase is not dramatic, the further increase in strength to values of 18 MPa is of great interest with regards to technical applications, and especially as the strain to rupture is barely influenced. The addition of more than 4 wt.% SWNT appears to be of minimal value, as the mechanical properties are not significantly influenced beyond this point.

Remarkably higher strengths and *Young*'s moduli may be obtained with strands of carbon nanotubes [30]. For example, Li *et al.* produced such strands with lengths of up to a few centimeters, and consisting of single- and double-walled nanotubes. The strand diameter ranged from 3 to 20 μm, and a typical scanning electron microscopy image of such a strand is shown in Figure 8.39. This micrograph not only shows clearly that the strand is composed of many individual nanotubes, but also indicates

Figure 8.39 Scanning electron micrograph of a strand consisting of single- and double-walled nanotubes. Note that the fibers are more or less perfectly aligned in the direction of the fiber axis. This micrograph also shows that the nanotubes are not simply straight but rather are bent in different directions (reprinted with permission from [30]), Copyright: Elsevier, 2007).

that the nanotubes although not perfectly aligned are, to a large extent, oriented in the direction of the strand. In order to produce such a strand it was not necessary to apply any polymer as binder.

The carbon nanotube strands exhibit interesting mechanical properties. For example, Figure 8.40 shows a typical stress–strain diagram obtained with such a strand, as it is depicted in Figure 8.39, with a length of 5 mm and a diameter of 5 μm. The maximum strength of 1.2 GPa and the *Young*'s modulus of 16 GPa are remarkably high values. The stress–strain diagram shown in Figure 8.40 highlights three

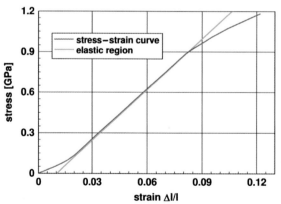

Figure 8.40 Stress–strain diagram of a strand consisting of single- and double-walled nanotubes. The specimen was 5 mm long, with a diameter of 5 μm. The first region is characterized by a reduced value stress/strain ratio; this may be caused by settling phenomena in the interaction with the tensile apparatus. The *Young*'s modulus should be calculated from the second, linear range, indicated in green color [30].

distinct regions. The first region is characterized by a relatively small value of stress/strain, which may be caused by settling phenomena in the interaction with the tensile apparatus. The following linear range, denoted by a green straight line, may be used to determine *Young*'s modulus. Finally, at about 0.9 GPa, a decrease in the slope of the stress–strain curve is observed; this may be caused by nonlinear elastic behavior of the nanotubes or, more probably, by slippage between the aligned nanotubes, and the fracture of a small number. The details of this mechanism are supported by scanning electron microscopy images.

Even higher strength and huge strains were obtained by Dalton *et al.* [31] with single-walled carbon nanotube composites bound together by an interphase region consisting of polyvinyl alcohol forming a coating on the nanotubes. The stress–strain diagram of such a composite specimen, produced in lengths up to 100 m and with a diameter of 50 μm, is shown in Figure 8.41. In this figure, for comparison, the authors included a stress–strain diagram for spider silk, the fiber known to have the greatest strength among natural products.

The data in Figure 8.41 show that this SWNT nanocomposite, which consists of 60 wt.% nanotubes, has an ultimate strength of 1.8 GPa, the *Young*'s modulus is 80 GPa, and the yield stress is approximately 0.7 GPa. Of particular interest is the huge plastic deformation which occurs after the yield stress is reached; this deformation, which resembles superplasticity, is possible because these fibers do not develop any necking. It may be assumed that slippage between the individual nanotubes within the fiber might contribute to this large plastic deformation. The mechanism involved may be essentially the same as for plastic deformation, as assumed for the material depicted in Figures 8.39 and 8.40. The ultimate stress of such a composite is two- or threefold that of an average steel, and lies within the range of values found in the very best quality steel wires.

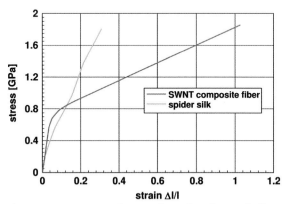

Figure 8.41 Stress–strain diagram of a single-wall nanotube fiber, bound together with polyvinyl alcohol at the surface of the carbon nanotubes. The diameter of the tensile specimen was 50 μm; the nanotubes content was 60 wt.%. The stress–strain diagram of spider silk is provided for comparison [31].

References

1 Karch, J., Birringer, R. and Gleiter, H. (1987) *Nature*, **330**, 556.

2 Siegel, R. and Fougere, G.E. (1995) *NanoStruct. Mater.*, **6**, 205–2016.

3 MacKenzie, J.K. (1950) *Proc. Phys. Soc.*, **63**, 2–11.

4 Sanders, P.G., Eastman, J.A. and Weertmann, J.R. (1997) *Acta Mater.*, **45**, 4019–4025.

5 Youngdahl, J., Sanders, P.G. and Eastman, J.R. (1997) *Scr. Mater.*, **37**, 809–813.

6 Nieman, G.W., Weertman, R.W. and Siegel, R.W. (1990) *Scr. Metall. Mater.*, **24**, 145.

7 Meyers, M.A., Mishra, A. and Benson, D.J. (2006) *Prog. Mater. Sci.*, **51**, 427–556.

8 Dash, W. (1957) in: *Dislocations and Mechanical Properties of Crystals* (ed. J. Fisher), J. Wiley, New York, p. 57.

9 Pande, C.S., Masumura, R.A. and Armstrong, R.W. (1993) *NanoStruct. Mater.*, **2**, 323–331.

10 Chang, H., Altstetter, C.J. and Averback, R.S. (1992) *J. Mater. Res.*, **7**, 2962–2979.

11 Chokshi, A.H., Rosen, A., Karch, J. and Gleiter, H. (1989) *Scr. Mater.*, **23**, 1679–1684.

12 Asby, M.F. and Verall, R.A. (1973) *Acta Mater.*, **21**, 149.

13 Chang, H., Altstetter, C.J. and Averback, R.S. (1992) *J. Mater. Res.*, **7**, 2962–2970.

14 Kadau, K., Germann, T.C., Lomdahl, P.S., Holian, B.L., Kadau, D. and Entel, P. (2004) *Met. Mater. Trans.*, **35A**, 2719.

15 Kim, H.S., Estrin, Y. and Bush, M.B. (2000) *Acta Mater.*, **48**, 493–504.

16 Jia, D., Wang, Y.M., Ramesh, K.T., Ma, E., Zhu, Y.T. and Valiev, R.Z. (2001) *Appl. Phys. Lett.*, **79**, 611–613.

17 McFadden, S.X., Valiev, R.Z. and Mukherjee, A.K. (2001) *Mater. Sci. Eng.*, **A319–321**, 849–853.

18 Mishra, R.S., Stolyarov, V.V., Echer, C., Valiev, R.Z. and Mukherjee, A.K. (2001) *Mater. Sci. Eng.*, **A298**, 44–50.

19 Betz, U., Padmanabhan, K.A. and Hahn, H. (2001) *J. Mater. Sci.*, **36**, 5811–5821.

20 Gao, H., Ji, B., Jäger, I.L., Arzt, E. and Fratzl, P. (2003) *Proc. Natl. Acad. Sci. USA*, **100**, 5597–5600.

21 Jordan, J., Jacob, K.I., Tannenbaum, R., Sharaf, M.A. and Jasiuk, I. (2005) *Mater. Sci. Eng. A*, **393**, 1–11.

22 Reynaud, E., Jouen, T., Gautheir, C. and Vigier, G. (2001) *Polymer*, **42**, 8759–8768.

23 Ash, B.J., Stone, J., Rogers, D.F., Schadler, L.S., Siegel, R.W., Benicewicz, B.C. and Apple, T. (2000) *Mater. Res. Soc. Symp. Proc.*, **661**.

24 Gilman, J.W., Jackson, C.L., Morgan, A.B., Harris, R., Jr. Manias, E., Giannelis, E.P., Wuthenow, M., Hilton, D. and Phillips, S.H. (2000) *Chem. Mater.*, **12**, 1866–1873.

25 Fu, X. and Qutubuddin, S. (2001) *Polymer*, **42**, 807–813.

26 Gloaguen, J.M. and Lefebvre, J.M. (2001) *Polymer*, **42**, 5841–5847.

27 Alexandre, M. and Dubois, P. (2000) *Mater. Sci. Eng.*, **28**, 1–63.

28 Yu, M.-F., Files, B.S., Arepalle, S. and Ruoff, R.S. (2000) *Phys. Rev. Lett.*, **84**, 5552–5555.

29 Chang, T.E., Jensen, L.R., Kisliuk, A., Pipes, R.B., Pyrz, R. and Sokolov, A.P. (2005) *Polymer*, **46**, 439–444.

30 Li, Y., Wang, K., Wei, J., Gu, Z., Wang, Z., Luo, J. and Wu, D. (2005) *Carbon*, **43**, 31–35.

31 Dalton, A.B., Collins, S., Muñoz, E., Razal, J.M., Ebron, V.H., Ferraris, J.P., Coleman, J.N., Kim, B.G. and Baughman, R.H. (2003) *Nature*, **423**, 703.

9
Nanofluids

9.1
Definition

Nanofluids are stable suspensions of nanoparticles in a liquid. In order to avoid coagulation of the particles, the particles must be coated with a second distance-holder phase which, in most cases, consists of surfactants that are stable in the liquid. The distance-holder must also overcome the tendency to form *van der Waals* bond clusters. Typically, nanofluids contain up to 10 vol.% of nanoparticles, and usually more than 10 vol.% of surfactant. Either oil or water is used as a carrier liquid, and the suspensions are designed in such a way that *Brownian* molecular movement thwarts the sedimentation of the particles.

9.1.1
Nanofluids for Improved Heat Transfer

The high heat capacity of nanoparticles, coupled with the possibility of producing stable suspensions, has many technical applications. Notably, such stable suspensions may show unprecedented combinations of two or more of the properties or features that are required in thermal systems, namely a high heat capacity, a good thermal conductivity, and good compatibility with technical systems. One typical application of nanofluids containing nanoparticles is as a coolant, since the addition of only a few volume percent of nanoparticles to a liquid coolant can significantly improve its thermal conductivity, yet have no negative influence on its heat capacity. In fact, the high heat capacity of nanoparticles can actually improve the heat capacity of a coolant. Figure 9.1 shows, graphically, the heat capacity ratio of a suspension consisting of ethylene glycol (which is widely used as a coolant, or may be added to aqueous cooling liquids) as the liquid and copper dispersed as the nanoparticles. The heat capacity ratio is defined as the heat capacity of the suspension compared to that of the base fluid without any additions. Consequently, the pure base liquid has a heat capacity ratio of 1. The data in Figure 9.1 show a slight improvement in the heat capacity ratio, such that the increase in heat capacity is greater than the volume content.

Nanomaterials: An Introduction to Synthesis, Properties and Application. Dieter Vollath
Copyright © 2008 WILEY-VCH Verlag GmbH & Co. KGaA, Weinheim
ISBN: 978-3-527-31531-4

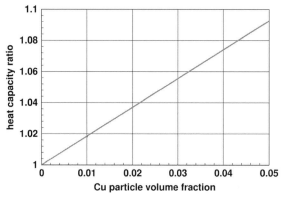

Figure 9.1 Heat capacity ratio of a nanofluid consisting of ethylene glycol and copper. The ratio is defined as the heat capacity of the nanofluid over that of pure ethylene glycol.

However, when considering the thermal conductivity a significant improvement is obtained. The related results from studies conducted by Keblinski *et al.* [1] and by Eastman *et al.* [2] are shown in Figure 9.2, where the ratio of the thermal conductivity is plotted versus the volume content of nanopowders. As in the above case, the thermal conductivity ratio is defined as the ratio of the thermal conductivity of the suspension over that of the base fluid. As shown by Masuda *et al.* [3], this improvement is not only limited to metallic nanoparticles but also is observed with ceramic nanoparticles (in this example, alumina).

The thermal conductivity of nanofluids depends heavily on the amount of nanoparticles dispersed in the liquid. Kwak and Kim [4] determined an improvement in thermal conductivity by adding CuO to ethylene glycol (see Figure 9.3). Here, the

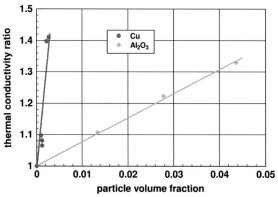

Figure 9.2 Thermal conductivity ratios of nanofluids consisting of ethylene glycol and copper or alumina, respectively. The ratio is defined as the thermal conductivity of the nanofluid over that of pure ethylene glycol. (Experimental data for copper-containing nanofluids from Keblinski *et al.* [1] and Eastman *et al.* [2]; data for alumina-containing nanofluids from Masuda *et al.* [3].)

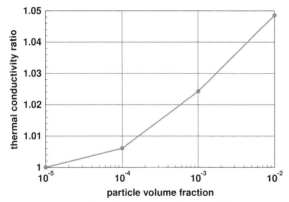

Figure 9.3 Thermal conductivity ratio of a nanofluid consisting of CuO in ethylene glycol as a function of the particle volume fraction (according to Kwak and Kim [4]).

improvement in thermal conductivity expressed as the thermal conductivity ratio, is plotted against the volume content of nanoparticles in the liquid. It is clear that, at least in a volume fraction range of up to 0.01, the thermal conductivity increases significantly, though less steeply than in the example shown in Figure 9.2.

When considering the technical applications of these fluids, it is not only the heat capacity and thermal conductivity but also the rheological parameters that are of vital importance. As an example, the dynamic viscosity η of nanofluids consisting of CuO nanoparticles in ethylene glycol is plotted as a function of the volume fraction c of nanoparticles [4] in Figure 9.4. The dispersed particles up to a volume fraction of approximately 0.001 have almost no influence on viscosity, but above this volume

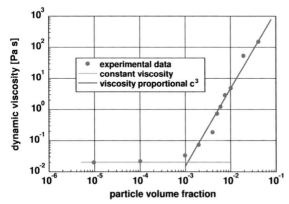

Figure 9.4 Dynamic viscosity of an ethylene glycol/CuO nanofluid as a function of the volume content of nanoparticles [4]. Remarkably, below a particle volume fraction of 10^{-3}, the viscosity is unchanged compared to pure ethylene glycol; however, above that value the viscosity increases with the third power of the particle content.

fraction the viscosity changes abruptly and increases following the proportionality $\eta \propto c^3$.

Such highly efficient coolants, as are obtained by the use of nanomaterials, may have a broad range of applications, especially in situations where the cooling channels are extremely narrow, such as in microtechnological applications. For applications in automobiles, the high thermal conductivity allows a significant reduction in the size of the cooling system. However, the inadequate long-term stability against sedimentation of these materials has, to date, impeded any broader technical applications.

9.2
Ferrofluids

9.2.1
General Considerations

One fascinating application of superparamagnetic particles – and one which already has been widely applied in technical products – is that of *ferrofluids*, which are a special type of nanofluids. A ferrofluid is a stable suspension of superparamagnetic particles in a liquid. In order to avoid magnetic coagulation of the particles, they must be coated with a second, distance-holder phase. Ferrofluid contains between 3 and 8 vol.% of magnetic nanoparticles, and usually more than 10 vol.% of the surfactant. Normally, oil or water is used as the carrier liquid. *Brownian* molecular movements thwart the sedimentation of the particles in the absence of any external magnetic field, while in the presence of a magnetic field such movements prevent demixing of the suspension. Magnetic sedimentation can be avoided if the magnetic moment of the particles is not too large. Characteristically, in the absence of an external magnetic field the net magnetic moment of a ferrofluid is nil. However, the particles adjust within a few milliseconds in the direction of an external magnetic field, and this leads to a net magnetic moment. As is typical for superparamagnetic systems, following removal of the magnetic field the magnetic moments of the particles randomize almost immediately, leading again to a net magnetic moment of nil. In a magnetic field gradient, the whole fluid moves to the region of highest flux, and consequently an external magnetic field can be used for the precise positioning and control of ferrofluids. This also allows the design of actuators which are based on ferrofluids. It is essential, however, that the ferrofluids are stable against the sedimentation of magnetic nanoparticles, even in strong magnetic fields.

In his book, *Ferrohydrodynamics*, Rosenzweig [5] has provided a complete theoretical basis for ferrofluids, and also explained the series of instabilities which bears his name. The most famous of these shows the surface of a ferrofluid in an inhomogeneous magnetic field forming spikes, rather than a flat or cambered surface (see Figure 9.5). The spikes which follow the gradient of the magnetic field are the consequence of an interaction of surface energy, and gravitational and magnetic energies. These are formed above a critical magnetic field, where the reduction in the

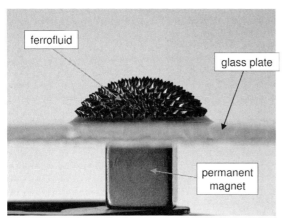

Figure 9.5 The appearance of a ferrofluid in an inhomogeneous magnetic field. The corrugated surface is the result of an interaction between surface, gravitational, and magnetic energies. This phenomenon is termed *Rosenzweig* instability [6] (wikipedia. com).

magnetic field energy is greater than the increase in surface and gravitational energies. The forces which act at magnetic fluids are proportional to the gradient of the magnetic field and the magnetization of the fluid. Therefore, the retention force of a ferrofluid may be adjusted by changing either the magnetization of the fluid or the external magnetic field.

9.2.2
Properties of Ferrofluids

Ferrofluids, as magnetic materials, may be used to transfer magnetic fields or to close magnetic circuits in a simple way, without the need for any complicated and shaped parts. Besides these possibilities, the variation of viscosity as a function of an external magnetic field represents one of the most striking properties of a ferrofluid. In Figure 9.6, the results of Patel *et al.* [7] are depicted, where the reduced viscosity of a ferrofluid consisting of Fe_3O_4 nanoparticles in kerosene is plotted as a function of the reduced magnetic field. In this context, the temperature-independent reduced magnetic field is defined as $\alpha = mH/kT$. The reduced viscosity is, in that case, defined as the viscosity at the value of the reduced magnetic field α divided by the viscosity at $\alpha = \infty$. The mean value of the log-normal-distributed particle sizes was approximately 13 nm. In order to estimate the actual field (in Tesla) at room temperature in Figure 9.6, the numbers shown at the abscissa must be divided approximately by 100. In this way, a clear characteristic of ferrofluids – that an extreme increase in viscosity occurs with relatively small magnetic fields – becomes apparent.

As shown in Figure 9.6, a magnetic field, when held constant over time, leads to an increase in the ferrofluid viscosity. However, the situation is more complex in an alternating magnetic field, where one observes a positive contribution at low

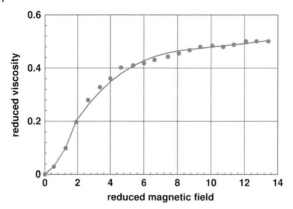

Figure 9.6 Reduced viscosity of a ferrofluid consisting of 13-nm Fe_3O_4 nanoparticles in kerosene (according to Patel *et al.* [7]). The temperature-independent reduced magnetic field is defined as mH/kT. In this case, at room temperature, the magnetic field (in Tesla) is approximated by dividing the reduced values given at the abscissa by 100. The reduced viscosity is obtained by dividing the experimentally measured viscosity values by that at an infinite reduced magnetic field.

frequencies of the magnetic field, and negative contribution at high frequencies. This is due to rotatory oscillations of the particles, caused by the alternating magnetic field. As there is no preference for any direction of rotation, in a first approximation, half of the particles rotate clockwise and the other half counter-clockwise. Therefore, from a macroscopic viewpoint, the angular velocity of the particles equals zero. However, any vortex results in a nonzero angular velocity of the particles, which in turn leads to a decrease in the effective viscosity; this is seen as a negative contribution to the viscosity.

The dependency of a ferrofluid's viscosity on the strength of an external magnetic field leads to a very interesting application, as an adjustable and "intelligent" shock absorber fluid. As ferrofluids can rapidly adopt (within milliseconds) the damping characteristics of shock absorbers, they can be used to replace older systems based on piezoelectric elements; typical applications are in high-performance CD and DVD player systems. A further advantage is that the dynamic control of the damping characteristics of shock absorbers (e.g., in cars) allows powers of up to kilowatt range to be controlled with an electrical power of only a few watts.

9.2.3
Applications of Ferrofluids

One of the first applications of ferrofluids in engineering was as a means for sealing off feed-throughs (see, for example, http://www.ferrotec.com.sg/category.asp? catid=9 or http://www.vacuum-guide.com/vacuum_components/vacuum_feed-through/mechanical_feedthrough_america.htm.) The general design of such a

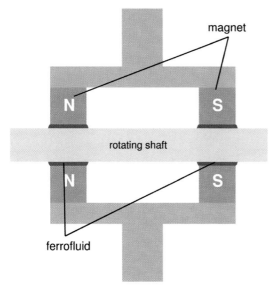

Figure 9.7 Design of a feed-through of a rotating shaft sealed by ferrofluids. The ferrofluids are kept in position by permanent magnets.

system is shown in Figure 9.7, in which this application, as most other successful uses, exploits the increase of viscosity within a magnetic field.

One of the commercially most successful applications of ferrofluids in consumer products is in loudspeakers, where the ferrofluid plays three important roles: (i) it centers the voice coil within the magnet; (ii) it acts as a coolant for the voice coil, by removing the heat caused by *Ohmic* losses; and (iii) it acts as damping medium. The design of a loudspeaker using a ferrofluid is shown schematically in Figure 9.8.

One other possible future application of nanofluids has been reported by Krauß *et al.* [8], who demonstrated the pumping of ferrofluids in a channel by an alternating current (AC) magnetic field. The magnetic particles in the fluid were cobalt-based.

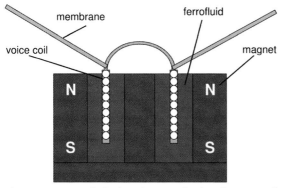

Figure 9.8 Design of a high-performance loudspeaker using a ferrofluid.

Figure 9.9 A magnetic pumping system for ferrofluids (according to Krauss *et al.* [8]). Coils 1 and 2 are connected to the same AC source. The electric currents show a phase difference of 90°, which leads to a rotating magnetic field at the location of the circular channel. (Reproduced with permission from [8], Copyright: American Institute of Physics.)

The set-up, which consists of a circular channel and two coils is shown in Figure 9.9. Here, coil 1 produces a vertical magnetic field, while coil 2 is wrapped around the channel and produces a magnetic field in azimuthal direction. Both coils are driven with an alternating current from the same source, though with a phase difference of 90°. The two coils together generate a rotating magnetic field, and changing the phase difference from $+90°$ to $-90°$ reverses the flow direction in the channel. The nanofluids used by the authors had a viscosity of 5.5×10^{-3} Pa s.

The velocity of the ferrofluid in the circular channel (see Figure 9.9), as driven by the AC magnetic fields, is shown in Figure 9.10. As might be expected intuitively, the velocity increases with increasing amplitude of the magnetic AC field. The data in this experiment were determined at a frequency of 1 kHz.

In this system, it is not only the amplitude but also the frequency of the field which has a significant influence. The velocity of the ferrofluid as a function of frequency at

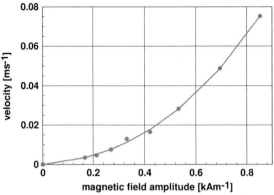

Figure 9.10 Velocity of the ferrofluid in the circular channel of Figure 9.9, as a function of the amplitude of the applied AC magnetic field at a fixed frequency of 1 kHz [8].

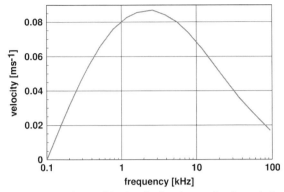

Figure 9.11 Velocity of the ferrofluid in the circular channel of Figure 9.9 as a function of the frequency of the applied AC magnetic field, and applying a field of fixed amplitude [8].

an amplitude of the AC magnetic field of $0.95\,\mathrm{kA\,m^{-1}}$ is shown in Figure 9.11; here, the maximum frequency was approximately 2.5 kHz is found. Beyond the maximum, the ferrofluid is increasingly unable to follow the frequency of the AC magnetic field. The data shown graphically in Figure 9.11 were calculated using a generalized function for the velocity as a function of the magnetic field amplitude and frequency as reported by the authors.

One other interesting phenomenon (among many) that might be used for sensing vibrations was reported by Kubasov [9], who showed that a vibrating ferrofluid in a static magnetic field could induce an electric voltage in a sensing coil. A simplified diagram of the set-up used is shown in Figure 9.12. In this system, 1.1 vol.% magnetite particles (each of diameter ca. 16 nm) suspended in kerosene was used as the ferrofluid. As compared to γ-Fe_2O_3, Fe_3O_4 has a significantly higher energy of anisotropy, and the particles were not superparamagnetic. An estimation of the relaxation times, assuming *Néel's* or *Brownian* superparamagnetism, resulted in $\tau_{\mathrm{Neel}} = 2.8 \times 10^{-4}\,\mathrm{s}$ and $\tau_{\mathrm{Brown}} = 3.4 \times 10^{-6}\,\mathrm{s}$. As $\tau_{\mathrm{Neel}} \gg \tau_{\mathrm{Brow}}$, the particles relax the

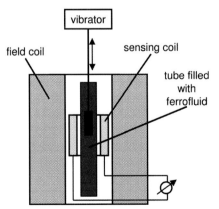

Figure 9.12 Design of a vibration sensor based on nanofluids in a magnetic field [9].

Brownian type (see Chapter 5) and as a whole follow the movement of the ferrofluid; this in turn causes a voltage to be induced in the sensing coil. The signal determined in the sensing coil consists of two harmonics: the frequency of the vibrator; and the second harmonics.

Further broad applications of ferrofluids are in the visualization of magnetic structures and domains. This is used extensively for the quality control of all types of magnetic storage devices, such as magnetic tapes, floppy disks, or magneto-optical disks. In materials sciences, the studies of magnetic domains in alloys, garnets, and minerals and the identification of small defects in steel and weldings are typical fields of application. In such cases, small external magnetic fields are often used to enhance the contrast.

Within this context, many economically extremely interesting applications may be identified. For example, in medical diagnostics ferrofluids are applied to increase the contrast of nuclear magnetic resonance (NMR) imaging. In NMR, the concentration of hydrogen is measured by monitoring the spins of protons; hence, in an NMR instrument a constant high magnetic field is superimposed by a small, high-frequency magnetic field. The spins of the hydrogen nuclei (the protons) are oriented either parallel or antiparallel to the magnetic field and, under the influence of the superimposed high-frequency field, they may "flip" into the other direction. In an NMR system, the resonance frequency of this flipping process is measured, and found to be directly proportional to the external magnetic field. Local variations of the amount of ferrofluid or the particle concentration cause changes in the magnetic field and, therefore, in the resonance frequency of the protons. This leads to variations of contrast which are more pronounced than those based on the concentration of protons alone. A typical example of the image obtained (in this case of a Novikoff hepatoma in rat liver) is shown in Figure 9.13 [10]. For comparison, images with and without ferrofluid addition are shown to highlight the striking effect of ferrofluids on contrast enhancement. In a more advanced version, the magnetic γ-Fe_2O_3 particles

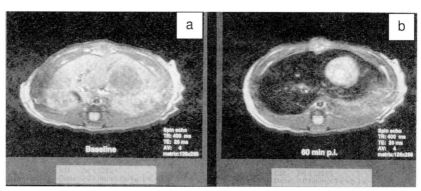

Figure 9.13 NMR tomography of Novikoff hepatoma in a rat liver [10]. (a) Without ferrofluid contrast enhancement. (b) With addition of γ-Fe_2O_3 nanoparticles as contrast enhancement. (Reprinted with permission from Deutsches Apothekerverlag [10].)

may be functionalized by using proteins that are characteristic for a certain organ or tumor, thus providing highly tissue-specific diagnoses.

References

1 Keblinski, P., Phillpot, S.R., Choi, S.U.S. and Eastman, J.A. (2002) *Int. J. Heat Mass Transfer*, **45**, 855–863.

2 Eastman, J.A., Choi, S.U.S., Li, S., Yu, W. and Thomson, L.J. (2001) *Appl. Phys. Lett.*, **78**, 718.

3 Masuda, H., Ebata, A., Teramae, K. and Hishinuma, N. (1993) *Netsu Bussei*, **4**, 227–233.

4 Kwak, K. and Kim, C. (2005) *Kor. Aust. Rheol. J.*, **17**, 35.

5 Rosenzweig, R.E. (1985) *Ferrohydrodynamics*, Cambridge University Press, Cambridge.

6 Maxwell, G.F. (2006) http://upload.wikimedia.org/wikipedia/commons/7/7c/, accessed July 2007. Ferrofluid_Magnet_under_glass.jpg.

7 Patel, R., Upadhyay, R.V. and Metha, R.V. (2003) *J. Colloid Interface Sci.*, **263**, 661.

8 Krauß, R., Reimann, B., Richter, R., Rehberg, I. and Liu, M. (2005) *Appl. Phys. Lett.*, **86**, 024102.

9 Kubasov, A.A. (1997) *J. Magn. Magn. Mater.*, **173**, 15.

10 Kresse, M., Pfefferer, D. and Lawaczeck, R. (1994) *Dt. Apotheker Zeitg.*, **134**, 3079–3089.

10
Nanotubes, Nanorods, and Nanoplates

10.1
Introduction

Nanotubes, nanorods, and nanoplates are frequently observed. Whilst nanotubes and nanorods are often referred to as one-dimensional nanoparticles, nanoparticles and fullerenes, in contrast, are generally denominated as zero-dimensional structures. Consequently, nanoplates could be considered as two-dimensional nanoparticles. Although, nanorods and nanoplates are often found as more or less spherical or facetted particles, their one- or two-dimensionality is clearly visible. Notably, as very few routes of synthesis are available for the preferential delivery of aggregates that are not zero-dimensional, interest has centered on these specially shaped nanoparticles and continues to be promoted by the wide range of interesting physical properties associated with these structures.

A typical example of nanorods (in this case, ZnO) is shown in Figure 10.1. These rods are over 5 μm long (most are about 15 μm long), with diameters ranging from 120 to 140 nm, and are clearly separated. (According to the definitions of nanomaterials, rods with linear dimensions over 100 nm are, strictly speaking, no longer nanomaterials. However, the perfection of this micrograph guaranteed its selection as an example.) The most important point is that a bulge is visible on one end of most particles, this being typical of the synthesis process via the gas phase.

ZnO nanoparticles and nanorods are of special interest because of their excellent luminescence properties in ultraviolet (UV) light. The intensity of the UV emission line that is found in the wavelength range from 380 to 390 nm of the nanorods as a function of the intensity of the excitatory light as shown in Figure 10.1 can be clearly seen in Figure 10.2.

The graph in which luminescence intensity is plotted against pumping power (= intensity of the excitatory light) (Figure 10.2) is of special interest, since above a pumping power of 600 kW cm^{-2} a stimulated emission is observed rather than only luminescence, as the ZnO nanorods act as lasers. In fact, this is one reason why nanorods are of special importance. In this example, the excitation power for conventional luminescence ranges from ca. 200 to 600 kW cm^{-2}, and this is demonstrated graphically in the insert of Figure 10.2.

Nanomaterials: An Introduction to Synthesis, Properties and Application. Dieter Vollath
Copyright © 2008 WILEY-VCH Verlag GmbH & Co. KGaA, Weinheim
ISBN: 978-3-527-31531-4

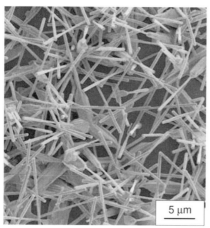

Figure 10.1 Secondary electron micrograph of ZnO nanorods [1]. These nanorods have lengths of about 15 μm, and diameters ranging from 120 to 140 nm. At one end, most of the nanorods show a bulge, which is typical of synthesis via a gas-phase route. (Reproduced with permission from [1], Copyright: Russian Academy of Sciences 2006.)

An example of nanoplates (in this case gold nanoplates) is shown in Figure 10.3a [2]. The size of these platelets is approximately 400 nm in the plane, and their thickness ranges from 25 to 60 nm. As can be seen from the hexagonal shape of the platelets, the nanoplates have a single orientation with the [111] direction perpendicular to the plane. This is also clearly visible in the electron diffraction

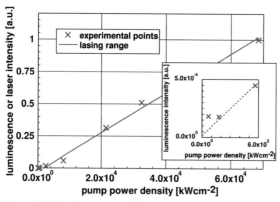

Figure 10.2 Luminescence and lasing intensity of the ZnO nanorods depicted in Figure 10.1 [1]. At low intensities of the exciting light, the nanorods show luminescence. At pumping powers above 600 kW cm^{-2}, an onset of laser action is observed. The range of conventional luminescence starts at ca. 200 kW cm^{-2}. (The range of conventional emission is depicted in the inset.)

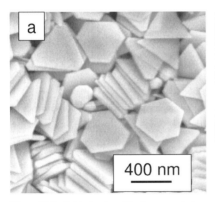
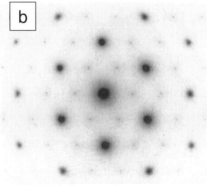

Figure 10.3 Gold platelets. The hexagonal shape is obtained by adding polyvinylpyrrolidone to the solution from which the platelets are precipitated [2]. (a) Electron micrograph of the gold platelets; these are about 400 nm wide, with thickness ranging from 25 to 60 nm. (b) An electron diffraction pattern of a gold platelet as shown in Figure 10.3a. The hexagonal symmetry of the diffraction pattern indicates that the [111] direction of the platelets was perpendicular to the faces of the platelet; in other words, the electron beam was exactly parallel to the [111] direction. (Reproduced with permission from [2], Copyright: American Institute of Physics 2005.)

pattern (Figure 10.3b), which shows the (111) reflexes as the most intensive. This example is of special interest because gold crystallizes in a cubic structure, and therefore cubes rather than platelets would be expected. However, the synthesis process applied in this case was in aqueous solution, and the size and shape of the gold platelets were controlled by the addition of polyvinylpyrrolidone in different quantities. The application of other organic agents may also lead to the formation of nanorods.

The nanoplates shown in Figure 10.3a are almost atomic flat. The root-mean-square roughness of the gold nanoplates, when measured with an atomic force microscope, is approximately 0.24 nm (this should be compared with the diameter of a gold atom, which is 0.29 nm). Plates, as described above, represent a precious commodity for nanotechnology, and they are applied for the manufacture of many small devices. The example in Figure 10.4 is of a nanogearwheel made from such gold platelets by using electron beam nanolithography.

A further example of nanoplates displaying a ceramic material is shown in Figure 10.5. Here, hexagonal platelets of $CuFe_2O_4$ [3] are crystallized in the cubic spinel structure. This copper ferrite is ferrimagnetic with a relatively low energy of anisotropy. As the magnetic properties of materials depend heavily on the anisotropy of the particles, the ability to produce magnetic particles in different shapes, especially for hard magnetic compounds, is essential. As in the case of gold platelets, the deviation from a cubic shape for the particles was achieved by using properly selected surfactants in the chemical process for synthesis. The change in particle shape from square (as expected for this cubic material) to hexagonal was achieved by adding sodium dodecylbenzenesulfonate and toluene to the solution from which the particles were precipitated. The addition of different amounts of polyethylene glycol (PEG) may change the shape from plate to rod.

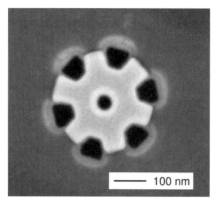

Figure 10.4 A nanometer-sized gearwheel made from a gold platelet as shown in Figure 10.3a [2]. This gearwheel has a diameter of 300 nm, and was produced using electron lithography. (Reproduced with permission from [2], Copyright: American Institute of Physics 2005.)

The hexagonal $CuFe_2O_4$ platelets have a lateral size close to 100 nm (details of the thickness were not provided). Apart from some contamination at the surface, the electron micrograph conveys the impression of extreme smoothness. This is not surprising, as any imperfection of the faces increases not only the surface area but also, therefore, the surface energy.

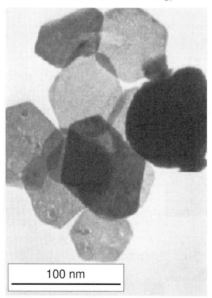

Figure 10.5 Electron micrograph of hexagonal $CuFe_2O_4$ platelets. Copper ferrite crystallizes in the cubic spinel structure; the hexagonal shape, deviating from the expected cubic form, is obtained by adding surfactants to the solution from which the ferrite is precipitated. (Reprinted with permission from [3], Copyright: Elsevier 2005.)

Initially, it may seem astonishing that nanotubes, nanorods, nanoplates and other particles with similar nonspherical shapes are thermodynamically stable. Even for facetted particles, intuitively, one expects the shape of the particles to be not too distant from that of a sphere. However, there are thermodynamically well-founded reasons for the existence of nanoparticles with shapes far from spherical. Besides the highly specific methods of synthesis that result in such nanostructures, three major reasons can be proposed for the existence of stable nanotubes and nanorods; these are discussed in the following paragraphs.

10.1.1
Conditions for the Formation of Rods and Plates

The first point for discussion is the influence of *surface energy*. For nonspherical nanostructures, this is especially important in the case of anisotropic (noncubic) structures. For reasons of simplicity, and without any loss of generality, tetragonal bodies with the sides a and c, and surface energies γ_a and γ_c, are assumed. The surface energy U_{surf} of such a prism is

$$U_{surf} = 4\gamma_a ac + 2\gamma_c a^2 \tag{10.1}$$

By assuming a constant volume V

$$V = a^2 c \Rightarrow c = \frac{V}{a^2} \Rightarrow U_{surf} = 4\gamma_a \frac{V}{a} + 2\gamma_c a^2$$

a minimum of surface energy is found by the condition

$$\frac{\partial U_{surf}}{\partial a} = -4\gamma_a \frac{V}{a^2} + 4\gamma_c a = -4\gamma_a c + 4\gamma_c a = 0$$

This leads to the important relationship

$$\frac{\gamma_a}{\gamma_c} = \frac{a}{c} \tag{10.2}$$

Equation (10.2) states that the ratio of the sides of a tetragonal prism is equal to the ratio of the surface energies. Lastly, this is the thermodynamic basis for the formation of nanorods or nanoplates. In the case of $\gamma_a = \gamma_c$, which is to be expected in cubic structures, one obtains $a = c$, a cube. The derivation is practically identical for hexagonal structure. However, it must be noted that, by attaching surface-active compounds, the surface energy of lattice planes may be modified in such a way as to influence the habitus of nanoparticles in a significant manner; this possibility is used widely in the synthesis of nonspherical nanoparticles.

The next question to be answered in this context is the geometry of the minimum surface energy of agglomerates. Again, using the example of a prism, the question to be asked is: Which of the configurations depicted in Figure 10.6 has the least surface energy?

The configuration according to Figure 10.6a has a surface energy of

$$U_a = 8\gamma_a ac + 2\gamma_c a^2 \tag{10.3}$$

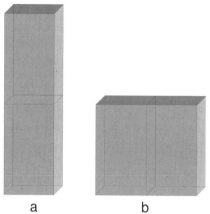

a   b

Figure 10.6 Two possibilities of combining two tetragonal prisms to a new form. (a) The prisms are connected at the end faces. (b) The prisms are connected at the lateral faces.

Analogously, for the arrangement depicted in Figure 10.6b:

$$U_b = 6\gamma_a ac + 4\gamma_c a^2 \tag{10.4}$$

By comparing Equations (10.3) and (10.4), one obtains for the stability condition of arrangement (a)

$$8\gamma_a ac + 2\gamma_c a^2 < 6\gamma_a ac + 4\gamma_c a^2 \Rightarrow \frac{\gamma_a}{\gamma_c} \frac{c}{a} < 1$$

or

$$\frac{\gamma_a}{\gamma_c} < \frac{a}{c} \tag{10.5a}$$

Equation (10.5a) is, lastly, equivalent to Equation (10.2). A system fulfilling Equation (10.5a) leads to the formation of rods. For the case drawn in Figure 10.6b, one obtains analogously:

$$\frac{\gamma_a}{\gamma_c} > \frac{a}{c} \tag{10.5b}$$

The agglomeration of particles of a compound fulfilling the condition in Equation (10.5b) will result in growing platelets.

Simply speaking, agglomerates of nanorods reduce their surface energy by increasing their aspect ratio and, in the case of nanoplates, the surface energy is reduced by decreasing the aspect ratio. In both cases, the character of being a rod or a plate will be enhanced.

The mechanisms described above are valid for "clean" surfaces only (these are surfaces that are not modified by contaminants or functionalization). As noted above, by correctly selecting surface-active molecules it is possible to grow rods or plates even from isotropic materials. In this context, it should be noted that even from gold, a cubic material, nanorods, and nanoplates are well known (see Figure 10.3).

10.1.2
Layered Structures

The second possibility of obtaining nanorods and nanotubes is related to layered structures, where the crystal structure is built from layers held together with *van der Waals* forces rather than by electrostatic attraction. The general arrangement of a particle crystallized in such a layered structure is shown schematically in Figure 10.7a, where the layers are independent. At the circumference of each layer, the bonds are not saturated (these "dangling bonds" are indicated in Figure 10.7b). In crystals of conventional size, the excess energy caused by these dangling bonds is negligible in comparison to the total energy. However, this is not the case for nanoparticles of layered compounds, where the contribution of the dangling bonds is significant, and therefore the system will attempt to saturate them. During synthesis, the most effective way of avoiding dangling bonds is simple curling of the sheets to form cylinders, or nanotubes.

Based on this explanation, it is clear that all compounds which crystallize in layered structures show a tendency to form nanotubes. Typical examples are BN, WS_2, MoS_2, WSe_2, $MoSe_2$ and, most importantly, carbon. The formation of nanotubes requires some time, but in situations where the time is insufficient the particles will seek other possibilities to saturate dangling points. One such approach is simply to join the ends of different particles, as shown in Figure 10.8. Here, three WS_2 particles consisting only of a small number of lattice planes are joined together to reduce the number of dangling bonds [4]. Clearly, other types of closing dangling bonds (e.g., fullerene-type particles) are also possible, and these will be discussed below.

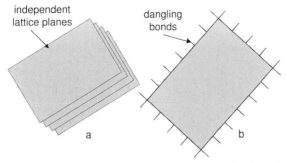

independent lattice planes

dangling bonds

a

b

Figure 10.7 The layout of a particle that crystallizes in a layered structure. (a) The particle set-up. (b) One layer of a particle as depicted in (a). The bindings at the circumference of the layer are not saturated. These dangling bonds (indicated by short lines) require additional energy; hence, there is a strong tendency for these dangling bonds to be saturated.

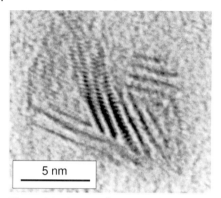

Figure 10.8 Three small WS$_2$ particles, each consisting of only a few lattice planes, bound together to saturate the dangling bonds at the circumference (Szabó, Vollath [4]).

10.1.3
One-Dimensional Crystals

The third possibility of obtaining nanotubes is to use compounds that crystallize in only one dimension. In theory, this is the most promising way to obtain long fibers, but unfortunately the importance of this route is negligible as the numbers of compounds coming into question is small. The most important class of one-dimensional compounds is in the class of silicates called *allophanes*. These are short-range, ordered aluminosilicates which follow the chemical formula Al$_2$O$_3$(SiO$_2$)$_x$(H$_2$O)$_y$, with $1.3 < x < 2$ and $2.5 < y < 3$. In most cases, allophanes crystallize in tubes with diameter ranging from 2 to 5 nm. To some extent, the aluminum in allophanes may be replaced by iron, magnesium, or manganese. These substitutions influence the diameter of the tube and the color of the material.

The most important compound in this context is *imogolite*, with an ideal composition of Al$_2$SiO$_3$(OH)$_4$. The ratio of silicon over aluminum is somewhat flexible, and can be used to adjust the tube diameter. Imogolite tubes with the ideal composition are very narrow, with internal diameters of 1 nm and external diameters of 2 nm.

The structure of imogolite (see Figure 10.9) is characterized by aluminum, silicon, oxygen, and (OH)$^-$ ions arranged in rings. This structure allows the addition of organic molecules (to "functionalize") at the surface.

Imogolite fibers synthesized using a wet-chemical process are shown in Figure 10.10 [6]. In this way, fiber bundles with different diameters can be created, and the extremely high aspect ratio is clearly apparent. The tubes may be up to a few micrometers in length, and both natural and synthesized imogolite tubes form bundles with diameters ranging from 5 to 30 nm. The surface area of imogolite has been determined experimentally as being in the range of $1000 \pm 100 \, \mathrm{m^2 \, g^{-1}}$. The *Mohs* hardness is quite low, ranging from 2 to 3. Although the geometry of the fibers suggests a possible use as a filler in composite with polymer matrix, the relatively poor strength of these fibers greatly limits the benefits of such composites.

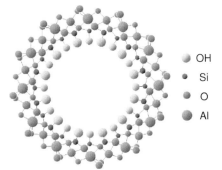

OH

Si

O

Al

Figure 10.9 The arrangement of ions in imogolite [5]. It is possible to attach organic compounds at the outer layer consisting of oxygen ions. The tube diameter can be adjusted by altering the aluminum : silicon ratio. (Reprinted with permission from [5], Copyright: The Korean Chemical Society 2006.)

It is possible to functionalize the surface of imogolite with organic molecules in order to add new properties to the material. A typical example, reported by Lee *et al.* [5], demonstrates the synthesis and functionalization of the surface with electrically conductive polypyrole. The effect of surface functionalization with polypyrole on electrical conductivity is demonstrated in Figure 10.11, which shows a clear increase from 1.21×10^{-7} S for the uncoated imogolite to 1.04×10^{-6} S for the polypyrole-coated fibers. Even when the resistance of such wires is relatively high, the increase in electrical conductivity by an order of magnitude promises many interesting technical applications. The huge current density of approximately 3×10^{10} A m^{-2} $(= 3 \times 10^4$ A mm$^{-2})$ is remarkable. Taking only the cross-section of the conducting polypyrole molecules into account, the current density is approximately one order of magnitude higher.

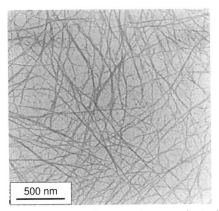

500 nm

Figure 10.10 Imogolite nanotubes as synthesized by Koenderink *et al.* [6]. The fibers form bundles with diameters ranging from 5 to 30 nm. The fibers may be up to a few micrometers in length. (Reprinted with permission from [6], Copyright: Elsevier, 2007.)

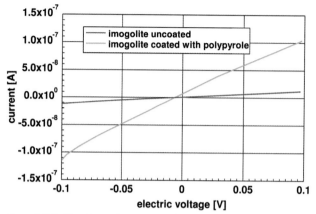

Figure 10.11 The electrical properties of imogolite coated with electrically conductive polypyrole. These data are plotted in comparison to the current/voltage dependency of the uncoated imogolite fibers [5]. Even when the current in the range of 10^{-7} A at 0.1 V is small, the current density is huge, due to the small diameter of the fibers (in the range of a few 10^{10} A m^{-2}).

When comparing the three possibilities of one- and two-dimensional nanostructures, those related to layered compounds are seen to be the most important.

10.2
Nanostructures Related to Compounds with Layered Structures

For small particles of compounds crystallizing in layered structures, a minimum free energy can be achieved by reducing the number of dangling bonds by forming tubes; nanotubes are observed especially with these types of compound. The most prominent representative of this class of compounds is graphite, although nanotubes consisting of boron nitride, the sulfides and selenides of molybdenum and tungsten, as well as many other compounds, have also been identified.

10.2.1
Carbon Nanotubes

In order to understand carbon nanotubes, it is essential first to discuss graphite and fullerenes as special modifications of carbon. The modifications of a substance differ in the ways in which the atoms are arranged and bond with each other, and so different modifications will have different physical and chemical properties. For example, graphite crystallizes in a layered hexagonal structure (see Figure 10.12) in which each carbon atom is bound covalently to its three neighbors. Therefore, only three of the four valences of the carbon atom are saturated. The fourth electron of the atoms remains unbound, and becomes delocalized across the hexagonal atomic

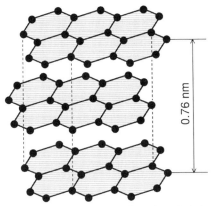

Figure 10.12 The structure of graphite. Each layer consists of interconnected hexagons with one carbon atom at each vertex. The bonding within the layers is covalent; in between the layers, the bondings are of the *van der Waals* type. At the circumference of each layer, the bondings are not saturated, as the number of neighbors is less than three.

sheets of carbon. As these electrons are mobile, graphite shows electrical conductivity within the layers, but perpendicularly to the layers graphite is an insulator. Within the layers are strong covalent bonds, whereas in between the layers are weak *van der Waals* bonds, and consequently it is possible to cleave pieces of monocrystalline graphite. These single layers of graphite are known as *graphene*, and because of its structure and bondings graphene is often denominated as an infinitely extended, two-dimensional aromatic compound. (The simplest aromatic compound is benzene; this consists of one hexagon of carbon atoms surrounded by six hydrogen atoms, each connected to one carbon atom.)

In the sense that graphene is a two-dimensional aromatic compound, fullerenes are three-dimensional aromatics. Instead of hexagons, fullerenes consist of a combination of hexagons and pentagons; Figure 10.13 depicts one pentagon surrounded by five hexagons. As such an arrangement leaves gaps between the hexagons, the closure of these (as indicated by an arrow in Figure 10.13) leads to the formation of a three-dimensional structure. This is the basic structural element of fullerenes.

By combining a larger number of these structures, spherical shapes are formed; the existence of such *polyeders* was first predicted by the mathematician *Euler*. The most common fullerene, and the first to be identified (by Kroto and Smalley, in 1996) [7], consists of 60 carbon atoms (this is written as C_{60}), with the molecular structure comprising 12 pentagons and 20 hexagons. Many other fullerenes exist where the number of vertices N (which, in the case of fullerenes is identical to the number of carbon atoms) present in polyhedrons consisting of hexagons and pentagons follows the simple formula: $N = 2i$, where i is an integer larger than 12. Therefore, although the smallest fullerene should consist of 24 carbon atoms, a

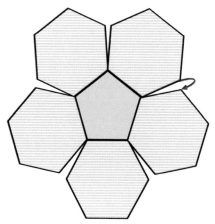

Figure 10.13 Five hexagons surrounding a pentagon. Closing the gaps between the hexagons leads to a three-dimensional structure, the basic element of fullerenes.

stable, smaller version comprising less than 60 carbon atoms has never been found. In addition to C_{60}, the most important other fullerenes are C_{70}, C_{76}, C_{78}, and C_{84}. The appearances of C_{60} and C_{70} are shown in Figure 10.14a and b. Clearly, C_{60} resembles a soccer ball, and therefore is often referred to as the "soccer ball molecule". The distribution of hexagons and pentagons can be clearly seen in both parts of Figure 10.14.

Even when fullerene molecules are quite stable, it is possible to attach metal atoms or other molecules at the surface, and this is of major importance in view of the applications of these molecules. Fullerenes also appear quite often in many layers; these aggregates are known as "nested fullerenes" or "onion molecules". The

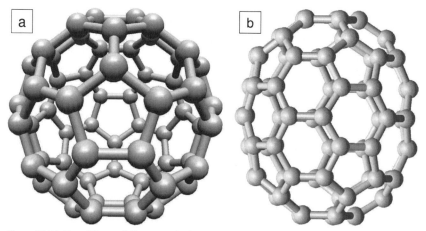

Figure 10.14 Two different fullerenes. The hexagons and pentagons, the constitutive elements of fullerenes, can be seen easily in both models [8]. (a) C_{60} fullerene; (b) C_{70} fullerene.

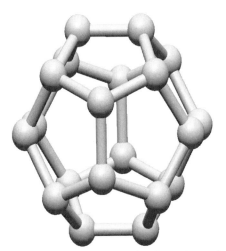

Figure 10.15 The C_{20} molecule. This is the smallest experimentally verified fullerene but, in contrast to the larger molecules, this is unstable [9]. (Reprinted with permission from Nature [9]) The geometry of this fullerene is also different; unlike larger molecules, which consist of hexagons and pentagons, C_{20} is composed only of pentagons [8].

smallest fullerene to be identified experimentally is C_{20}, which comprises only pentagons [9]. In contrast to the larger fullerenes, C_{20} is unstable; a schematic representation of the molecule is shown in Figure 10.15.

It may be easily conceived that single graphite layers (graphene) reduce the energy stored in the dangling bonds by forming tubes. There are, however, alternative possibilities for these planes to form coils, and this determines the properties of the carbon nanotubes. The structure of a graphene sheet is shown in Figure 10.16. Such a layer is described using a coordinate system with the unit vectors $\vec{e}_1$ and $\vec{e}_2$. The coordinates in this system are given, for some points, in Figure 10.16. A vector in this system describing a nanotube is termed the "chirality vector" $\vec{c} = n\vec{e}_1 + m\vec{e}_2$, where n and m are integers which describe the length of the coordinates in the directions $\vec{e}_1$ and $\vec{e}_2$. Furthermore, in order to describe carbon nanotubes, the convention $0 \le |m| \le n$ was adopted. The tube axis is perpendicular to the chirality vector. Two types of chirality vector provide the nanotubes with a very special arrangement of the carbon atoms; these are the "zig-zag line" $\vec{c} = (n, 0)$ and the armchair line $\vec{c} = (n, n)$. In both cases, n is an arbitrary integer.

Based on the chirality vector, it is possible to calculate the diameter d of a nanotube, which is given by

$$d = \frac{\sqrt{3}}{\pi} a_{C-C}(n^2 + m^2 + nm)^{0.5} = 0.0783(n^2 + m^2 + nm)^{0.5} \, [\text{nm}] \qquad (10.6)$$

where $a_{C-C} = 0.14$ nm is the distance between two neighboring carbon atoms. In reality, the range of diameters of carbon nanotubes is limited. Experimentally,

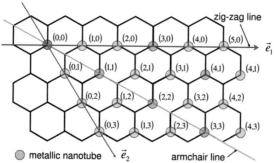

Figure 10.16 Geometry of a graphene sheet using a coordinate system based on the unit vectors $\vec{e}_1$ and $\vec{e}_2$. To illustrate the system of coordinates, the values of the coordinates are given for some of the vertices. Any vector in this system can serve as the chirality vector; two special forms – leading to the "armchair" and the "zig-zag" line – are indicated.

single-wall carbon nanotubes are observed with diameters ranging from 1.2 to 1.4 nm. However, nanotubes with significant larger diameters may also be produced. The chiral angle – the angle between the $\vec{e}_1$ axis and the chirality vector $\vec{c}$ – is given by

$$\delta = \arctan\left[\sqrt{3}\,\frac{m}{2n+m}\right] \tag{10.7}$$

The chiral angle for the armchair line is $30°$, and that for the zig-zag line is $0°$. Carbon nanotubes with a chirality vector fulfilling the condition $\frac{2n+m}{3} = q =$ integer show metallic electrical conductivity. In Figure 10.16, these vertices are indicated with red dots. The system of coordinates of a graphene sheet is re-displayed in Figure 10.17, where the chirality vector of the nanotube (4,1) and the tube axis of the corresponding nanotube are indicated. Since for this tube $q = 3$, this chirality vector describes a metallic nanotube.

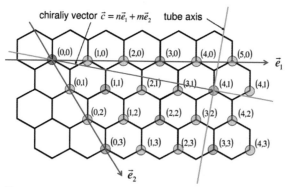

Figure 10.17 Graphene sheet showing the chirality vector (4,1) in red. Perpendicular to the chirality vector is the direction of the tube axis. A nanotube with a chirality vector (4,1) is metallic.

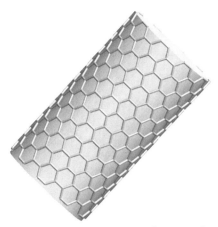

Figure 10.18 An armchair-type carbon nanotube with the chirality
vector (10,10). By using Equation (10.6), the diameter is calculated
as 1.35 nm [10].

After rolling the graphene sheet to form a tube, a (10, 10) nanotube with a diameter
of 1.35 nm has the appearance as depicted in Figure 10.18 [10]. This nanotube is of
the armchair type.

Nanotubes are closed with fullerene halves. An excellent electron micrograph
showing the caps at nanotubes is shown in Figure 10.19 [11], where the nanotube
consists of four walls. As the contrast at the caps, compared to the body of the
nanotubes, is significantly reduced, electron microscopy examinations of these end-
caps are very difficult to perform.

The formation of nanotubes is not limited to single graphene layers and, as for
fullerenes, both "multiwall" and "single wall" nanotubes may be observed. The
multiwall nanotubes consist of a series of coiled graphene layers, and can be depicted
perfectly using electron microscopy (see Figure 10.19). Previously, many excellent
electron micrographs of multiwall nanotubes comprising up to six layers have
been prepared. However, the actual structure can be demonstrated to better effect
in calculated examples. A typical calculated example consisting of four layers is
shown in Figure 10.20 where, for the individual nanotubes, the chirality vectors were
chosen as (7,0), (10,0), (13,0), and (16,0); this indicates that all four nanotubes are of

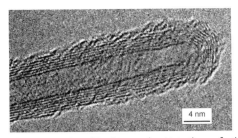

Figure 10.19 A multiwall nanotube. Note the perfectly depicted
end-caps (Ritschel, Leonhardt; reproduced with permission [11];
Copyright: IFW Dresden 2007).

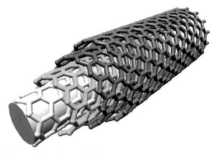

Figure 10.20 A calculated example of a multiwall carbon nanotube of the zig-zag type with the chirality vectors (7,0), (10,0), (13,0), and (16,0). The outer diameter of this multiwall nanotube is 1.25 nm (Weber [10]).

the zig-zag type. The outer diameter of this multiwall nanotube was 1.25 nm [10]. Most of the production processes involved deliver primarily multiwall nanotubes, with single wall nanotubes being the exception.

Carbon nanotubes have many fascinating applications, and those employing their electrical conductivity or high strength are also discussed in Chapters 7 and 8, respectively. In addition, a small diameter combined with an extreme stiffness means that carbon nanotubes are ideal materials for the tips used in scanning force or scanning tunnel microscopes.

Potentially, a major application of nanotubes is as electron emitters. Electron emission in an electrical field requires a sharp tip, and the sharper the tip the lower the electrical voltage required for electron emission. This is because the electrical field at the tip controls electron field emission. Although single-wall nanotubes have the sharpest tip occurring in nature, in reality multiwall nanotubes are used as they are more readily available. The emission current of carbon nanotubes as a function of the applied voltage is shown in Figure 10.21 [12]. Two types of nanotube were applied, namely closed and open. The open nanotubes were obtained by removing the end-caps in an oxidizing medium. The closed nanotubes start emission at 120 V, whereas the open tubes start significantly later, at 240 V. At 170 V, the emission current is 10^{-7} A. Assuming a diameter of ca. 1.5 nm, this leads to an electrical current density in the range of 5.7×10^{10} A m^{-2} ($= 5.7 \times 10^{4}$ A mm^{-2}). Compared to macroscopic metallic electrical conductors, this is an "astronomically" high current density, and the number would be even higher if the current density were related to the actual material-containing cross-section of the nanotube. When considering the voltages shown in Figure 10.21, it must be borne in mind that the actual voltage of the onset of the emission depends heavily on the geometry of the experiment. Therefore, the voltages shown in this figure must not be taken as standard values. However, despite these geometric influences, the voltages shown in Figure 10.21 are lower than those necessary for tungsten tips.

When considering technical applications in displays, it is necessary to analyze the stability of the emission as well as the basics of the emission itself. The field emission current of carbon nanotubes determined over a broader range of voltages is shown in

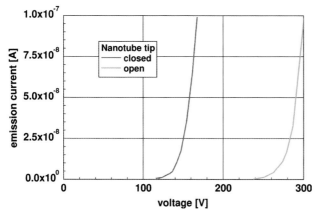

Figure 10.21 Electron field emission characteristics of carbon nanotubes. Here, the electron emission of closed and open nanotubes is compared [12].

Figure 10.22 [13]. For these experiments, bundles of nanotubes were embedded in an electrically nonconductive polymer matrix, such that the emission measured stemmed from more than one nanotube. In the range of lower voltages, where the emission increases exponentially with the applied voltage, the emission showed fluctuations of 50% and more. However, at higher voltages, characterized by a reduced dependency of the emission on the applied voltage, the emission was stable for hours – that is, over a time range which would be applicable for technical devices.

Electron field emission is of major economic importance in terms of field emission displays. Compared to tungsten tips, carbon nanotubes have the advantage of a

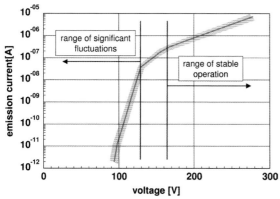

Figure 10.22 Field emission current of carbon nanotubes determined on bundles of carbon nanotubes embedded in a matrix of electrically nonconducting polymer [13]. Within the range of lower voltages, significant fluctuations of more than 50% are observed.

higher stability of the emission, and a better oxidation resistance. This is because, during field emission, the tungsten cathode tips become very hot, which in turn causes distortion of the tip geometry and, on occasion, even local melting. Both events lead not only to deformation but also to a reduction in the electrical field at the tip, which is required for emission. Additionally, in contrast to carbon nanotubes, the resistivity of metals increases with temperature, and this leads to higher *ohmic* losses – which results in even further increases in temperature. This feedback cycle may cause the emission tips to be destroyed. Since it is unavoidable that some residual oxygen will be present in the vacuum of a field emission device, the heated tungsten cathode oxidizes, and this results in a further-reduced field emission.

The set-up of a display using carbon nanotube as emitters is shown in Figure 10.23. The carbon nanotubes are grown at the surface of the cathode, which consists of addressable points, in most cases printed on an insulating carrier plate. Close above the carbon nanotubes, a grid is located that produces the electrical field at the tips of the carbon nanotubes and also accelerates the electrons. In most experimental devices, the distance between the nanotubes and the grid is less than 0.1 mm. The electrons fly to the anode, which is at the same electrical potential as the grid, and also patterned with luminescent material to produce light of the desired colors.

The cathode of a demonstration device is shown in Figure 10.24 [14]. Here, the patterned cathode and spots covered with carbon nanotubes are clearly visible. The cathode has different rows for the three basic colors of red, green, and blue. If finally developed, devices such as this may, in time, replace conventional television sets and computer monitors.

One further interesting application of carbon nanotubes exploits their electrical conductivity and large length-to-diameter ratios. Because of the huge aspect ratio, the amount of carbon nanotubes required to achieve electrical conductivity is very small (see Section 7.4), and it is possible to produce an optical transparent coating with relatively good electrical conductivity. Such coatings are necessary as contacts for organic light-emitting diodes (OLEDs) and organic solar cells. OLEDs may be used for any type of display, ranging from television sets to computer monitors. The electrical conductivity of such a carbon nanotube/polymer composite as a function of the applied voltage is shown in Figure 10.25 [15].

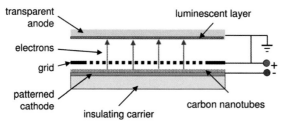

Figure 10.23 The general set-up of a display based on the field emission of carbon nanotubes. The electrons emitted by the carbon nanotubes are accelerated by the grid and move to the anode, which is covered with an electroluminescent layer.

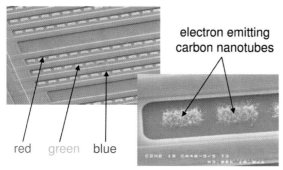

Figure 10.24 A first design for commercial cathode parts for a display system based on carbon nanotubes as electron field emitted (reproduced with permission from [14], Copyright: Samsung Advanced Institute of Technology, 2001).

In this example, the amount of carbon nanotubes (single-walled) added to the composite was 0.1 wt.%, and PFO [poly(2,7-9,9-(di(oxy-2,5,8-trioxadecane))fluorene)] was used as the polymer. In Figure 10.25, the conductivity of PFO, which is in the range of $10^{-13}\,S\,cm^{-1}$, is also plotted. The authors correlated the gradual increase in conductivity at about 2.5 V, by charge injection from the single-walled carbon nanotubes to the PFO polymer. The most important point, as shown in Figure 10.26, was the optical transmission of the composite. For comparison, the transmittance of a sputtered indium tin oxide (ITO) thin film is also provided (ITO is the standard material used in electro-optical devices).

When comparing the optical transmission of PFO with 0.1 wt.% single-wall nanotubes with that of ITO, except for the wavelength range below ca. 450 nm,

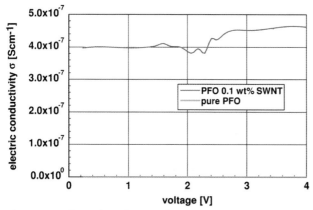

Figure 10.25 Electrical conductivity of an optically transparent PFO [poly(2,7-9,9-(di(oxy-2,5,8-trioxadecane))fluorene]–0.1 wt.% single-wall carbon nanotube composite. The increase in conductivity at about 2.5 V is explained by a charge injection from the single-wall carbon nanotubes to the PFO polymer [15].

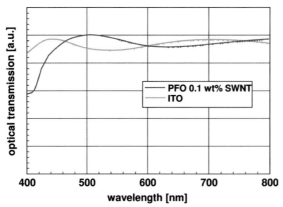

Figure 10.26 Optical transmission of an electrically conductive PFO [poly(2,7-9,9-(di(oxy-2,5,8-trioxadecane))fluorene]–0.1 wt.% single-wall carbon nanotube composite in comparison to an indium tin (ITO) thin film [15]. In contrast to ITO, the carbon nanotube composite may be applied onto flexible substrates.

there are no significant differences. The reduction in optical transmission in the blue regime of the optical spectrum is a property of the selected polymer. When compared with ITO, several advantages of the PFO composites become apparent:

- They are more easily fabricated than the ITO layers, which are mostly sputtered. PFO nanocomposite layers are produced by spin coating or printing, which is a significantly cheaper method of production.

- They are much more flexible than the brittle ceramic ITO layers, and so can be; hence, the PFO electrically conductive composites may also be applied to flexible substrates.

It is also possible to fill the interior of carbon nanotubes with metals or other compounds. As in the case of coated nanoparticles, this strategy allows two different properties to be combined within one particle. These filled carbon nanotubes may have many exciting applications. An electron micrograph of a carbon nanotube filled with CuI is shown in Figure 10.27, where the different layers of the multiwall nanotube can be clearly seen. The lattice of the filler, CuI, is also visible, as is the perfect filling of the nanotube. From this figure it is clear that the CuI filler in the nanotube is monocrystalline, as might be expected since the formation of a grain boundary requires additional energy. Filling a carbon nanotube with iron leads to a ferromagnetic part with a high shape anisotropy; hence, such aggregates show a large hysteresis in their magnetization curve.

Figure 10.28 represents a typical example of a magnetization curve of such a composite, where the remanent magnetization as a function of the external field is shown for an iron-filled carbon nanotube. This multiwall nanotube had a diameter of approximately 60 nm and a length of a few micrometers. The red lines in Figure 10.28 describe the magnetization during the change of the external field. The experiment was started at an external magnetic field of –1000 mT, and changed gradually up to

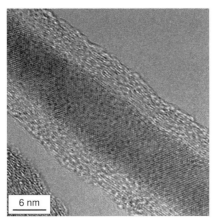

Figure 10.27 A multiwall carbon nanotube filled with CuI. The different layers of the wall and the lattice structure of the filling are clearly visible in this high-resolution electron micrograph (Hampel, Leonhardt; reproduced with permission [16a]; Copyright: IFW Dresden 2007.)

400 mT. In Figure 10.28, the arrows **1a** and **1b** follow this trace. After reaching the maximum, the magnetic field was again reduced to zero (indicated by arrow **2**). While increasing the magnetic field from −1000 to 400 mT, at approximately 250 mT the magnetization switches from a negative to a positive value. A red arrow indicates this switching point, **A**. If the experimental data drawn with red lines are extrapolated to negative magnetic fields, a field is found where the magnetization of the composite

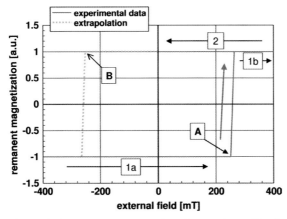

Figure 10.28 Remanent magnetization of an iron-filled carbon nanotube with a length of a few micrometers and 60 nm diameter. To obtain this data, the magnetic field was changed from −1000 to 400 mT. At an external field of approximately 250 mT, the magnetization of the iron filling changed its direction suddenly; this point is indicated as **A**. After reducing the external field, the changed direction is maintained. By extrapolation, the direction of magnetization changes back at −250 mT [16b].

may change its direction again. This point, obtained by extrapolation, is indicated as **B**. An iron-filled carbon nanotube acts as a magnetic switch with clearly defined switching points. As the magnetization loop shown in Figure 10.28 is rectangular, the switching occurs at exactly defined values of the external magnetic field. The energy product of this aggregate (remanent magnetization × coercitivity) is quite large.

10.2.2
Nanotubes and Nanorods from Materials other than Carbon

At a very early stage, Tenne *et al.* [17] showed that, in general, all compounds which crystallize in layered structures may form nanotubes and fullerene-like structures. The first noncarbon nanotubes consisted of MoS_2 and WS_2, but such structures were later observed with the selenides of molybdenum and tungsten [18]. Despite minor differences, these compounds are built according to the same scheme and, like graphite, they crystallize in layered structures with each layer being built up of three sublayers consisting either of metal (Me) or nonmetal ions (X):

$$X - Me - X \qquad X - Me - X \qquad X - Me - X \qquad X - Me - X \quad \ldots \ldots$$

Within each triple package of layers, there is covalent bonding; in between the packages the bonding is of the *van der Waals* type. Hence, the packages can be shifted against each other, and this is the reason why MoS_2 and WS_2 are, like graphite, used on a technical basis as solid lubricants. Another potentially important compound which forms nanotubes and fullerene-like structures is boron nitride (BN). Although this compound crystallizes in the same structure as graphite, it has no free electrons and therefore it is an insulator, and the color is white.

An electron micrograph of WS_2 nanotubes with diameters ranging from 15 to 20 nm is shown in Figure 10.29 [19]. The insert in the figure shows one of the nanotubes at a higher magnification, such that the four layers of the multiwall nanotube are clearly visible. When comparing these micrographs with those of carbon nanotubes, there is a significantly better contrast; this is due to the higher atomic number of tungsten compared to carbon. Like carbon, these compounds form not only nanotubes but also fullerene-like structures.

It is of interest to note that these ball-shaped, fullerene-like structures often consist of many layers. Insofar, they are zero-dimensional in analogy to multiwall nanotubes. A typical example is shown in Figure 10.30, where a $ZrSe_2$ particle is crystallized in an "onion" form [20].

Similar particles to those in Figure 10.30 were also observed in connection with multiwall fullerenes. Such carbon particles seem to collapse under the influence of surface tension. In fact, there is the opinion that in the case of carbon the pressure at the center of such an onion particle can become so high that the graphite-like structures become unstable and may under electron irradiation transform into the diamond structure [21]. On the basis of this experimental evidence, the existence of nonspherical structures, such as that depicted in Figure 10.31, is difficult to understand.

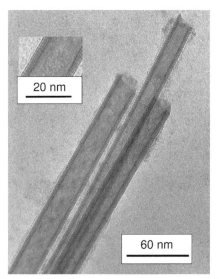

Figure 10.29 Three multiwall WS$_2$ nanotubes with diameters ranging from 15 to 20 nm [19]. The insert shows one of the tubes at higher magnification; four walls are visible. (Reproduced with permission from R. Tenne.)

A MoS$_2$ a multiwall, fullerene-like particle – referred to as an "onion crystal" – is shown in Figure 10.31 [19]. The special point about this particle is that the center is not circular, but rather is triangular. When comparing this shape with that of collapsed carbon nanotubes in which diamond was formed at the center, it is difficult to appreciate that this shape is indeed stable.

So, the question to be answered here is the application of these materials. The fullerene-like structures of MoS$_2$ and WS$_2$ have high potential as solid-state

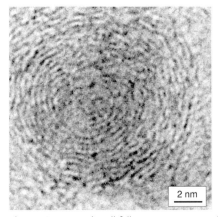

Figure 10.30 A multiwall fullerene, "onion crystal" consisting of ZrSe$_2$ [20].

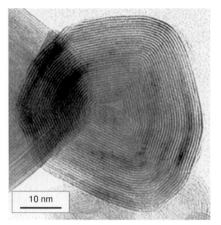

10 nm

Figure 10.31 A multiwall fullerene particle consisting of MoS$_2$ [19]. The deviations from sphericity visible in this micrograph are difficult to understand.

lubricants, and conventional MoS$_2$ powder is widely used for such purpose as the different *van der Waals* bonded layers glide easily upon each other. These compounds also minimize metal wear as the platelets act as distance holders – that is, as spacers – between the two metal surfaces and thus eliminate any direct contact. These properties are of special importance in the case of extreme loads, where oils and greases which are normally used as lubricants may be squeezed out of the contact area. However, with time, the dangling bonds at the borders of the platelets react with the metal surfaces and, at least to some extent, lose their lubricant properties. Fullerene-like MoS$_2$ nanoparticles have no open bonds, and therefore do not react with metal surfaces. In contrast to platelets, these spherical particles behave as "nano-ball bearings" [22], and do not lose their lubricant properties until they are lost to the system (e.g. they leave with the oil that is squeezed out) or are oxidized. The improving properties of WS$_2$ additions on friction coefficient, compared to a lubricant without such additions, are shown graphically in Figure 10.32. These are results of friction experiments in a ball-on-flat device as a function of the load. Paraffin oil, with additions of 0.5 and 1 wt.% WS$_2$ fullerene-like particles, was applied as a lubricant. The most important point is the load where the friction coefficient suddenly increases, as this deflection point describes the load capability of a friction pair. The critical load to seizure of the metal surface increases with the addition of only 0.5 wt.% WS$_2$, from 270 to 430 N. The advantage of a further increase in WS$_2$ content was minimal, as higher concentrations appear to have only a minor influence. These results vindicate the statement that such fullerene-like particles outperform conventional lubricants. Similar positive effects were observed using WS$_2$ particles as an addition to metal-working fluids applied to high-precision shaping techniques such as drilling or lathing. The advantages of such an application are two-fold:

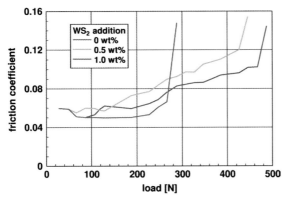

Figure 10.32 Tribological properties of paraffin oil with WS_2 fullerene-like particle addition. Effect on friction coefficient compared to a lubricant without such additions, determined with a ball-on-flat device as a function of the load. The load where the friction coefficient suddenly increases is described as the load capability of a friction pair. The critical load to seizure increases with the addition of WS_2 [22].

- Due to reduced friction, the applied forces and temperature are each reduced. When narrow-tolerance parts are involved, a controlled temperature is essential in order to obtain the intended dimensions.
- The metal-working fluid requirements are reduced, and the lifetime of the cutting tools is extended.

Reduced friction was observed when impregnating fullerene-like particles into self-lubricating solid films. The initial attempts to fill nickel–phosphorus electroless coatings with fullerene-like WS_2 particles were successful, and improved the properties of these self-lubricating coatings. Some typical results of a tribological test versus time are shown in Figure 10.33, where the friction coefficient of a pure nickel–phosphorus layer is compared to that of a layer of the same material, but containing some dispersed WS_2 fullerenes. During the first two cycles, there was almost no difference between the friction of the layers, with or without WS_2 filling. Clearly, the WS_2 particles were fixed in the matrix. However, after some wear the filled layer released some of the WS_2 particles, after which a change was observed from a sliding to a rolling friction. As a result, the coefficient of friction did not show any further increase, unlike that for the pure nickel–phosphorus layer.

10.2.3
Synthesis of Nanotubes and Nanorods

Carbon nanotubes and fullerenes may be produced when an electric spark is passed between carbon electrodes. However, as small amounts may also be found in any

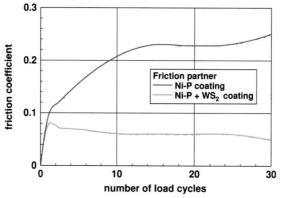

Figure 10.33 Time evolution of the friction coefficient of a nickel–phosphorus coating, with and without WS_2 fullerene particle addition [22]. As the WS_2 particles are embedded in the coating, friction is reduced after some wear, when the WS_2 particles have been released.

soot, one of the well-proven methods of producing nanotubes and fullerenes is that of *laser ablation*. The process of nanotube formation in an electric arc seems to be quite complicated, and there are indications that, at arc temperatures above 5000 K, a liquid carbon phase is involved. The micrograph in Figure 10.34 shows carbon nanotubes, together with some beads conveying the impression of frozen droplets [23].

In an electric DC arc at the surface of the electrodes, carbon is known to melt. Subsequently, as the temperature drops, the liquid carbon becomes supercooled and begins to crystallize. The crystallizing nanotube grows through the liquid layer at the cathode, dragging along small droplets of supercooled carbon. These droplets, as well as the carbon nanotubes, are visible in Figure 10.34. These explanations are backed up by the fact that the droplets remain amorphous, and the nanotubes are unintentionally coated with a thin amorphous layer. As mentioned above, these procedures result in the production of carbon nanotubes, fullerenes and often large amounts of soot. As soot is more susceptible to oxidation as compared to carbon nanotubes and

Figure 10.34 Carbon nanotubes produced in an electric arc. During this process, the arc causes the graphite to melt. The outside of the droplets cools at a faster rate, leading to the formation of glassy particles. The liquid carbon inside the drops cools so slowly that it becomes supercooled, and later crystallizes as nanotubes, dragging out some liquid carbon as droplets (Reprinted with permission from Georgia Institute of Technology [24] 2007.)

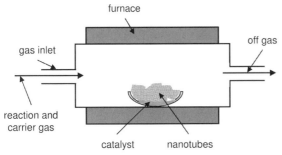

Figure 10.35 The synthesis of carbon nanotubes in a tubular furnace. The essential points in this process are the selection of an appropriate catalyst and a well-suited gaseous precursor.

fullerenes, it is removed by careful oxidation at elevated temperatures, ranging from 1000 to 1100 K.

The yield of carbon nanotubes is significantly improved by adding metal catalysts to the carbon electrodes. However, in order to obtain larger amounts of nanotubes, more sophisticated processes of synthesis are required. The most common approach utilizes a conventional tubular furnace at temperatures of about 1300 K, at which carbon nanotubes are obtained in the presence of a catalyst. The process is shown, schematically, in Figure 10.35.

Some typical precursors and catalysts used to obtain nanotubes include:

- For carbon nanotubes, a mixture of methane (CH_4) and hydrogen are used as the reaction gas, which is diluted with argon. As a catalyst, iron, nickel, and alloys of these metals, for example with molybdenum, are currently used.

- For MoS_2 or WS_2 nanotubes, the process begins with the oxides. Sulfur is delivered by H_2S in an argon, nitrogen, and hydrogen mixture.

- For the synthesis of GaN, iron is used as catalyst. However, in this case, the catalyst is supplied as $Fe(C_5H_5)_2$ vapor. Ga-dimethyl amide ($Ga_2[N(CH_3)_2]_6$) is well-proven as a precursor for gallium. The reaction is performed in an atmosphere of ammonia.

A typical product obtained by such a process resembles a disordered ball of wool, as shown in Figure 10.36. This is a typical example of a carbon nanotube product made from methane as precursor and an iron-containing catalyst [25]. By correctly selecting the catalyst it is possible to adjust the diameter of the nanotubes.

Instead of using randomly placed catalyst particles, it is possible to prepare clear-cut patterns of catalyst particles, either by printing or vapor deposition. With regards to the application (e.g., as a field electron emitter), this allows nanotubes to be grown exactly at the places where they are needed. This ability is of particular importance in the production of field emission displays.

So, the question arises of: How does the catalyst work? In an effort to provide an answer to this problem, a model of the growth of nanotubes, supported by a catalyst, is shown in Figure 10.37.

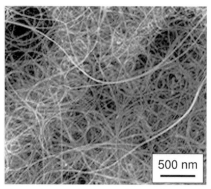

Figure 10.36 Carbon nanotubes made in a tubular furnace as shown in Figure 10.35. Methane was used as the carbon precursor, and iron as catalyst. The reaction temperature was approximately 1300 K (Hampel, Leonhardt; reproduced with permission [25]; Copyright: IFW Dresden 2002.)

The production of nanotubes or nanorods begins with a droplet of liquid catalyst. The precursor is gaseous, and the precursor and catalyst are selected in such a way that the precursor molecules dissociate at the surface of the catalyst droplet. For example, in order to obtain carbon nanotubes, methane (CH_4) dissociates at the surface of an iron or nickel droplet; the released carbon is then dissolved in the metal droplet, and the hydrogen leaves the system. At this stage, it is important to note that the selection of catalysts is quite critical. To obtain carbon nanotubes, the process begins with a solution of carbon in the catalyst metal particle. In order to avoid too-high temperatures, it is advantageous to seek eutectic systems. After some time, the material dissolved in the liquid catalyst reaches saturation and precipitates at the surface of the properly selected substrate. There is no doubt, that this process occurs more rapidly if the catalyst particles are liquid, as solid-state diffusion is significantly slower than diffusion in a liquid. Once the process has started, there is a steady transport of dissociated material to the precipitate; the nucleus becomes larger, and the nanotube or nanorod grows. When considering a further process, for example to

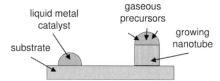

Figure 10.37 Model of the catalytic action during the synthesis of nanotubes or nanorods. Initially, the precursor for the product reacts with the catalyst metal. It is advisable to select a metal as precursor that forms a liquid phase together with the material for the nanotube or nanorod. After exceeding the maximum solubility, the nanorod or nanotube is precipitated. The one-dimensional product will grow as long as the precursor is supplied.

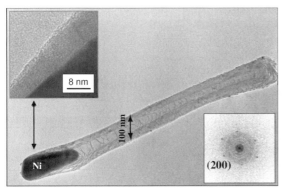

Figure 10.38 Electron micrograph of a carbon nanotube. The frozen nickel droplet, which was used as catalyst, is visible at one end. This micrograph clearly confirms the model of catalytic action as depicted in Figure 10.37. The electron diffraction pattern (inset, lower right) shows the hexagonal structure of the graphene sheet, the structural element of carbon nanotubes [26] (Copyright: Wiley-VCH 2007.)

obtain GaN nanorods, iron may be used as the catalyst. Both, iron and gallium form a low-melting phase with a melting point below 1200 K. Interestingly, the process described above is not simply a hypothetical model, but has been proved in many instances by electron microscopy. A transmission electron micrograph of a larger multiwall carbon nanotube, with a droplet of nickel (used as catalyst) embedded at the tip of the nanotube, is shown in Figure 10.38 [26]. At the top left of the figure is a higher-magnification insert, showing the different layers of the carbon nanotube. In the electron diffraction pattern shown in the insert at bottom right, the hexagonal structure of the graphene sheets – the constitutive part of the carbon nanotubes – is clearly visible.

A further quite spectacular example, which demonstrates the growth of a germanium nanorod, starting from a metal catalyst particle, is shown in Figure 10.39 [27]. This series of electron micrographs was taken *in situ* during the growth of the nanorod. The temperature for this process ranged between 1000 and 1200 K, where the germanium precursor GeJ$_2$ is gaseous. Additionally, within this temperature range in the gold–germanium binary system a liquid phase is expected, as the eutectic temperature is 630 K. The gold nanoparticle at the start of the process is shown in Figure 10.39a. Following the dissolution of some germanium in the gold particle, the particle melts (see Figure 10.39b) and this results in an increased diameter. After saturation of the gold droplet with germanium, the nanorod begins to grow (Figure 10.39c–e). Finally, a long nanorod with a catalyst droplet at the tip, is obtained (Figure 10.39f).

The process of growing nanotubes or nanorods is quite slow, and it may take days to obtain fibers with lengths of a few microns. The length of a single-wall carbon nanotube grown *in situ* in an electron microscope as a function of time is shown in Figure 10.40 [28]. Here, the length of the carbon nanotubes was measured in

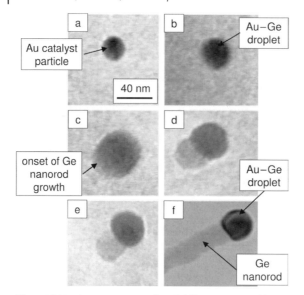

Figure 10.39 Electron micrographs of different stages of growth of a germanium nanorod [27]. (a) The gold catalyst particle as a starting point. (b) The gold particle dissolves germanium from the vaporized precursor GeJ$_2$. (c) The gold droplet has exceeded its solubility for germanium; hence, the onset of germanium precipitation occurs and the nanorod begins to grow. (d,e) The germanium nanorod continues to grow. (f) On completion of the experiment, the long germanium nanorod is capped with a droplet of a gold–germanium alloy. (Reproduced with permission from [27], Copyright: American Chemical Society 2001.)

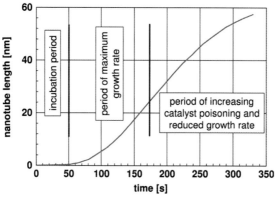

Figure 10.40 Carbon nanotube length versus time of synthesis. Nanotube growth begins after an incubation period. Most likely, this is the time required to saturate the nickel catalyst with carbon from the precursor. The rate of nanotube growth then increases until the catalyst becomes poisoned [28].

real-time, the temperature was 920 K, C_2H_2 was used as a precursor, and metallic nickel served as the catalyst. In Figure 10.40, three regimes of particle growth may be realized. During the first 50 s, there is no detectable growth in particle length, and obviously this is the time required for the nickel particle to be saturated with carbon and to nucleate the nanotube itself. Following this incubation period, the regime of rapid growth begins, during which the growth rate is controlled by carbon diffusion from the surface of the catalyst particle to the growing nanotube. In the third regime, the growth rate is steadily decreasing, even when all experimental variables are left constant. This reduction in the growth rate is attributed to an increasing coverage of the catalyst surface by strongly adsorbed carbon atoms, and this leads to a poisoning of the catalyst. When considering this growth mechanism, it is intuitively clear that the diameter of the nanotubes depends heavily on the size of the catalyst particles. As can be seen from Figure 10.39, the nanorod diameter is somewhat smaller than that of the catalyst particles; this relationship is also valid for the carbon nanotubes.

A further interesting process for the production of nanotubes, especially in relation to oxides, begins from highly anisotropic metal embryos. For example, Zn metal particles on a substrate form hexagonal prisms; this is due to the significant differences in surface energy between the hexagonal base plane and the lateral surfaces. At an elevated temperature (in the range of 700–800 K), and in a slightly oxidizing atmosphere, the less-stable lateral surfaces of the prisms begin to oxidize. Growth then occurs in the direction perpendicular to the hexagonal base plane. Assuming a continuing vapor transfer of Zn, the oxide sheath layer will continue to grow, forming a ZnO nanotube, the interior edge length of which corresponds to the outer edge length of the original Zn embryo particle. As the vapor pressure of the pure metal is significantly higher than that of the oxide, within a short time the material of the starting embryo will evaporate and be used for the growth of the nanotube. These three stages of the ZnO nanotube growth process are shown schematically in Figure 10.41.

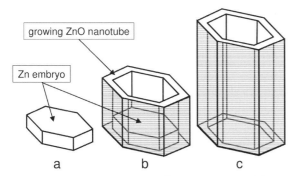

growing ZnO nanotube

Zn embryo

a　　　　　b　　　　　c

Figure 10.41 Growth mechanism of ZnO nanotubes starting from a metallic embryo, according to Xing *et al.* [29]. (a) The Zn metal embryo forms a body with minimum surface energy. (b) The lateral surfaces of the embryo begin to oxidize. The oxide skin grows, and the material for growth is spent from the vaporized precursor and the evaporating metallic embryo. (c) When the metal reservoir of the embryo is exhausted and any further supply of precursor has ceased, the nanotube has reached its final size.

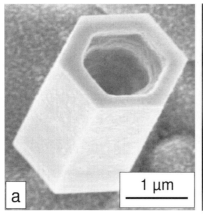
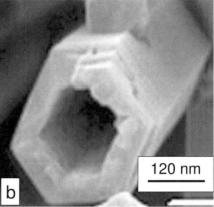

Figure 10.42 Electron micrographs of two hexagonal ZnO nanotubes. These nanotubes are grown using a mechanism as shown in Figure 10.41. (a) A micrograph of a larger ZnO nanotube. Note the perfect flat outer surfaces, whereas the inner surface is somewhat corrugated [30]. (b) An electron micrograph of a small ZnO nanotube. It is interesting to note that this nanotube may have grown as a spiral (Reprinted with permission from [31]; Copyright: Elsevier 2006.)

In Figure 10.41a, the metallic Zn embryo is shown, followed by a schematic of the growing oxide tube (Figure 10.41b). Finally, the metal reservoir from the metallic embryo is consumed. If there is no further supply of vaporized zinc, then the nanotube growth will stop. Electron micrographs of such zinc oxide nanotubes are shown in Figure 10.42, where the two ZnO nanotubes have entirely different dimensions. The nanotube in Figure 10.42a [30], which is rather a micro than a

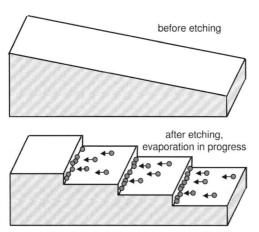

Figure 10.43 Process of producing wires of noble metals by evaporating the metal onto the surface of a cut-and-etched surface of a single crystal. The metal atoms migrate to the edges which stem from the etching process, where they form nanowires.

nanotube, has an edge length of more than 1 μm and a wall thickness of approximately 200 nm. The nanotube shown in Figure 10.42b is entirely different [31], having an edge length of approximately 200 nm and a wall thickness of less than 50 nm. When examining Figure 10.42b in detail, it can be seen that this nanotube, like a whisker, may have grown as a spiral.

One more physical process to produce nanowires is worthy of mention, and this involves primarily noble metals. The process starts from a graphite or silicon substrate that is cut at a small angle against the lattice planes. After etching, a series of regular steps are observed at the surface. When noble metal atoms are evaporated onto the surface of the substrate, they migrate to the edges of the steps, as shown in Figure 10.43. It is possible to release these wires from the substrate and to handle them subsequently.

References

1 Georgobiani, A.N., Gruzintsev, A.N., Kozlovskii, V.I., Makovei, Z.I., Red'kin, A.N. and Skasyrskii, Ya.K. (2006) *Neorganicheskie Materialy*, **42**, 830–835.

2 Yun, Y.J., Park, G., Ah, C.S., Park, H.J., Yun, W.S. and Haa, D.H. (2005) *Appl. Phys. Lett.*, **87**, 233110–233113.

3 Du, J., Liu, Z., Wu, W., Li, Z., Han, B. and Huang, Y. (2005) *Mater. Res. Bull.*, **40**, 928–935.

4 Szabo, D.V. and Vollath, D. (1998) unpublished results.

5 Lee, Y., Kim, B., Yi, W., Takahara, A. and Sohn, D. (2006) *Bull. Kor. Chem. Soc.*, **27**, 1815–1824.

6 Koenderink, G.H., Kluijtmans, S.G.J.M. and Philipse, A. (1999) *J. Colloid Interface Sci.*, **216**, 429–431.

7 Kroto, W., Heath, J.R., O'Brien, S.C., Curl, R.F. and Smalley, R.E. (1985) *Nature*, **318**, 162–163.

8 Weber, S. (2007) http://www.jcrystal.com/steffenweber/pb/swpb2.pdf.

9 Prinzbach, H., Weiler, A., Landenberger, P., Wahl, F., Wörth, J., Scott, L.T., Gelmont, M., Olevano, D. and Issendorff, B.V. (2000) *Nature*, **407**, 60–63.

10 Weber, S. http://www.jcrystal.com/steffenweber/pb/swpb1.pdf.

11 Ritschel, M. and Leonhardt, A. (2007) IFW-Dresden, Germany, unpublished results.

12 Bonard, M., Kind, H., Stockli, T. and Nilsson, L.-O. (2001) *Solid State Electron.*, **45**, 893–914.

13 Collins, P.G. and Zettl, A. (1996) *Appl. Phys. Lett.*, **69**, 1969–1972.

14 Samsung Advanced Institute of Technology (2001), Suwong, Korea.

15 Zhang, T., Simens, A., Minor, A. and Liu, G. (2006) Carbon Nanotube-Conductive Polymer Composite Electrode for Transparent Polymer Light Emitting Device Application, PMSE.

16a Hampel, S., Leonhardt, A. (2007) IFW-Dresden, Germany, unpublished results.

16b Winkler, A., Mühl, T., Leonhardt, A., Kozhuhorova, R., Menzel, S. and Mönch, J. (2007) IFW-Dresden, Annual Report.

17 Tenne, R., Margulis, L., Genut, M. and Hodes, G. (1992) *Nature*, **360**, 444–446.

18 Vollath, D. and Szabo, D.V. (1998) *Mater. Lett.*, **35**, 236–244.

19 Tenne, R. (2005) http://www.weizmann.ac.il/ICS/booklet/20/pdf/reshef_tenne.pdf.

20 Vollath, D. and Szabo, D.V. (2000) *Acta Mater.*, **48**, 953–967.

21 Banhardt, F. and Ajayan, P.M. (1996) *Nature*, **382**, 433–435.

22 Rapoport, L., Fleischer, N. and Tenne, R. (2005) *J. Mater. Chem.*, **15**, 1782–1788.

23 de Heer, W.A., Poncharal, P., Berger, C., Gezo, J., Song, Z., Bettini, J. and Ugarte, D. (2005) *Science*, **307**, 907–910.

24 de Heer, W. (2007) http://www.gatech. edu/news-room/release.php?id=516, Georgia Tech Communications and Public Affairs, 404-385-2966, david.terraso@ icpa.gatech.edu.

25 Hampel, S. and Leonhardt, A. (2002) IFW-Dresden, Germany, unpublished results.

26 Rybczynski, K.J., Huang, Z., Gregorczyk, K., Vidan, A., Kimball, B., Carlson, J., Benham, G., Wang, Y., Herczynski, A. and Ren, Z. (2007) *Adv. Mater.*, **19**, 421–426.

27 Wu, Y. and Yang, P. (2001) *J. Am. Chem. Soc.*, **123**, 3165–3166.

28 Lin, M., Tan, J.P.Y., Boothroyd, C., Tok, E.S. and Foo, Y.-L. (2006) *Nano Lett.*, **6**, 3449–3452.

29 Xing, Y.J., Xi, Z.H., Zhang, X.D., Song, J.H., Wang, R.M., Xu, J., Xue, Z.Q. and Yu, D.P. (2004) *Solid State Commun.*, **129**, 671–675.

30 Gao, P.X., Lao, C.S., Ding, Y. and Wang, Z.L. (2006) *Adv. Funct. Mater.*, **16**, 53–62.

31 Liu, J. and Huang, X. (2006) *J. Solid State Chem.*, **179**, 843–848.

11
Characterization of Nanomaterials

When considering characterization methods, it is very important to distinguish between methods that deliver values averaged over a large ensemble of particles, and those that provide information about a limited number of particles. The most important among the latter group of methods are those that are microscopic in nature. The behavior of nanomaterials is controlled by their global properties, which provide indications of how an ensemble behaves. However, in order to understand why an ensemble behaves in a certain way, it is necessary to utilize microscopic methods.

11.1
Global Methods for Characterization

11.1.1
Specific Surface Area

As each particle is characterized by its surface, and the surface is directly related to the particle size, then measuring the surface will provide an indication on particle size and the state of agglomeration. The specific surface A is usually given per gram, and is the number of particles per gram multiplied by the surface area per particle. Assuming a spherical particle, A is given by

$$A = \frac{6}{\pi D^3 \rho} \pi D^2 = \frac{6}{D\rho} \tag{11.1}$$

where D is the particle diameter and ρ the materials density. In addition, Equation (11.1) shows that any statement on particle size – or, more generally speaking, on an individual property – requires assumptions to be made about the particle shape.

Assuming spherical particles, Figure 11.1 displays the specific surface (in $m^2\,g^{-1}$) for alumina. (Even when $m^2\,g^{-1}$ is not a SI unit, it will be used in this context, because it is the only unit generally accepted for specific surfaces.) In this case, a density of $3.5 \times 10^3\,kg\,m^{-3}$ was assumed. Even when the assumption of a constant density over the whole range of particle sizes is possibly not correct, this figure provides at least an

Nanomaterials: An Introduction to Synthesis, Properties and Application. Dieter Vollath
Copyright © 2008 WILEY-VCH Verlag GmbH & Co. KGaA, Weinheim
ISBN: 978-3-527-31531-4

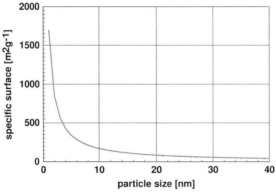

Figure 11.1 Specific surface of spherical alumina particles as a function of the particle size (a reduction of the surface due to clustering was not assumed).

insight into the surfaces of the nanoparticle. Certainly, clusters of particles necessarily have smaller surfaces, and the ratio of the expected surface over the measured surface indicates the degree of clustering.

Provided that it is possible to cover the surface of a specimen with a well-defined number of gas molecules N, which is proportional to the surface, in the case of a monolayer, the surface of a specimen is given by

$$A = Na_M \tag{11.2}$$

where a_M is the area covered by one gas molecule. Equation (11.2) assumes that, between the particle surface and the gas molecules, there are attractive forces which overcome any disordering effects of thermal motion. Such an attractive interaction results from the *van der Waal's* interaction. This process is called *physisorption*; when there is chemical interaction between the surface and the adsorbate, the process is known as *chemisorption*. The limit between physisorption and chemisorption is defined, somewhat arbitrarily, at an enthalpy of interaction of approximately $50\,kJ\,mol^{-1}$.

Brunauer *et al.* [1] first described a method to measure the specific surface of powders by gas adsorption at the surface, and this is now used as the standard procedure to determine specific surfaces. This so-called BET method is named after the first letters of the names of its inventors (Brunauer, Emmett and Teller). The BET theory is based on the *Langmuir* adsorption isotherm but, more realistically, is expanded beyond monolayer coverage. *Langmuir* assumes that at a surface, there is a fixed number of possible sites for gas adsorption; additionally – and this is very important – a further layer of gas molecules is not allowed until all possible sites for the first layer are occupied. To derive *Langmuir's* formula for adsorption, one starts with the following reaction:

$$M_G + V \rightarrow N - V \tag{11.3}$$

where M_G is the quantity of gas to be adsorbed, N is the number of possible sites for adsorption, and V is a vacant site for adsorption at the surface. Equation (11.3) assumes that there is no interaction between the adsorbed molecules, and that the reaction is not influenced by the coverage; or, the enthalpy of adsorption is independent of the coverage. The equilibrium constant K of Equation (11.3) is

$$K = \frac{(N-V)}{VM_G} \tag{11.4}$$

The relative amount of adsorbate is $\Theta = \frac{N-V}{N}$, and the number of adsorbed gas molecules M_G is proportional to the gas pressure p; therefore, one may assume $M_G = \alpha p$. The number of vacancies is $V = N(1-\Theta)$; hence, one obtains for the equilibrium constant

$$K = \frac{\Theta}{(1-\Theta)\alpha p} \quad \text{or modified} \quad b = \frac{\Theta}{(1-\Theta)p} \tag{11.5}$$

Experimentally accessible is the amount of gas adsorbed at the surface; therefore, the following expression derived from Equation (11.5) is used

$$\Theta = \frac{bp}{1+bp} \tag{11.6}$$

Equation (11.6) is the famous *Langmuir* adsorption isotherm. For large values of the gas pressure p, Θ approaches asymptotically 1. This is independent of the temperature. The factor b is determined by measuring the adsorption isotherm at different temperatures; it is a function of the enthalpy of adsorption ΔH_{ads}

$$b = \exp\left(\frac{\Delta H_{ads}}{RT}\right) \tag{11.7}$$

From Equation (11.7) it is obvious that b increases with decreasing temperature. To obtain the asymptotic value of Θ at not too-high temperature, it is advised that these measurements be made, if possible, at low temperatures. In most cases, nitrogen is selected as the adsorbate, and therefore it is not too difficult to make the measurements at the temperature of boiling nitrogen. Figure 11.2 displays *Langmuir*'s adsorption isotherms for different values of the temperature-dependent equilibrium constant b. This figure also demonstrates, drastically, the advantage of applying low temperatures for adsorption measurements.

At first glance, Equation (11.6) seems to be ready for evaluation and for determining specific areas. However, one of the crucial assumptions – the adsorption of one complete monolayer – is a too far-reaching simplification. Brunauer *et al.* [1] expanded *Langmuir*'s theory of adsorption isotherms by taking the possibility of multiple layers of adsorbed gas atoms into account. As did *Langmuir*, these authors assumed that there is no interaction between the layers of adsorbed molecules and, most importantly, that *Langmuir*'s theory is applicable to each layer. Equation (11.8) displays the series of reactions assumed for the BET process.

$$M_G + V_S \rightarrow L_1$$
$$M_G + V_{L1} \rightarrow L_2$$
$$M_G + V_{L2} \rightarrow L_3 \tag{11.8}$$
$$\ldots\ldots$$
$$\ldots\ldots$$

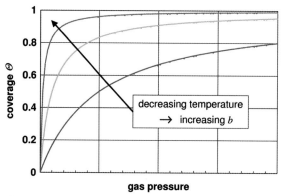

Figure 11.2 *Langmuir*'s adsorption isotherm as a function of the gas pressure. Decreasing temperatures lead to increasing values of the temperature-dependent parameter *b*. At low temperatures, it is easier to obtain saturation; hence, adsorption experiments are performed at reduced temperatures.

In Equation (11.8), V_S represents a vacancy of the adsorbed layer at the surface, and V_{Li} for a vacancy in the ith layer L_i of adsorbed molecules. The situation represented by the reaction in Equation (11.8) is shown in Figure 11.3.

After some mathematical manipulation, the BET isotherm is given by

$$\Theta = \frac{c\frac{p}{p_0}}{\left(1-\frac{p}{p_0}\right)\left[1-(1-c)\frac{p}{p_0}\right]} \tag{11.9}$$

(For details of the lengthy derivation of this formula, please consult a textbook on physical chemistry.) Figure 11.4 shows the coverage Θ as a function of the ratio p/p_0 for different values of the BET constant c.

In Equation (11.9), p is the gas pressure, p_0 the saturation pressure of the adsorbate, and Θ is now defined as the ratio $\Theta = N_{ads}/N_{mono}$, the ratio of the total number of adsorbed molecules over the number of molecules in the monolayer (first layer).

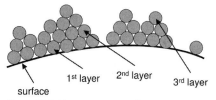

surface   1st layer   2nd layer   3rd layer

Figure 11.3 The assumption of Brunauer, Emmett, and Teller [1] about the adsorbed layers at a surface. Basically, these authors assumed a multilayer system, without the need to form completely filled layers.

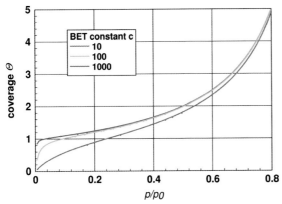

Figure 11.4 Coverage of surface according to Brunauer, Emmett, and Teller as a function of the pressure ratio p/p_0. The parameter for the curves is the BET constant, c [1].

Therefore, Θ may achieve values larger than 1. In a first approximation the BET constant c is given by

$$c = \frac{\exp\left(\dfrac{\Delta H_{ads}}{RT}\right)}{\exp\left(\dfrac{\Delta H_{vap}}{RT}\right)} \tag{11.10}$$

where ΔH_{vap} is the enthalpy of vaporization of the adsorbate. For $\Delta H_{ads} \gg \Delta H_{vap}$ leading to large values of c, the BET isotherm degenerates to $\Theta|_{c \to \infty} = \frac{1}{1 - \frac{p}{p_0}}$. Lastly, this is again *Langmuir's* adsorption isotherm, and is observed when the adsorption to the surface is significantly stronger than the condensation to the liquid. This case is found if an unreactive gas, such as nitrogen or krypton, is adsorbed at a polar surface.

Experimentally, the coverage is measured by determining the amount of gas adsorbed at the surface. Measuring the gas volumes in adsorption or desorption should lead to comparable results. The BET isotherm describes physical reality for a coverage Θ in the range from approximately 0.8 to 2 very well, and the pressure range of validity is restricted from $0.05 < p/p_0 < 0.35$. Even when that range is quite limited, it is large enough to obtain reliable experimental results. To evaluate BET experiments, one plots the BET function

$$\text{BET function} = \frac{\frac{p}{p_0}}{V_{ads}\left(1 - \frac{p}{p_0}\right)} \tag{11.11}$$

in the appropriate pressure region versus p/p_0, where V_{ads} is the volume of the adsorbed gas. This plot gives a straight line, as shown in Figure 11.5 for the data calculated in Figure 11.4.

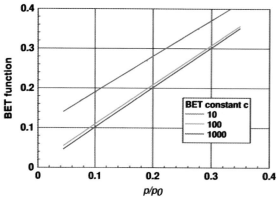

Figure 11.5 BET function according to Equation (11.11) plotted versus the pressure ratio p/p_0. The parameter for the curves is the BET constant, c.

According to the BET theory, the axis intercepts in this plot give the parameter for the surface. (For the detailed mathematics, the reader is referred to a higher-level textbook of physical chemistry.) From

$$\text{BET function} = \frac{\frac{p}{p_0}}{V_{ads}\left(1 - \frac{p}{p_0}\right)} = \frac{1}{V_{ads}\left(\frac{p_0}{p} - 1\right)} = \frac{p}{p_0}\frac{1}{cV_{mono}} + \frac{1}{V_{mono}}$$

(11.12)

The intercept of the straight line defined by Equation (11.12) with the abscissa $\frac{p}{p_0} = 0$, gives $\frac{1}{V_{mono}}$, the amount of gas adsorbed at the surface in one monolayer, the slope $\frac{1}{(cV_{mono})}$ may be used to calculate the BET constant c.

The BET experiment delivers the amount of gas adsorbed at the surface. The surface area of the material is calculated from the number of gas molecules in a monolayer at the surface $N_s = \frac{V_s}{V_M}N_A$, where V_s is the gas volume adsorbed at the surface and V_M the volume of one mole of gas, both under standard temperature and pressure conditions, and N_A Avogadro's number. The specific surface of a specimen A with the weight m is then given by

$$A = \frac{N_s\sigma}{m}$$

(11.13)

where σ is the area covered by one molecule; in the case of nitrogen, this value is $0.158\,\text{nm}^2$.

Technically, many powders with huge surface areas are produced, and some examples are listed in Table 11.1. In the case of nanoparticles, the values of the specific surface areas range from 45 to 400 $\text{m}^2\,\text{g}^{-1}$. However, in interpreting these values great care must be taken as they are related to a constant weight and not to a certain volume of material. Values of the surface per cm^3 are listed in Table 11.1. The densities used for calculation are those of bulk material, which is a rough approxi-

Table 11.1 BET surfaces and surface per unit volume of different commercial nanopowders.

	BET-surface $(m^2 g^{-1})$	Surface per unit volume $(m^2 cm^{-3})^a$	Composition
AEROSIL® 90[1]	90 ± 15	200	SiO_2
HDK® T40[2]	400 ± 40	880	SiO_2
AEROXIDE® Alu C[1]	100 ± 15	400	Al_2O_3
AEROXIDE® TiO_2P25[1]	50 ± 15	200	TiO_2
VP Zirkonoxid 3-YSZ[1]	60 ± 15	340	ZrO_2/Y_2O_3
NanoTek® Indium–tin oxide[3]	30–60	330	In_2O_3, SnO_2 90:10 wt.%
NANOCAT® SFIO[4]	ca. 250	1300	Fe_2O_3

The materials listed in this table are products of the following companies: [1]DEGUSSA; © 2001, Degussa Aerosil & Silane; [2]Wacker Chemie AG, München, Germany; [3]Nanophase Technologies Corporation Romeoville, USA; [4]MACH I King of Prussia, USA.
[a] Roughly estimated values.

mation. Among the examples shown in Table 11.1, the NANOCAT® SFIO iron oxide is the most finely dispersed powder.

If an attempt is made to correlate particle sizes with BET surfaces, it becomes apparent that, in most cases, the experimentally determined surface is significantly smaller than that calculated using Equation (11.1). The reason behind this phenomenon is the clustering of the particles.

11.2
X-Ray and Electron Diffraction

Both, X-ray and electron diffraction techniques are used to study the crystal structure of specimens, and it is also possible to obtain information on the particle size in this way. The physical background of diffraction is found in the wave nature of electrons and X-rays. Provided that these waves are in an appropriate range of wavelength relative to the lattice structure of the specimen, a diffraction pattern is obtained that is typical of the material in question. However, when considering nanocrystalline materials, the diffraction lines may be so much broadened that an unequivocal assignment to a certain structure is impossible. Typical of this problem is the differentiation between cubic and tetragonal phases; therefore, great care must be taken in the interpretation of diffraction patterns of nanoparticles.

Diffraction experiments may be conducted in transmission, usually in electron diffraction, or in reflection, as it is used primarily in X-ray diffraction techniques. The basic principles of diffraction on a three-dimensional lattice are shown in Figure 11.6, where the incoming waves are scattered at each atom in the lattice of the specimen. The scattered waves form a spherical wave that interferes, and the interference pattern thus formed carries the information about the arrangement of atoms in the lattice.

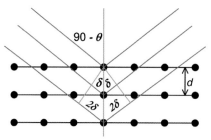

Figure 11.6 Basic geometry of diffraction. The waves incoming under an angle of θ against the surface are "reflected" at the same angle. Between two consecutive lattice planes, the "reflected" waves have a phase difference of 2δ.

When examining the incoming wave (as represented by their path in Figure 11.6), a difference becomes apparent in the path length between the wave scattered in the 0th lattice plane, representing the plane directly at the surface, and the wave scattered in the 1st plane. The difference in the path length δ is a function of the distance between the lattice planes d and the angle between the incoming wave and the lattice planes θ. This path difference is $δ = d \sin θ$. The scattered wave, leaving the specimen under the same angle, has the same difference in the path length, and therefore the total difference of the path length is $δ_{total} = 2d \sin θ$. The path difference between the 0th and the 2nd plane is, therefore, exactly doubled; or more generally, the difference of the path length to the nth plane is $2nδ$. An interference maximum is observed when the difference in the path length is exactly an integer multiple of the wavelength. This may be the path difference between two parallel adjacent lattice planes or between planes in the distance nd. This leads to the condition for an interference maximum

$$n\lambda = 2d \sin θ \quad n = 1, 2, 3, \ldots \tag{11.14}$$

where d is the distance between two arbitrary planes in the lattice and n is the order of the diffraction line.

In order to understand the appearance of complex diffraction patterns, it is necessary to explain the geometry and the notations in crystallography. This explanation uses the simplest case of a primitive cubic lattice, but for more details the use of more specialized books on diffraction techniques and crystallography is advised. Figure 11.7 shows a cubic cell and three lattice planes indicated in blue; the lattice planes are denoted by *Miller* indices.

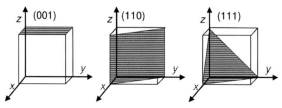

Figure 11.7 Lattice planes in a cubic lattice. These planes are characterized by their *Miller* indices.

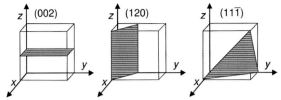

Figure 11.8 *Miller indices of some lattice planes in a cubic structure.*

The *Miller* indices are the reciprocal values of the intercept of the lattice planes with the axes of the coordinate system. The coordinate system is normalized in such a way that it achieves the value 1 at the lattice constant *a*. Generally, one uses the letters *h* for the value in the *x* direction, *k* in *y*, and *l* in the *z* direction. Due to this normalization, *Miller* indices are integers, and they are written in round brackets (*hkl*). The minus sign for negative values is written above the number. Figure 11.8 displays three cases of lattice planes with nonelementary indices. In the first case, the intercept of the lattice plane in question is at *z* = 0.5, and therefore, the *Miller* indices are (002). Similarly, in the second case (120), and in the third case, the intercepts between the lattice plane and the coordinate systems are at *x* = *y* = 1, *z* = −1. Therefore, the Miller indices are (11$\bar{1}$).

In a cubic lattice, the spacing of two lattice planes with the indices (*hkl*) and the lattice constant *a* is given by

$$d_{(hkl)} = \frac{a}{(h^2 + k^2 + l^2)^{0.5}} \tag{11.15}$$

Using Equation (11.15), the interference condition writes:

$$n\lambda = 2\frac{a}{(h^2 + k^2 + l^2)^{0.5}} \sin \theta \tag{11.16}$$

In Equation (11.16), again, *n* is the order of the diffraction line. To avoid too-large numbers for the *Miller* indices, conventionally one incorporates the order of diffraction into the indices in the following way:

$$\lambda = \frac{2a}{(n^2h^2 + n^2k^2 + n^2l^2)^{0.5}} \sin \theta = \frac{2a}{n(h^2 + k^2 + l^2)^{0.5}} \sin \theta \tag{11.17}$$

This equation allows one to calculate the lattice constant *a* and the *h*, *k*, *l* values from diffraction patterns. The next question is directed to the angular resolution. This is important, because sometimes, for example, in order to differentiate between cubic and tetragonal phases it is essential to measure small differences. This means that one must seek the conditions where the expression

$$\frac{\frac{\partial d}{\partial \theta}}{d} \Rightarrow \text{maximum}$$

is fulfilled. Simple calculations show that maximal resolution goes with cot θ. Or, in other words: one obtains maximal resolution by selecting experimental conditions leading to a diffraction angle close to 180° or π. As in electron diffraction, the

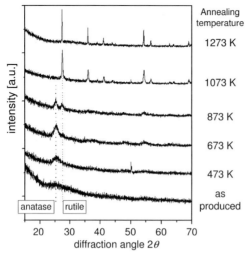

Figure 11.9 X-ray diffraction pattern from an annealing experiment. The measurements were performed at room temperature after annealing for 6 h. At 1273 K the annealing time was 20 h (reproduced with permission from [2]; Copyright: American Institute of Physics).

diffraction angles usually are extremely small, this must be interpreted as instruction to seek diffraction lines at the maximum possible diffraction angle θ. According to Equation (11.17), this corresponds to lattice planes with small values of *d*, although it must be borne in mind that the intensity of diffraction lines at higher angles, usually, is very small. This creates, especially in the case of nanoparticles, severe problems as the diffraction patterns of small particles usually are connected to low intensities.

Figure 11.9 displays a typical application; this demonstrates grain growth and phase transformation in nanoparticulate titania during annealing [2]. In the "as-produced" state, the particle size is in the range between 2 and 3 nm. Even when the specimen is crystallized, the X-ray diffraction spectrum shows just a broad shoulder in the range, where the intensive diffraction line of the anatase phase is expected. With increasing annealing temperature, the diffraction peaks becomes more pronounced. Additionally, after annealing at 873 K new diffraction lines appear which indicate the onset of a phase transformation from the anatase to the rutile phase. At the maximum annealing temperature, 1273 K, the diffraction lines of rutile (now the only ones) are very sharp, indicating large particle sizes of the rutile.

The example shown in Figure 11.9 intuitively suggests the possibility of particle size measurement by analysis of the width of the diffraction lines. In fact, this is a topic of broad research in the physics of X-ray diffraction. The first approach to this problem by *Scherrer* has, until now, been the most useful, even when, in the original report, some correction factors were not taken into account [3]. The famous *Scherrer* formula is

$$D = \frac{\kappa \lambda}{b \cos \theta} \tag{11.18}$$

In Equation (11.18) D is the crystallite size vertical to the analyzed lattice plane with the *Miller* indices (*hkl*). In this context, extreme care must be taken as the particle may be larger since it might be an agglomerate of many crystallites. Here, θ is the diffraction angle and b the width of the diffraction line at half intensity (in a 2θ-intensity plot). The constant factor κ depends on the crystal structure and habitus, and is found to be in the range between 0.89 and 1.39. For cubic materials, a value of 0.94 is often selected. λ is the wavelength of the X-rays applied in the experiment. In the case of small nanoparticles, the determination of the line width b is not problematic. When the particle size comes into the range of 100 nm and more, the instrumental influences on the line width must be taken into account. The instrumental line broadening, determined at perfectly crystallized coarse-grained material is – as simplest assumption – subtracted from the measured line width to obtain the line width for crystallite determination. The crystallite sizes determined from the diffraction profiles shown in Figure 11.9 are displayed in Figure 11.10.

In the example shown in Figures 11.9 and 11.10, the interpretation was not too difficult. However, the situation is significantly more confusing in the case of tetragonal–cubic transformations. Because of the broad diffraction lines, it is often (perhaps in most cases) not possible to decide unequivocally if the structure is cubic or tetragonal, or if there is a superposition of the spectra of both structures. As an example, this problem is faced in the case of zirconia or $BaTiO_3$, but a decisive statement based on diffraction data alone is impossible in such cases. Additional information that is helpful may be obtained from density measurements or extended X-ray absorption fine structure (EXAFS) results. A typical example is depicted in Figure 11.11 [4].

Figure 11.11 displays the (200) and (002) diffraction lines of tetragonal $BaTiO_3$ as a function of the annealing temperature, which is lastly correlated to the particle size. In the case of the highest annealing temperature, leading to the largest particles, the two diffraction lines of the tetragonal phase are clearly visible. However, with

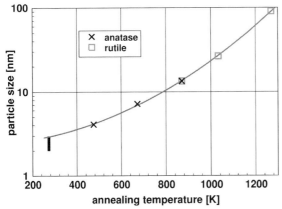

Figure 11.10 Size of titania particles determined by application of the *Scherrer* formula from the X-ray diffraction patterns displayed in Figure 11.9 [2].

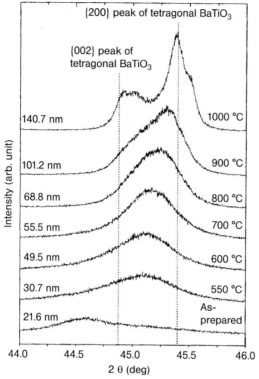

Figure 11.11 (200) and (002) diffraction lines of tetragonal BaTiO$_3$, which was annealed at different temperatures. The two diffraction lines are well separated after annealing at 1000 °C. At lower temperatures, it was not entirely clear whether the specimen consisted of the tetragonal, the cubic phase, or a mixture of both (reproduced with permission from [4]; Copyright: Elsevier).

decreasing annealing temperature the separation of the two lines becomes poorer; this is caused by the increasing width of the diffraction lines. In any case, the slight asymmetric line profile indicated the presence of the tetragonal phase, as the (002) peak has about double the intensity of the (200) peak. Additionally, it cannot be excluded that in the range of the small particles, and low annealing temperature, a significant fraction of the material remained cubic. A detailed analysis of the line profile by the authors led to the conclusion that there should be a content of both phases; however, a clear result was not achieved.

When comparing X-ray diffraction and electron diffraction, there is one essential difference – X-rays applied for structural analyses usually are in wavelength range between 0.05 and 0.22 nm, although most common is the use of Cu Kα radiation with a wavelength of 0.154 nm. In order to obtain optimal conditions for analysis of the line width with respect to particle size, the application of a monochromator is often recommended. In the case of electron diffraction, the energy of the electrons is selected to be in a range from 100 to 200 keV. This is equivalent to a wavelength range from 2.5×10^{-3} to 3.7×10^{-3} nm. Additionally, electron diffraction is based on registration with photo plates or CCD devices, whereas X-ray diffraction applies goniometer readings, as depicted in Figures 11.9 and 11.11.

A typical electron diffraction pattern of nanoparticles, together with the corresponding electron micrographs, is shown in Figure 11.12 [2]. These micrographs and diffraction patterns are taken from the same study as in Figures 11.9 and 11.10. In Figure 11.12a, the powder is seen in the as-produced state, with particle sizes ranging

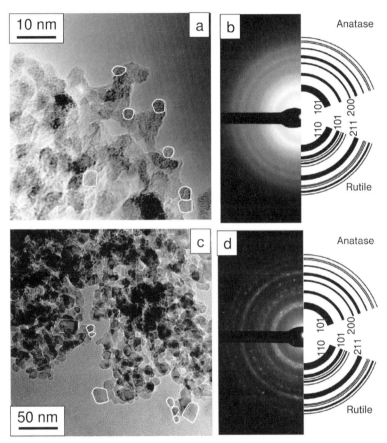

Figure 11.12 Morphology and electron diffraction pattern of titania in the as-produced state and after annealing at 873 K [2]. (a) Powder in the as-produced state; some of the grains are highlighted to show their real dimensions. (b) Electron diffraction pattern of the material displayed in panel (a). Note the very broad and weak lines that are typical of small grain sizes. The diffraction pattern fits only to the anatase state. (c) The same powder as shown in panel (a), but after annealing at 873 K. Note the significant grain growth; the relatively narrow size distribution, visible in panel (a) has also become broad. (d) Electron diffraction pattern of the specimen displayed in panel (c). As annealing led to grain growth, the diffraction lines became more intense and clearer. A few exaggerated grown particles also gave rise to a diffraction pattern consisting of isolated points. (Reprinted with permission from [2], Copyright: American Institute of Physics 2006.)

from 2 to 3 nm. For easier observation, some of the particles are highlighted. In contrast to the X-ray diffraction spectrum displayed in Figure 11.9, where only one weak shoulder is visible, the electron diffraction pattern in Figure 11.12b shows six clearly visible lines; this is a quite often-observed phenomenon that, especially for small particles, electron diffraction patterns show more details as compared to X-ray diffraction patterns. In connection with the electron diffraction pattern, the theoretically expected diffraction pattern is also shown. Figure 11.12c shows the same specimen after annealing at 873 K; now, this specimen consists of a mixture of the anatase and rutile phases. The process of grain growth during annealing led to a relatively broad particle size distribution in the range from 5 to 20 nm, with a majority of the particles found in the range below 10 nm. As the intensity of the diffraction lines is weighted by the particle volume, the average particle size determined by X-ray diffraction was 13 nm.

The electron diffraction pattern displays an additional interesting feature, namely that within the diffraction rings small bright spots can be seen. These stem from a limited number of larger particles that give diffraction patterns of that high intensity. It should be noted that the uniform rings come from a large number of randomly oriented particles. However, one large particle provides only one set of diffraction points; a series of larger particles in the electron beam results in spotted rings, as shown in Figure 11.12d.

Today, many instruments have the ability to focus the electron beam used for diffraction to such a large extent that it is possible to obtain the diffraction pattern of one particle only. An example of this technique, known as selected area diffraction (SAD), is shown in Figure 11.13 [5]. Figure 11.13a shows an electron micrograph of a $CuFe_2O_4$ platelet, a ferrite which crystallizes in the cubic spinel structure, while Figure 11.13b shows an electron diffraction pattern of exactly this particle. This precise correlation is important when determining the crystal structure.

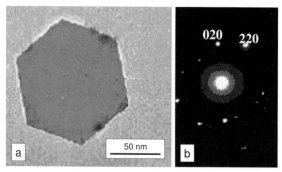

Figure 11.13 A micrograph of a $CuFe_2O_4$ platelet, and its electron diffraction pattern [5]. (a) An electron micrograph of the platelet. (b) An electron diffraction pattern of the same particle. The numbers indicate the lattice plane belonging to the diffraction signal. (Reproduction with permission from [5]; Copyright: Elsevier 2005.)

11.3
Electron Microscopy

11.3.1
General Considerations

In order to study the shape, size, and structure of nanoparticles, electron microscopy is the best-suited technique. Today, however, as electron microscopy is considered a broad science in its own right, within this chapter we will outline a few basic facts, aiming to avoid the impression that simply by reading these few pages it would be possible to interpret electron micrographs. Electron microscopy is, indeed, a task for "specialists".

Roughly speaking, electron microscopy functions like optical microscopy, except that the difference is simply the application of electron waves instead of electromagnetic waves. Both microscopes consist of an illumination system, a specimen holder, an objective system followed by a projection system, and finally a registration device. The latter may be either a photographic plate or an electronic camera system. The need to change from optical microscopy to electron microscopy lies in the length of the applied waves. According to *Abbe*, the minimum feature that can be seen with a conventional optical system is limited by diffraction to the half of the applied wavelength λ. In other words, it is impossible to resolve structural details smaller than half of the wavelength. Nowadays, optical microscopes come close to the theoretical possible values of the resolution, but in technical reality in electron microscopes this resolution limit is by far not achieved.

The resolution of an optical system is limited by the numerical aperture (NA), which is the ratio of the radius of the lens (which usually is limited by an aperture diaphragm) over the focal distance. For practical purposes, one considers the minimum distance x_{min} of two points that are separated in the image

$$x_{min} = \frac{\lambda}{NA} \qquad (11.19)$$

Equation (11.19) states that the distance of two points which are distinguishable in the image plane decreases with increasing value of the numerical aperture; lastly, with increasing diameter of the optical system. The theoretical background to Equation (11.19) is, according to *Abbe*, that optical resolution is limited by the highest transmitted diffraction order. Therefore, the highest resolution is obtained if all diffraction orders contribute to the image. As the numerical aperture is finite in any optical instrument, not all diffraction orders are transmitted, and image blurring and distortions are unavoidable. On the other hand, a small numerical aperture increases the depth of the field. Whilst this is advantageous in cases where maximum resolution is not necessary, it is a major disadvantage if micrographs are to be prepared at different depths of the specimen for three-dimensional reconstruction.

Because of the incomplete correction of an electron optical system, usually, the numerical aperture of an electron microscope is less than 10^{-2}. Therefore, in order to obtain a certain resolution it is necessary to apply electron waves with an extremely short wavelength.

In light optical systems, the shortest wavelengths that can be applied are 400 nm. Some recent developments have also used shorter wavelengths in the UV-region, although to date very few commercial microscopes using UV are available commercially. However, the minimum feature of 200 nm resolvable in optical microscopy is by far too large for nanomaterials.

The wavelength of electrons is selected by the acceleration voltage. In the case of electron microscopy, this is the operating voltage of the electron microscope. According to *de Broglie*, the wavelength λ associated to a particle with the mass *m* is given by

$$\lambda = \frac{h}{mv} \tag{11.20}$$

In Equation (11.20), $h = 6.63 \times 10^{-34}\,\text{J s}^{-1}$ is *Planck's* constant, and *v* is the velocity of the electrons. From the energy balance

$$U = eV = \frac{mv^2}{2} \Rightarrow v = \left(\frac{2eV}{m}\right)^{\frac{1}{2}} \tag{11.21}$$

the speed *v* needed in Equation (11.20) may be calculated. In Equation (11.21), $e = 1.602\,\text{C}$ is the electric charge, $m = 9.11 \times 10^{-31}\,\text{kg}$ the mass of the electrons, and V the acceleration voltage of the system. In electron microscopy, voltages above 100 kV are applied. At these high energies, the velocity of the electrons come into a range, where mass is increased by relativistic phenomena. Using the *Lorentz* transformation, the mass *m* of a particle (in this case, an electron) traveling with the speed *v* and the mass $m_0 = 9.11 \times 10^{-31}\,\text{kg}$ at $v = 0$ ("rest mass"), is calculated by

$$m = \frac{m_0}{\left[1 - \left(\frac{v}{c}\right)^2\right]^{\frac{1}{2}}} \tag{11.22}$$

where $c = 2.998 \times 10^8\,\text{m s}^{-1}$ is the speed of light. After inserting in Equation (11.20), one obtains

$$\lambda = \frac{h}{\left[2m_0eV\left(1 + \frac{eV}{2m_0c^2}\right)\right]^{\frac{1}{2}}} \tag{11.23}$$

The relativistic increase of the electron mass reduces the wavelength of the electrons. Figure 11.14 displays the wavelength of electrons as a function of the acceleration voltage. In this graph, the wavelength of the electrons is calculated with and without relativistic correction.

Now, it is possible to estimate the necessary voltage for electron microscopes. In order to identify points at a distance of 0.5 nm apart, and the numerical aperture of the electron microscope is 5×10^{-3} (which is a reasonable value for electron microscopes), an electron energy of at least $10^5\,\text{eV}$ is needed. To compensate for other problems, electron microscopes used in materials science studies apply voltages ranging from 150 to 300 kV, although for very special purposes instruments with

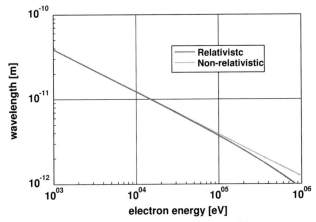

Figure 11.14 Wavelength of electrons as a function of the acceleration voltage. The wavelength is plotted with and without the relativistic increase in electron mass.

acceleration voltages of up to 1 MV have been built. However, it must be noted that the resolution power of these high-powered instruments is not significantly better.

Today, the lenses for electron microscopes apply magnetic fields and these so-called magnetic lenses show rotational symmetry. When the first electron microscopes were built, instruments using electrostatic lenses were also available commercially.

When electrons pass a specimen they are scattered, and a distinction should be made at this point between two cases:

- *elastic scattering*, where the energy of the electrons is not altered
- *inelastic scattering*, where the electrons lose energy.

Electron optical systems are connected to a significant chromatic aberration, and provide a sharp image only at exactly one energy of the electrons; therefore, inelastic scattered electrons blur the image. A major part of these electrons are scattered over a wider angle, and therefore they are caught by diaphragms within the system. Problems with chromatic aberration begin already at the illumination of the system with electrons. A modern field emission gun for electrons has an inherent energy spread of typically 0.7 eV, but to obtain maximum resolution even this is too much. By applying an electron monochromator this energy spread can be reduced to values less than 0.2 eV.

The second severe source of image blurring in electron microscopy is that of *spherical aberration*. This is caused by the fact that the focal length of rays close to the axis of the microscope is different when compared to that of rays further away from the optical axis. This leads to imperfect, delocalized pictures, and also limits the resolution of the system. This occurs because the electrons coming from one point of an object are not imaged into a single point, but rather into a small disk, thus blurring the image. The way to avoid rays with a larger distance from the optical axis is to reduce the numerical aperture. Whilst both chromatic and spherical aberration will limit the resolution of electron microscopes, conventionally the best electron microscopes can separate points within a distance of 0.15 to 0.2 nm.

Since 1936 (*Scherzer*), it has been acknowledged via a theoretical, well-based theorem, that spherical aberration in electron optics with rotationally symmetric electron lenses is not correctable. However, in a groundbreaking report, Rose [6] showed that, by combing the electron optical lenses with multipole electron optical elements, spherical aberration could be reduced by several orders of magnitude. This paved the road to "sub-Ångström" resolution electron microscopy (an Ångström is a non-SI length unit, equal to 0.1 nm).

The set-up of a modern electron microscope is shown in Figure 11.15. The electrons are emitted from a field emission point source and accelerated to the demanded energy, in most cases in the range between 200 and 300 keV. The next element, the monochromator, selects electrons of a very narrow energy band to reduce chromatic errors in the optical system. The condensor system (*Köhler* illumination system) focuses the electrons at the specimen. Although, in Figure 11.15 the lenses are shown as light optical lenses, in reality electron microscopes apply magnetic lenses, or more general, magnetic electron optical elements.

The picture of the specimen, which is formed by the elastic and inelastic scattering of electrons within it, is enlarged with the objective lens, after which the electron beam is limited by the objective lens diaphragm. By correctly selecting the size and position of this diaphragm it is possible to adjust the contrast and resolution within certain limits, and to change between bright- and dark-field microscopy. Before the

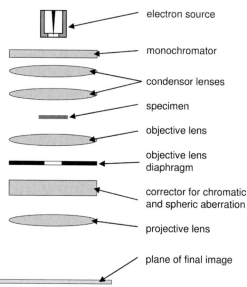

electron source

monochromator

condensor lenses

specimen

objective lens

objective lens
diaphragm

corrector for chromatic
and spheric aberration

projective lens

plane of final image

Figure 11.15 The set-up of a modern transmission electron microscope. Even when electron lenses are magnetic systems, for reasons of simplification, they are drawn like optical lenses.

image is magnified to its final size, it passes a correction system, where spherical and chromatic errors are minimized. This second correction for chromatic deviations is necessary, because by inelastic scattering, some of the electrons lose energy and are no longer within their intended energy range. Finally, the magnified image of the specimen becomes visible on the fluorescence screen, from where it may be documented using either photographic or electronic means.

11.3.2
Interaction of the Electron Beam and Specimen

The specimens used in transmission electron microscopes are thin foils, usually with thicknesses ranging between a few nanometers and 100 nm, at maximum. The electrons derived from the illumination system are scattered in the specimen. In contrast to elastic scattering, inelastic scattered electrons have lost energy. In view of the high-resolution imaging, the inelastic scattered electrons lead to chromatic aberration in the image, whereby the quality of the images will be reduced. On the other hand, the energy loss of the electrons represents a "fingerprint" for the elements in the specimen, and therefore the energy distribution of the inelastic scattered electron may also be used for elemental analysis of the specimen. Moreover, with appropriate instrumentation an elemental mapping of the specimen can be achieved, with extremely high lateral resolution. Together, these techniques are summarized by the acronym EELS (= electron energy loss spectroscopy). The largest portion of the inelastic scattered electrons are spread over a much larger angle as compared to the elastic scattered electrons, and are stopped at the objective lens diaphragm.

Elastic electron scattering is essentially proportional to $(Z/V)^2$. This means that the interaction of the electrons with the specimen increases with increasing atomic number Z, and decreases with increasing electron beam energy V. Therefore, it is difficult to "see" structural details from light elements in the vicinity of those of heavy elements. For example, in the case of zirconia, ZrO_2, it is impossible to obtain an image of the positions of the oxygen ions, $Z = 8$; $Z^2 = 64$, next to the zirconium ions, $Z = 40$, $Z^2 = 1600$, as the elastic scattering of zirconium is 25-fold that of the oxygen ions. The probability for inelastic scattering depends heavily on the electronic structure of the elements in the specimen. In a first approximation, it may be said that inelastic scattering increases with atomic number Z, and therefore all methods which apply processes connected to the inelastic scattering of electron (such as EELS or X-ray analysis) function better with heavy elements than with light elements.

Unlike optical microscopy, which in order to function depends primarily on contrast due to absorption, the situation is completely different for electron microscopy. Here, high-resolution images are formed by the interference of elastic scattered electrons, leading to a distribution of intensities that depends on the orientation of the lattice planes in a crystal relative to the electron beam. Therefore, at certain angles the electron beam is diffracted strongly from the axis of the incoming beam, whilst at other angles the beam is almost completely transmitted. In the case of high-

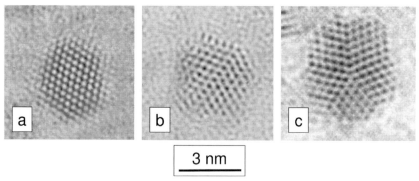

Figure 11.16 High-resolution electron micrographs of a fluctuating gold particle [7]. (a) A gold particle; (b) single twinned gold particle; (c) a five-fold twinned gold particle. (Reproduction with permission from [7]; Copyright: Springer 2002.)

resolution imaging, this allows the arrangement of atoms within a crystal lattice to be deduced. The lattice images of different gold nanoparticles are shown in Figure 11.16; these pictures are taken from a series of fluctuating particles that differ in both shape and twinning [7].

As shown in Figure 11.16a–c, at the highest resolution, the micrographs of the crystallized specimen show a "picture" of the lattice. Such micrographs can be readily obtained provided that the electron beam is exactly in the direction of a crystallographic orientation. In this way, the electrons "see" the atoms which are exactly in a row, although this is useful only if the crystallographic orientation has low *Miller* indices. Since, in general, this is not the case, the microscope must be capable of tilting the specimen, but without losing the point of observation. This is a quite a difficult task in the design and manufacture of the specimen holder.

As mentioned above, the micrographs shown in Figure 11.16 are not "shadow" images of the lattice, and therefore the interpretation of dark and bright points is not straightforward. This means that, without a detailed analysis, it cannot be said whether a dark point is showing the position of a column of atoms, or a hole. In addition, artifacts often also disturb lattice images; for example, in the case of high-Z specimen, due to an enhanced absorption of higher-order diffraction (because of the larger scattering angle, there will be a longer path through the specimen), the lattice image may be a function of the specimen thickness.

One very important point here in the interpretation of electron micrographs is the comparison of bright- and dark-field micrographs of the same spot of specimens. Different arrangements leading to such conditions are shown in Figure 11.17.

In Figure 11.17, it can be seen how the objective diaphragm limits the number of diffraction orders used to obtain the enlarged image. The main difference between the arrangement applied for bright- and dark-field illumination is found in the position of the objective lens diaphragm. For dark-field imaging, the diaphragm is shifted out of the optical axis; therefore, the information connected to the 0th diffraction order is blocked. The same result is obtained (but significantly better) by tilting the illumination system. This is not a mechanical tilting, but rather is tilting

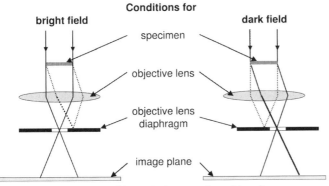

Figure 11.17 Comparison of the optical system of conditions for bright- and dark-field imaging. For dark-field imaging the 0th diffraction order is blocked and therefore does not contribute to image formation.

with a magneto-optical element, and allows variation in both tilting angle and rotation.

In the case of nanoparticles, dark-field electron microscopy is applied to obtain a first overview in the analysis of unknown materials. By varying the tilting angle and rotation, it can be seen quite rapidly whether a material contains crystalline grains, or not. In dark-field electron microscopy, those crystallized particles which are in correct orientation to the electron beam appear bright; a typical example is shown in Figure 11.18, which compares bright- and dark-field electron micrographs. Here,

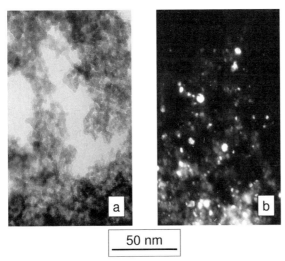

Figure 11.18 Comparison between (a) bright-field and (b) dark-field electron micrographs. The specimen is alumina with small zirconia precipitates (Sickafus, Vollath [8]). Note that only the large alumina particles are crystallized; the tiny bright spots indicate the crystallized zirconia precipitates.

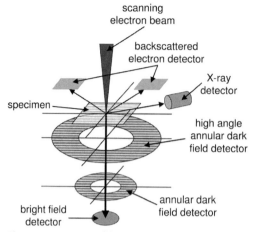

Figure 11.19 System of detectors in connection with a scanning transmission electron microscope.

the specimen was alumina with some precipitated zirconia. Up to a particle size of approximately 8 nm, alumina does not crystallize; however, the zirconia phase is precipitated in particles with dimensions ranging from 2 to 3 nm (see also Figure 1.1 in Section 1.1).

One further possibility of imaging in an electron microscope is connected to electron beam scanning. Originally, the scanning mode of the transmission electron microscope was primarily to produce elemental distribution images, by using the characteristic X-rays excited in the specimen. However, scanning transmission electron microscopy acquired an entirely new quality after the first design of aberration-free condenser lenses, as these systems allowed the electron beam to be focused on spots that were small enough to lie within the range of atomic resolution. Compared to transmission electron microscopy of similar resolution, these systems have the advantage that the contrast of the features in the image is not changed, for example from bright to dark, simply by changing the focus position. The possible signals that may be connected to a scanning transmission electron microscope are shown schematically in Figure 11.19.

11.3.3
Localized Chemical Analysis in the Electron Microscope

Inelastic scattered electrons represent the tools for chemical analysis in an electron microscope. Inelastic interaction of the incoming electrons with the specimens initiates a series of processes, characteristic of the composition and the electronic structure. This is shown schematically in Figure 11.20.

The primary electrons coming from the condenser system hit the specimen, while the largest fraction passes the specimen after elastic scattering or diffraction. The

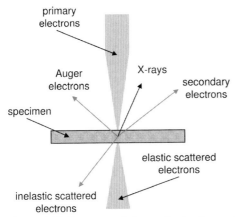

Figure 11.20 Electrons and photons leaving the specimen after the passing of energy-rich electrons for image formation.

electrons interacting with the specimen transfer parts of their energy to electrons in the specimen, and this results in the emission of:

- secondary electrons
- X-rays
- *Auger* electrons, and
- inelastic scattered electrons passing the specimen.

The sequence of processes starts with the excitation of an atom of the specimen. The absorbed energy is generally emitted as X-ray photon. The energy of the primary electron is reduced for the amount of energy necessary to excite the atom, and therefore the signals of the emitted X-rays and the energy loss of the inelastic scattered electrons are equivalent. As both carry the same information, both are characteristic of the specimen. In the case of light elements (those with low Z), the emitted X-ray photon pushes an electron with lower energy out of the specimen. This electron, which is called the "*Auger* electron" again has an energy characteristic of the atom. As a general rule, it can be said that the higher the energy of the primary X-ray photon (which is equivalent to a larger atomic number Z), the lower the probability of emission of an *Auger* electron. The phenomenon leading to the emission of *Auger* electrons is also called the "Internal Photo Effect".

Inelastic scattered electrons have lost part of their energy in the specimen, and a major part of this energy is converted to X-rays. There are two processes responsible for this phenomenon:

1. Deceleration of the electrons in the electric field of the atomic nucleus. This leads to the emission of "*Bremsstrahlung*" (radiation of deceleration). Although bremsstrahlung is characteristic of the deceleration process, it cannot be used to identify the decelerating target. Bremsstahlung has a continuous spectrum; the maximum energy of the emitted photons is equal to that of the incoming electrons.

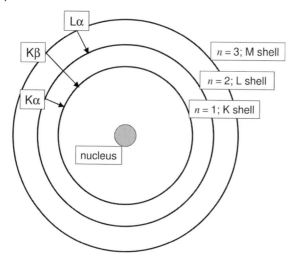

Figure 11.21 The schematic structure of an atom. The different electron shells with their names, and the most important electron transitions leading to the emission of X-rays, are indicated. The denomination of the X-rays is also indicated.

2. Ionization of the inner shells of the atoms; this process leads to the emission of characteristic X-rays, which are used for qualitative and quantitative analysis of the target.

The ionization of an inner shell leads to an electron vacancy in an inner shell, and this is re-filled by the transfer of an electron from another, more outer lying, shell. This filling process is connected to an emission of the now "superfluous" energy as an X-ray photon which is characteristic of the emitting atom. Depending on the shell, where the first ionization occurred, and the shell delivering the electron to fill this vacancy, characteristic X-rays are grouped into different series. This system is shown graphically in Figure 11.21.

A compilation of some X-ray emission series for elements where the largest principal quantum number n is 4 is provided in Table 11.2. In the periodic system of the elements, this system of emission series is continued up to a maximum principal quantum number of 7.

This system of emission lines allows an unequivocal determination of the atoms in the specimen. The wavelength of the emitted X-rays follows *Moseley's* law:

$$v = \kappa R_Y (Z - C)^2 \tag{11.24}$$

In Equation (11.24), v is the frequency of the emitted X-ray line, Z the atomic number of the emitting element, and R_Y is the *Rydberg* constant ($R_Y = 1.097 \times 10^{-7}\,\mathrm{m}^{-1}$), and κ and C are constants which depend on the line system. The values of κ and C for the K and L series are listed in Table 11.3.

All emission lines also carry information on the valency state of the atoms. In most cases, this line shift is so small that it cannot be evaluated in the X-ray detection

Table 11.2 Names and main quantum numbers of electron shells of atoms and the most important transitions leading to X-ray emission (the names of the emitted X-ray lines is also given).

Shell of the initial electron vacancy	Principal quantum number	Shell donating the electron to fill the vacancy	Principal quantum number	Designation of the emission lines series[a]	Difference of principal quantum numbers
K	1	L	2	$K\alpha$	1
K	1	M	3	$K\beta$	2
K	1	N	4	$K\gamma$	3
...	1				
L	2	M	3	$L\alpha$	1
L	2	N	4	$L\beta$	2
...	2				
M	3	N	4	$M\alpha$	1
...	3				

[a] This notation, which is preferred by physicists, is different to that known as the *Siegbahn* notation used by spectroscopists.

system of an electron microscope, and only those emission lines with energies significantly below 1 keV may be evaluated in this respect. The emitted X-rays may be analyzed using either a wavelength- or an energy-dispersive system. Wavelength-dispersive systems use single crystals as diffractive elements, while energy-dispersive systems apply semiconducting sensors, delivering an electric signal that is, in terms of its amplitude, proportional to the energy of the absorbed X-ray photons. The latter systems are most common in connection with electron microscopy.

A typical emission spectrum of a specimen consisting of zirconia particles is shown in Figure 11.22. This X-ray spectrum shows, as expected, the K lines of zirconium and oxygen and the L lines of zirconium. Although, in addition, there is a relatively strong signal of copper and carbon, these are *artifacts*! The copper lines stem from the copper mesh used as the specimen carrier, while the carbon signal is from the carbon film used as an electron transparent substrate for the zirconia particles.

Inelastic scattered electrons have lost their energy by interaction with the specimen. Therefore, the energy-loss of the electrons is – to a large extent – equivalent to the spectrum of the emitted X-rays. In other words, like emitted X-rays, the energy loss of the electrons can be used for characterization of the specimen material.

Table 11.3 Constants to calculate the frequency of X-rays using *Moseley*'s law given in Equation (11.25).

	κ	C
K series	0.75	1
L series	0.139	7.4

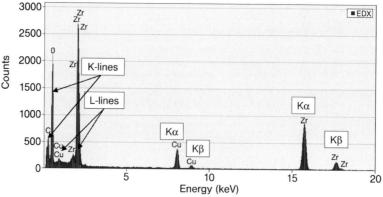

Figure 11.22 X-ray spectrum acquired with an energy-dispersive system of a specimen consisting of zirconia (ZrO_2) particles (Szabó [9]).

Experience has taught that the spectrum of the energy loss of the electrons contain additional information characterizing the specimen.

The typical EELS spectrum is shown in Figure 11.23, where the most important features are the zero loss peak stemming from the elastic scattered electrons, the plasmon peak, and the absorption edge with its fine structure, characterizing the elements, and their neighborhood.

In Figure 11.23, the highest peak at zero loss contains only elastic scattered electrons. The next prominent feature is the plasmon peak which, with energy loss of less than approximately 50 eV, is most prominent in metals; however, though not fully understood, it is also found in ceramics and polymers. Due to the chromatic error of the electron lenses, these electrons limit the resolving power of the electron microscope.

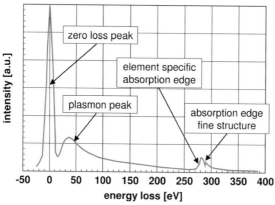

Figure 11.23 A typical EELS spectrum. The most important features in such a spectrum are the zero loss peak stemming from the elastic scattered electrons, the plasmon peak, and the absorption edge with its fine structure, characterizing the elements, and their neighborhood.

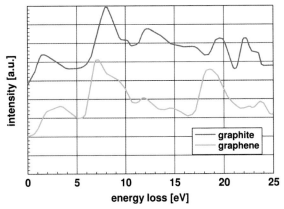

Figure 11.24 Comparison of the plasmon peaks of graphite and a single graphene layer. The spectra are background-corrected and stacked. (Reproduced with permission from [10]; Copyright: Elsevier 2001.)

The plasmon peak contains a wealth of information about valency states and binding in the specimen. A typical example is shown in Figure 11.24, where the plasmon peak of graphite and a single graphene layer are compared [10]. Here, the two spectra are background-corrected and stacked. It is important to note the significant differences between these two spectra which are, in theory, well understood.

The next group of features in the EELS spectrum are the absorption edges, which are specific for the elements in the specimen. On the side of higher energy losses (ranging up to 50 eV from the absorption edge), these structures contain information on binding, symmetry, bond length, and coordination numbers. The EELS spectrum shown in Figure 11.23 demonstrates only one absorption edge in the range of low electron loss energies, but depending on the composition of the specimen, absorption edges may also be found in the range of many keV. A typical spectrum of an absorption edge is shown in Figure 11.25, which shows the K absorption edge of carbon with the background having already been removed. These two spectra are for single- and multiwall carbon nanotubes. As in the case of the plasmon peak fine structure, which compares the spectra of a graphene sheet and graphite, Figure 11.25 shows that a more detailed structure stems from the multiwall nanotube as compared to that of a single wall nanotube [10].

The EELS technique has a very wide field of applications. For example, as described above, it may be used to determine energy loss spectra for a detailed analysis of the specimen. However, by using an imaging system it is possible to select a certain range in the energy loss spectrum for spatial mapping. This process delivers element-specific images, with a mapping of the distribution of the selected element. Compared with element mappings obtained from the characteristic X-rays, it can be said that the resolution of the elemental mappings obtained by EELS is significantly better. Also, depending on the instrumentation, EELS imaging may function in either standard or scanning transmission mode.

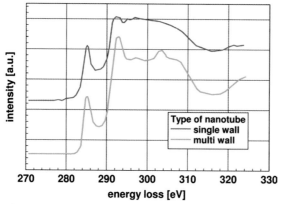

Figure 11.25 Carbon K-absorption edge of single-wall and multiwall nanotubes. The spectra are background-corrected and stacked. (Reproduced with permission from [10]; Copyright: Elsevier 2001.)

11.3.4

Scanning Transmission Electron Microscopy using a High-Angle Annular Dark-Field (HAADF) Detector

As mentioned in the description of electron microscopes, the introduction of aberration-corrected electron optical systems has led to the development of scanning transmission electron microscopes with lateral resolving power at least equal to that of the transmission electron microscope. Such systems produce images of outstanding quality, with maximum resolution significantly below 0.1 nm. Comparative electron micrographs taken from a grain boundary of a SrTiO$_3$ specimen are shown in Figure 11.26 [11].

Whereas, the HAADF image in Figure 11.26a clearly shows the lattice of the specimen, the "normal" bright-field electron micrograph in Figure 11.26b does not show any further detail. Collecting the inelastic scattered electrons at a high-scattering angle brings an additional advantage of an increased sensitivity on the atomic number (Z-sensitivity). Although this does not allow an actual identification of the elements to be made, in those cases where the structure of the specimen is well known a correlation is possible. Figure 11.27 shows such a micrograph [12] where the specimen resembled that shown in Figure 11.26 of SrTiO$_3$; however, compared with the upper grains in Figure 11.26, this micrograph has been rotated by 45°.

The positions of the atom columns in the lattice are shown in Figure 11.27. In addition, due to the Z-dependency of electron scattering, the atomic species Sr $Z = 38$ and Ti $Z = 32$ appear with significantly different contrast. However, because of their low Z number the oxygen atoms are invisible. When comparing Figure 11.27 with Figure 11.26a, the same differences in contrast are apparent but are less distinct than in the bright-field electron micrograph of Figure 11.26b.

Additionally, when compared to high-resolution transmission electron microscopy, HAADF imaging is more advanced for determining inhomogeneities and defect

 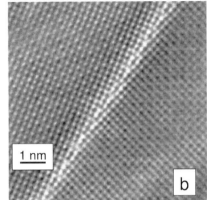

Figure 11.26 Comparison of transmission electron micrographs of SrTiO₃ taken in the high-resolution mode and with a HAADF system [11]. (a) HAADF micrograph. Note the arrangements of the atoms in a squared lattice. The intensity of atoms in the center of the squares (Ti; $Z = 22$) is less than that at the edges (Sr; $Z = 38$). This is caused by different Z numbers for these atoms; Z-contrast. (b) A high-resolution electron micrograph. In this case, the difference in Z is not easily visible. (Reprinted with permission from [11]; Copyright: Springer 2005.)

structures in the interiors of specimens at atomic resolution. In fact, when using HAADF Lupini and Pennycook [12] showed that it is possible to see even a single impurity atom in an otherwise perfect crystal. Certainly, although the examples shown in Figures 11.26 and 11.27 are selected from the very best specimens under optimal conditions, they clearly demonstrate the high potential of this imaging method.

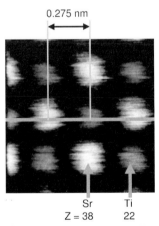

Figure 11.27 HAADF transmission electron micrograph of SrTiO₃. In this case, the Z sensitivity of this imaging system provides a perfect contrast between the strontium and titanium atoms [12].

References

1 Brunnauer, S., Emmett, P.H. and Teller, E. (1938) *J. Am. Chem. Soc.*, **60**, 309–319.

2 Schlabach, S., Szabó, D.V., Vollath, D., de la Presa, P. and Forker, M. (2006) *J. Appl. Phys.*, **100**, 024305.

3 Scherrer, P. and Nochrichten, G. (1918) *Math. Phys.*, 98–100.

4 Suzuki, K. and Kijima, K. (2006) *J. Alloys Comp.*, **419**, 234–242.

5 Du, J., Liu, Z., Wu, W., Li, Z., Han, B. and Huang, Y. (2005) *Mater. Res. Bull.*, **40**, 928–935.

6 Rose, H. (1990) *Optik*, **85**, 19–24.

7 Ascencio, J.A., Mendoza, M., Santamaria, T., Perez, M., Nava, I., Gutierrez-Wing, C. and Jose-Yacaman, M. (2002) *J. Clust. Sci.*, **13**, 189–197.

8 Sickafus, K.E. and Vollath, D. (1992) unpublished results.

9 Szabó, D.V. (2007) private communication.

10 Stéphan, O., Kociak, M., Henrard, L., Suenaga, K., Gloter, A., Tencé, M., Sandré, E. and Colliex, C. (2001) *J. Electron Spectrosc. Relat. Phenom.*, **114–116**, 209–217.

11 Ayache, J., Kisielowski, C., Kilaas, R., Passerieux, G. and Lartigue-Korinrk, S. (2005) *J. Mater. Sci.*, **40**, 3091–3100.

12 Lupini, A.R. and Pennycook, S.J. (2003) *Ultramicroscopy*, **96**, 313–322.

Index

a

aberration, chromatic/spherical   329
ablation process, laser   83–86
absorbance   158
– differential   197
– gold–silver alloy   174
– optical   179
absorption
– light quanta   13
– optical   150
– UV   147–149
absorption spectra, photochromic materials   196
additive technologies   3
adiabatic coagulation   27
adsorption
– enthalpy   315
– Langmuir isotherm   314–317
Aerosil®, index of refraction   148
aerosol process, flame   92–103
agglomerates, fractal   94–95
agglomeration   71
– coated particles   104–105
– in laser ablation process   85
– platelets   284
alloy, gold–silver   174
alumina particles, spherical   314
aluminum
– nanocrystalline   247
– surface energy   42
anisotropic lattices   26
anisotropy
– energy of   114–116
– magnetic   114–115, 129
annealing   57, 179
– X-ray diffraction pattern   322
annealing temperature   198

b

ballistic conductivity   212–213
band energy   153
band gap   154
barrier, energy   13
beam, electrons   331–334
BET (Brunauer, Emmett and Teller) method   314–319
BET surfaces, nanopowders   319
biologically functionalized nanoparticles   11
bleaching process   204
Bloch walls   109–112, 141–142
blocking temperature   113, 119
blue shift   148–158, 166, 175, 181–182, 204–207
Bohr magneton   116, 124
Bohr radius, excitons   150
Bose–Einstein statistics   43
"bottom-up" approach   1–3
boundaries, grain   22
Bragg's law   320
bremsstrahlung   335
"bricks-and-mortar" structure   257
bright-field imaging   332–333
Brownian molecular movement   270
Brunauer, see BET method

antiferroelectric–paraelectric transformation   58
antiferromagnetism   110–111, 122
aperture, numerical   327
arc, electric   304
armchair-type carbon nanotubes   292
array, gas sensing elements   18
Ashby–Verall mechanism   246–248
atoms, energy levels   152
Auger electrons   335

Nanomaterials: An Introduction to Synthesis, Properties and Application. Dieter Vollath
Copyright © 2008 WILEY-VCH Verlag GmbH & Co. KGaA, Weinheim
ISBN: 978-3-527-31531-4

bulk metallic materials   236–253
Burgers vector   242
burnable organic liquids   102–103

c
cadmium sulphide (CdS) nanorods
– photoluminescence spectrum   223
– surface   224
cadmium telluride (CdTe) nanoparticles
    175–179
calibration curves, gas sensors   19
cancer cells, luminescence staining   163
capacity, heat, *see* heat capacity
carbon, supercooled   304
carbon black   92
carbon nanotube-based composites   228–229,
    262–265
carbon nanotubes   288–300
– armchair-type   292
– debundling   230
– electrical conductivity   216–221
– electron field emission characteristics   295
– single-wall   263, 265
– zigzag type   291–292, 294
catalyst particles   305–306
cavity, resonant microwave   87
CCD devices   325
CD player systems   272
cell separation, magnetic   133
cells, cancer   163
ceramic-coated ceramic nanoparticles   104
ceramic core   164–165, 180–183, 185–186
ceramic materials, mechanical properties
    236–253
ceramic/polymer nanocomposites,
    photoluminescence   164
ceria (CeO$_2$) nanoparticles, facetted   26
I–V characteristics   213–215, 217–220
– imogolite   288
charge carrier transport   200
charge transfer   189
chemical analysis, localized   334–340
chemical reactions
– gas-phase   34–35, 71–108
chemical vapor synthesis processes   79–83
chemisorption   314
chemistry, colloid   71
chirality vector   291
chromatic aberration   329
Clausius–Clapeyron law   33
clusters, fractal   5
coagulation   71
– adiabatic   27
– and phase transformations   54–55
– gold particles   28

– relaxation time   37
coated particles
– agglomeration   104–105
– gas-phase synthesis   103–107
coating, nanoparticles   9–10
– nanoparticles   19
coercitivity   110–112, 137, 300–303
– superparamagnetism   119
collision parameter   74, 78
– influence of electrical charge   77
collision probability   72
colloid chemistry   71
coloring/bleaching process   204
composites, *see* nanocomposites
condensation   71
condensation process, inert gas   78–79
condensor   330
conductance   211
– cadmium sulphide (CdS)   225
conductivity
– ballistic   212–213
– diffusive   211–212
– electrical, *see* electrical conductivity
– thermal   268–269
confinement, quantum   149–161
contraction, lattice   32
contrast enhancement
– ferrofluidic   276
– NMR   134
cooling, rapid   86
copper ferrite (CuFe$_2$O$_4$) platelets   281–282
core, ceramic, *see* ceramic core
corona discharge   101
correlation length   137
coverage of surface   317
crystal
– one-dimensional   286–288, 301
crystal field, magnetic   132
crystal model, linear   42
crystallization, metal nanoparticles   51
crystallographic planes   25–26
cubic–tetragonal transformation   323
Curie temperature   59
current   211
– density at failure   216
current–voltage characteristics   213–215,
    217–220
– imogolite   288
curvature, surfaces   35–36
curvature-dependence, vapor pressure
    36

d
dark-field imaging   332–333
DeBroglie relationship   150, 328

debunding, carbon nanotubes   230
defoliated platelets   258–259
deformation
– mode   235
– nanocrystalline aluminum   247
density distribution, electrons   151
diamagnetism   109
diamond structure   300
dielectric constant   59
differential absorbance   197
differential transmittance   202–203
diffraction
– selected area   326
– X-ray/electron   319–326
diffusion coefficient, volume   245
diffusion scaling law   13–19
diffusive conductance   211–212
dimensionless surface/volume ratio   21
dipole–dipole interactions   121–122
discharge, corona   101
dislocation climbing/gliding   240–241
displays, field emission   295–296
distortion, pseudotetragonal   58
domains, magnetic   109–110
drift movement of electrons   211
DVD player systems   272
dynamic viscosity   269

e
EELS, *see* electron energy loss spectroscopy
elastic scattering   329
electric arc   304
electric quadrupole splitting   128
electrical charge, influence on collision
   parameter   77
electrical conductivity
– carbon nanotubes   216–221
– gold nanowire   213–214
– nanocomposites   225–230
– nanotubes and nanorods   211–216
– PFO coating   297
electrical current, *see* current
electrical properties, nanoparticles   211–231
electrical resistance, carbon nanotubes
   216–220
– resistance quantum   213
electrochromic materials   194, 200–204
– transmittance   201–203
electrodes, plate and needle   100–102
electroluminescence   188–194
electromigration   37–38
electron beam, interaction with specimen
   331–334
electron energy loss spectroscopy (EELS)   331,
   338–339

electron field emission characteristics, carbon
   nanotubes   295
electron microscopy   326–331
– localized chemical analysis   334–340
– scanning transmission   340–341
electron shells   336–337
electrons
– density distribution   151
– diffraction   319–326
– free   153
– mean free path length   88
– oscillations   169–171
– wavelength   329
elongated gold nanoparticles   170
embryos   63–65
emission spectra   154–155
– CdTe quantum dots   177
Emmett, *see* BET method
energy
– Fermi   153, 188
– of anisotropy   114–116
– solid–liquid interface   51
– strain   31–32
– surface, *see* surface energy
energy bands   153
energy barrier   13
energy levels
– atoms   152
– nucleus   128, 131
ensemble   67
enthalpy
– adsorption   315
– free, *see* free enthalpy
– Gibb's free   41
– monoclinic–tetragonal transformation   54
– of melting   42, 50
entropy   41
equations, *see* laws and equations
exchange-coupled magnetic nanoparticles
   136–142
excimer lasers   84
excitation, lumophores   183–184
excitons   149–150
– Bohr radius   150
extended X-ray absorption fine structure
   (EXAFS)   323

f
facetted nanoparticles   26
Faraday effect   204–206
feed-throughs   272–273
Fermi energy   153, 188
ferrimagnetism   111
ferrites   115, 120–121, 123, 126–127,
   129–130, 140–142, 206–207

ferroelectric material   56–57
ferrofluids   135, 270–277
– applications   272–277
– contrast enhancement   276
– viscosity   271–274
ferrohydrodynamics   270
ferromagnetism   109–110
FET, *see* field-effect transistor
Fick's laws   13–14
field
– inhomogeneous magnetic   271
– magnetic crystal   132
– reduced magnetic   117
field-effect transistor (FET)   220–221
field emission characteristics, carbon
      nanotubes   295
field emission displays   295–296
filled polymer composites   253–265
films
– thick-film layer   18
– thin, *see* thin films
FITC, *see* fluorescein isothiocyanate
flame aerosol process   92–103
flames, silica-producing   94
fluctuations
– structural   64–68
– thermal   125
fluids, nano-, *see* nanofluids
fluorescein isothiocyanate (FITC)   161–162
fluorescence   163, 181
formation, nanorods and nanoplates   283–285
formulae, *see* laws and equations
fractal agglomerates   94–95
fractal clusters   5
Frank–Reed source   241–242
free enthalpy   65–66
– Gibb's   41
– nanocrystalline bodies   248
– of formation   29
free path length, mean   72–73
– mean   88
free electrons   153
frequency, "kitchen"   89
– industrial   92
friction coefficient   303–304
fullerenes   279, 288–291, 293, 300, 302–305
– multiwall   300–302
– nested   291

g

gallium nitride (GaN) nanorods   304, 306
gap, band   154
gas condensation process, inert   78–79
gas-phase reactions   34–35

gas-phase synthesis   71–108
– coated particles   103–107
– flame aerosol process   92–103
– inert gas condensation process   78–79
– laser ablation process   83–86
– microwave plasma process   86–92
– physical/chemical vapor synthesis
      processes   79–83
gas pressure, reaction chamber   84
gas sensors
– general layout   15
– nanoparticles   15–17
gases, kinetic theory   73
gearwheel, nanometer-sized   282
germanium nanorods   307–309
Gibb's free enthalpy   41
gold nanoparticles
– coagulation   28
– elongated   170
– melting temperature   68
– phase diagram   63
– spheroidal   169
gold nanorods   172
gold nanowire, electrical conductivity   213–214
gold platelets   281
gold–ruby–glass   7, 172–174
gold–silver alloy, absorbance   174
grain boundaries, polycrystalline material   22
grain growth   27–29
grain size   238–251
– zirconia (ZrO$_2$) powder   5
graphene   217–220, 289, 291–293, 307
– electron microscopy   339
graphite, structure   289

h

HAADF, *see* high-angle annular dark-field
      detector
Hall–Petch relationship   239–240, 243–245
hard magnetic materials   136–138
hardness
– Vickers   244, 246–247
heat capacity
– nanoparticles   42–45
– sintered metallic nanocrystalline materials
      44
heat capacity ratio   267–268
heat release rate   261–262
heat transfer, nanofluid improved   267–270
hexamethyl disiloxane (HMDSO)   93, 95
high-angle annular dark-field (HAADF)
      detector   340–341
high-power lasers, pulsed   83
holes   153, 189, 195

homogenization time, relative   14
Hooke's law   234–235
hydrostatic pressure   24, 30–32
hydroxide layer   90
hypostoichiometric state   195
hysteresis   110–111, 116–117, 127

i
illumination hole   198–199
imaging, bright-/dark-field   332–333
imogolite $(Al_2SiO_3(OH)_4)$ 286–288
improved heat transfer   267–270
index of refraction, adjustment   145–149
indium tin oxide (ITO)   186–187, 190,
   297–298
inelastic scattering   329, 334–335
– HAADF Detector   340
inert gas condensation process   78–79
– up-scaling   80
inhomogeneous magnetic field   271
instability, thermal   113
interface energy, solid–liquid   51
inverse Hall–Petch relationship   244–245
ionization, partial   78
iron oxide (Fe_2O_3) powder
– gas-phase synthesis   81–82
– nanoparticulate   6
iron oxide particles, superparamagnetic   107
ITO, *see* indium tin oxide

k
Kelvin formula   33
Kerr effect   204
kinetic theory of gases   73
"kitchen frequency"   89

l
Landau's order parameter   60–61
Langevin's formula   116, 118, 123, 125
Langmuir adsorption isotherm   314–317
Langmuir's formula   314
laser ablation process   83–86
lasers   83–84
lateral limits, structuring processes   4
lattices
– anisotropic   26
– contraction   32
– spin   111
laws and equations
– Bragg's law   320
– Clausius–Clapeyron law   33
– DeBroglie relationship   150, 328
– diffusion scaling law   13–19
– Fick's   13–14
– Hall–Petch relationship   239–240, 243–245
– Hooke's law   234–235
– Kelvin (Thomson) formula   33
– Langevin's formula   116, 118, 123, 125
– Langmuir's formula   314
– Moseley's law   336
– Ohm's law   211, 213–215
– Pauli's principle   152
– Planck's equation   212
– Scherrer formula   322–323
– Tauc relationship   155–157
– Thomson's law   47
layered silicate   260
layered structures   3, 285–286
– compounds   288–311
light
– absorption   13
– scattering   146–147
limits, lateral   4
linear crystal model   42
liquids, burnable organic   102–103
localized chemical analysis   334–340
loudspeaker, ferrofluidic   273
luminescence   279–280
– cancer cell staining   163
– electro-   188–194
– nanocomposites   180–187
– photo-   160, 164–166
– spectra   182, 186–187
lumophores   161–169
– excitation   183–184

m
magnetic anisotropy   114–115, 129
magnetic cell separation   133
magnetic crystal field   132
magnetic domains   109–110
magnetic field
– inhomogeneous   271
– reduced   117
magnetic materials
– hard/soft   136–138
– ultrasoft   140
magnetic moment, molecular   110
magnetic nanoparticles, exchange-coupled
   136–142
magnetic particle size   124
magnetic properties, nanoparticles   109–143
magnetic pumping system   274
magnetic refrigeration   135
magnetic susceptibility   125–132
magnetization
– remanent   298
– saturation   111, 120–121, 137, 185

– thermal fluctuation   125
magnetization curve   110–111, 116–119, 121–124, 140–142
magneto-optic applications   204–207
mass increase, relativistic   328
mass susceptibility   126
matrices, transparent   169–180
matrix, nanocomposites   6–9
mean free path length   72–73
– electrons   88
mechanical properties, nanoparticles   235–266
melting, nanoparticles   60–64
melting enthalpy   42, 50
melting temperature
– gold nanoparticles   68
– reduction   47–51
metal nanoparticles, crystallization   51
metallic materials
– mechanical properties   236–253
– sintered nanocrystalline   44
metallic nanoparticles, transparent matrices   169–180
microscopy, electron   326–331
microwave cavity, resonant   87
microwave plasma process   86–92
Miller indices   320–321, 323, 332
molar surface energy   22
molecular magnetic moment   110
molecular movement, Brownian   270
molybdenum sulphide (MoS$_2$) nanotubes   300–303
monoclinic phase   29–30, 29–30
monoclinic–tetragonal transformation   52–53
– enthalpy   54
Moseley's law   336
Mössbauer spectrum   127–130, 132
mother-of-pearl   254, 257–258
multilayer system   191–192
multiwall nanotubes   216–220, 293–294, 298–301, 307
– EELS spectra   340

n
nacre   254, 257–258
nanocomposites   5–11
– carbon nanotube-based   262–265
– carbon nanotube–epoxy   228–229
– electrical conductivity   225–230
– filled polymer   253–265
– matrix   6–9
– photoluminescence   164–166
– platelet-filled polymer-based   257–262

– special luminescent   180–187
nanocrystalline aluminum, deformation   247
nanocrystalline materials
– sintered metallic   44
– yield stress   237
nanofluids   267–277
– improved heat transfer   267–270
– thermal conductivity   268–269
nanomaterials
– and nanocomposites   5–11
– characterization   313–342
– consequences of particle size   11–19
– surfaces   21–40
nanometer-sized gearwheel   282
nanomotors   37–39
nanoparticles   5
– biologically functionalized   11
– cadmium telluride (CdTe)   175–179
– ceria (CeO$_2$)   26
– coated   103–107
– coating   9–10, 19
– electrical properties   211–231
– elongated gold   170
– exchange-coupled magnetic   136–142
– gas-phase synthesis   71–108
– gas sensors   15–17
– heat capacity   42–45
– iron oxide (Fe$_2$O$_3$) powder   6
– magnetic properties   109–143
– magneto-optic applications   204–207
– mechanical properties   235–266
– melting   60–64
– melting temperature   68
– metallic and semiconducting   169–180
– optical properties   145–209
– phase diagram   63–64
– phase transformations   41–69
– semiconducting   169–180
– spheroidal gold   169
– structure   55–60
– surface   11–12
– suspensions, *see* nanofluids
– thermal phenomena   12–13
– thermodynamics   41–42
– zirconia (ZrO$_2$)   27–28
– , *see also* particles
nanoplates   279–311
– formation   283–285
nanopowders, BET surfaces   319
– , *see also* powder
nanorods   3, 5
– cadmium sulphide (CdS)   223
– electrical conductivity   211–216
– formation   283–285

– gallium nitride (GaN) 304, 306
– germanium 307–309
– gold 172
– nanotubes and nanoplates 279–311
– non-carbon 300–303
– photoconductivity 222–225
– synthesis 303–311
– zinc sulphide (ZnS) 191–193
nanostructures, compounds with layered structures 288–311
– , *see also* layered structures
nanotubes 3, 5
– carbon, *see* carbon nanotubes
– electrical conductivity 211–216
– multiwall 216–220
– nanorods and nanoplates 279–311
– non-carbon 300–303
– relation to nanorods and nanoplates 279–311
– single-wall carbon 263, 265
– synthesis 303–311
– tungsten sulphide (WS$_2$) 300–303
– zinc oxide (ZnO) 309–311
nanowire, gold 213–214
Nd-YAG lasers 84
needle electrodes 100–102
Néel-superparamagnetism 127, 129
nested fullerenes 291
non-carbon nanotubes/-rods 300–303
nuclear magnetic resonance (NMR) 133–134, 276
nucleation 71
nucleus
– energy levels 128, 131
numerical aperture 327

o
Ohm's law 211, 213–215
one-dimensional crystals 286–288
one-dimensional nanoparticles 279
"onion" crystal 301
"onion" molecules 291
optical absorbance 179
optical absorption 150
optical properties
– nanoparticles 145–209
– relation to quantum confinement 149–161
optical transmission, PFO 297–298
optoelectronics 222
order parameter, Landau's 60–61
organic light-emitting diodes (OLEDs) 296
organic liquids, burnable 102–103
oscillations, electrons 169–171

oxide/PMMA nanocomposites, photoluminescence 165–166

p
paraelectric–antiferroelectric transformation 58
paramagnetism 109–110
partial ionization 78
particle-filled polymers 253–256
"particle in a box" problem 150–151
particle size
– elementary consequences 11–19
– influence on thermodynamic behavior 41–68
– magnetic 124
particle size distribution, zirconia (ZrO$_2$) powder 75
– zirconia (ZrO$_2$) powder 91
particles
– pseudocrystalline 63–64
– , *see also* nanoparticles
path length
– mean free 72–73, 88
Pauli's principle 152
percolation threshold 225–226, 230
perovskite structure 56–57
PFO 297–298
phase, monoclinic 29–30
– sphalerite 178
– tetragonal 29–30
– wurtzite 178
phase diagram, gold nanoparticles 63
phase transformations
– and coagulation 54–55
– nanoparticles 41–69
phonons 169
photochromic materials 194–200
– absorption spectra 196
photoconductivity, nanorods 222–225
photoluminescence 160
– nanocomposites 164–166
– spectrum of CdS nanorods 223
physical vapor synthesis processes 79–83
physisorption 314
Planck's equation 212
planes, crystallographic 25–26
plasma process, microwave 86–92
plasmon peaks 339
plasmons 153, 169–171
plasticity, super-, *see* superplasticity
plate and needle electrodes 100–102
platelet-filled polymer-based nanocomposites 257–262
platelets

– copper ferrite (CuFe$_2$O$_4$)   281–282
– gold   281
PMMA, *see* polymethylmethacrylate
polarization plane   204–205
poly(2,7-9,9-(di(oxy-2,5,8-trioxadecane))
   fluorene), *see* PFO
polycrystalline material, grain boundaries   22
polymer nanocomposites
– filled   253–265
– photoluminescence   164
polymerization, PMMA   165
polymers, particle-filled   253–256
polymethyl methacrylate (PMMA)   10,
   106–107
– index of refraction   147–148
– nanocomposite photoluminescence
   165–166
– special luminescent nanocomposites
   180–186
– stress–strain diagram   256
porosity   236–238
powder
– BET surfaces   319
– iron oxide (Fe$_2$O$_3$)   6, 81–82
– titania (TiO$_2$)   96, 98
– zirconia (ZrO$_2$)   75, 91
– zirconia (ZrO$_2$) powder   5
pressure
– gas   84
– hydrostatic   24, 30–32
– vapor   34–35
probability, particle collisions   72
pseudocrystalline particles   63–64
pseudotetragonal distortion   58
pulsed high-power lasers   83
pumping system, magnetic   274
Q-value   142

q
quadrupole splitting, electric   128
quantum confinement, relation to optical
   properties   149–161
quantum dots   161–169
– CdTe   177
– electroluminescene   190
quenching   86

r
rapid cooling   86
reaction chamber, gas pressure   84
reactions, chemical, *see* chemical reactions
reduced magnetic field   117
reduction of melting temperature   47–51
Reed, *see* Frank–Reed source

refraction, index of   145–149
refrigeration, magnetic   135
relative homogenization time   14
relativistic mass increase   328
relaxation time, coagulation   37
remanence   110–112
remanent magnetization   298
resistance, electrical, *see* electrical resistance
resolution, sub-Ångström   330
resonant microwave cavity   87
rupture   234

s
SAD, *see* selected area diffraction
saturation magnetization   111, 120–121, 137,
   185
scaling law, diffusion   13–19
scanning transmission electron microscopy
   340–341
scattering of light   146–147
Scherrer formula   322–323
selected area diffraction (SAD)   326
self-lubricating solid films   303
self-organization processes   7
semiconducting nanoparticles, transparent
   matrices   169–180
sensitivity, gas sensors   17
sensors
– gas   15–17
– vibration   275
separation, magnetic cell   133
shear stress   242
silica-producing flame   94
silicates, layered   260
single-wall carbon nanotubes (SWNT)   263,
   265, 293
– EELS spectra   340
sintered metallic nanocrystalline materials,
   heat capacity   44
sintering   35–36
size, particle, *see* particle size
small particle size, elementary consequences
   11–19
"soccer ball" molecule   290
soft magnetic materials   137–138
solid films, self-lubricating   303
solid–liquid interface energy   51
sonication   229
specific surface area   313–319
spectra
– absorption   196
– EELS   338–339
– emission   154–155, 177
– luminescence   182

– Mössbauer   127–130, 132
– photoluminescence   223
sphalerite phase   178
spherical aberration   329
spherical alumina particles   314
spheroidal gold nanoparticles   169
spin-canting phenomena   130
spin lattices   111
splitting, electric quadrupole   128
statistics, Bose–Einstein   43
strain, *see* stress–strain diagram
strain energy   31–32
strand of carbon nanotubes   264–265
stress
– surface   23–24
– yield, *see* yield stress
stress–strain diagram   233–235, 238, 248,
   259–260, 263–264
– filled polymer composites   254–256
– superplastic materials   251–253
structural fluctuations   64–68
structure
– layered, *see* layered structures
– nanoparticles   55–60
– perovskite   56–57
structuring processes, lateral limits   4
sub-Ångström resolution   330
subtractive technologies   3
supercooled carbon   304
superferromagnetism   112
superparamagnetic materials   113–135
– applications   132–135
– ferrofluids   270–277
– iron oxide particles   107
– luminescence spectra   186
– magnetic susceptibility   125–132
superparamagnetism
– Néel-   127, 129
superplasticity   251–253
surface area, specific   313–319
surface energy   22–39, 283
– molar   22
– solid and liquid aluminum   42
– technical consequences   33–39
surface stress   23–24
surfaces
– cadmium sulphide (CdS) nanorods   224
– coverage   317
– curvature   35–36
– nanomaterials   21–40
– nanoparticles   11–12
– surface/volume ratio   12, 21
susceptibility
– magnetic   125–132

– mass   126
suspensions of nanoparticles, *see* nanofluids
SWNT, *see* single-wall carbon nanotubes
synthesis
– gas-phase   71–108
– nanotubes/-rods   303–311

t
Tauc relationship   155–157
technical consequences, surface energy
   33–39
technologies, additive/subtractive   3
Teller, *see* BET method
temperature
– annealing   198
– blocking   113, 119
– melting   47–51, 68
tetragonal–cubic transformation   323
tetragonal–monoclinic transformation
   52–54
tetragonal phase   29–30
thermal conductivity, nanofluids   268–269
thermal fluctuation, magnetization   125
thermal instability, superparamagnetic
   materials   113
thermal phenomena, nanoparticles   12–13
thermodynamics, nanoparticles   41–42
thick-film layer   18
thin films   3, 5
Thomson formula   33
Thomson's law   47
threshold
– percolation   225–226, 230
titania (TiO_2) powder   96, 98
"top-down" approach   1–3
transformation
– monoclinic–tetragonal   52–54
– paraelectric–antiferroelectric   58
– tetragonal–cubic   323
transistor, field-effect   220–221
transmission, optical   297–298
transmission electron microscopy, scanning
   340–341
transmittance
– differential   202–203
– electrochromic materials   201–203
transparent matrices, metallic and
   semiconducting nanoparticles   169–180
transport, charge carriers   200
tribological properties   303–304
tungsten sulphide (WS_2) nanotubes   300–303
twinning   62, 332
two-dimensional nanomaterials, *see* thin films
two-dimensional nanoparticles   279

u

ultimate strength   234
ultrasoft magnetic material   140
unit cell volume   56
up-scaling, inert gas condensation process
   80
UV-absorbing materials   147–149

v

van der Waals bonds   289, 300, 302
van der Waals forces   285
vapor deposition   305
vapor pressure   34–35
vapor synthesis processes, physical and
   chemical   79–83
Verall, *see* Ashby–Verall mechanism
Verdet constant   205
vibration sensor   275
Vickers hardness   244, 246–247
viscosity
– dynamic   269
– ferrofluids   271–274
voltage   211
– *I–V* characteristics   213–215, 217–220
volume
– expansion   33
– surface/volume ratio, *see* surface/volume
   ratio
– unit cell   56
volume diffusion coefficient   245

w

Waals, *see* van der Waals
wavelength of electrons   329
wurtzite phase   178

x

X-ray diffraction   319–326

y

yield stress   233–234, 243
– nanocrystalline aluminum   247
– nanocrystalline copper and palladium   237
Young's modulus   234, 236–239, 260–265
yttria-doped zirconia   252–253

z

zero-dimensional nanomaterials, *see*
   nanoparticles
zigzag type carbon nanotubes   291–292, 294
zinc oxide (ZnO) nanotubes   309–311
zinc sulphide (ZnS) nanorods   191–193
zirconia (ZrO$_2$)
– index of refraction   148
– yttria-doped   252–253
zirconia (ZrO$_2$) nanoparticles   27–28
zirconia (ZrO$_2$) powder
– grain size   5
– particle size distribution   91
– size distribution   75
zirconium nitride (ZrN) particles   92